Electricity and Magnetism

Electricity and Magnetism

Volume 1

B. I. BLEANEY
Formerly Fellow of St. Hugh's College, Oxford

and

B. BLEANEY
Dr. Lee's Professor Emeritus of Experimental Philosophy, University of Oxford

THIRD EDITION

OXFORD UNIVERSITY PRESS

Oxford University Press, Walton Street, Oxford OX2 6DP
Oxford New York Toronto
Delhi Bombay Calcutta Madras Karachi
Kuala Lumpur Singapore Hong Kong Tokyo
Nairobi Dar es Salaam Cape Town
Melbourne Auckland Madrid
and associated companies in
Berlin Ibadan

Oxford is a trade mark of Oxford University Press

First edition 1957
Second edition 1965
Third edition 1976
Reprinted (with corrections) 1978
Reprinted (with corrections) 1983
Reprinted 1985, 1987
First published as a two volume edition 1989
Reprinted 1990, 1991, 1993

British Library Cataloguing in Publication Data
Bleaney, B. I. (Betty Isabelle)
Electricity and magnetism.—3rd ed.
Vol. 1
1. Electricity 2. Magnetism
I. Title II. Bleaney, B. (Brebis)
537
ISBN 0–19–851172–8

Library of Congress Cataloging in Publication Data
(Data available)

Printed in Great Britain by
J. W. Arrowsmith Ltd, Bristol

Preface to the third edition

De manera que acordé, aunque contra mi voluntad, meter segunda vez la pluma en tan estraña lavor é tan agena de mi facultad, hurtando algunos ratos á mi principal estudio, con otras horas destinadas para recreacioń, puesto que no han de faltar nuevos detractores á la nueva edicion.

1499 *Fernando de Rojas*

So I agreed, albeit unwillingly (since there cannot fail to be fresh critics of a new edition), again to exercise my pen in so strange a labour, and one so foreign to my ability, stealing some moments from my principal study, together with other hours destined for recreation.

For the third edition of this textbook the material has been completely revised and in many parts substantially rewritten. S.I. units are used throughout; references to c.g.s. units have been almost wholly eliminated, but a short conversion table is given in Appendix D. The dominance of solid-state devices in the practical world of electronics is reflected in a major change in the subject order.

Chapters 1–9 set out the macroscopic theory of electricity and magnetism, with only minor references to the atomic background, which is discussed in Chapters 10–17. A simple treatment of lattice vibrations is introduced in Chapter 10 in considering the dielectric properties of ionic solids. The discussion of conduction electrons and metals has been expanded into two chapters, and superconductivity, a topic previously excluded, is the subject of Chapter 13. Minor changes have been made in the three chapters (14–16) on magnetism. The discussion of semiconductor theory precedes new chapters on solid-state devices, but we have endeavoured to present such devices in a manner which does not presuppose a knowledge in depth of the theory. The remaining chapters, on amplifiers and oscillators, vacuum tubes, a.c. measurements, noise, and magnetic resonance, bring together the discussion of electronics and its applications.

The authors are grateful to many colleagues in Oxford and readers elsewhere for helpful comments on previous editions which have been incorporated in the present volume. In particular we are indebted to Dr. G. A. Brooker for numerous and detailed comments and suggestions; to Drs F. V. Price and J. W. Hodby, whose reading of new material on electronics in draft form resulted in substantial improvement of the presentation; to Drs F. N. H. Robinson and R. A. Stradling for several helpful suggestions; and to Messrs

C. A. Carpenter and J. Ward for the considerable trouble taken in producing Fig. 23.3. We are indebted to Professors M. Tinkham and O. V. Lounasmaa for generously sending us material in advance of publication; and to Professor L. F. Bates, F.R.S., Drs R. Dupree, and R. A. Stradling for their kindness in providing the basic diagrams for Figs 15.6, 6.15, and 17.9. We wish to thank Miss C. H. Bleaney for suggesting the quotation which appears above.

Clarendon Laboratory,
Oxford
February 1975

B. I. B.
B. B.

Acknowledgements

THE authors are indebted to the following for permission to use published diagrams as a basis for figures in the text: the late Sir K. S. Krishnan; G. Benedek; R. Berman; A. H. Cooke; B. R. Cooper; G. Duyckaerts, D. K. Finnemore; M. P. Garfunkel; R. V. Jones; A. F. Kip; C. Kittel; D. N. Langenberg; D. E. Mapother; K. A. G. Mendelssohn; R. W. Morse; D. E. Nagle; N. E. Phillips; H. M. Rosenberg; J. W. Stout; R. A. Stradling; W. Sucksmith; W. P. Wolf; American Institute of Physics; American Physical Society; Institution of Electrical Engineers; Institute of Physics and the Physical Society (London); Royal Society (London); Bell Telephone Laboratories; G. E. C. Hirst Research Centre.

Note added in 1989

The opportunity has been taken of dividing this textbook into two volumes.

Volume 1: Chapters 1 to 9 inclusive, covering the basic theory of electricity and magnetism.

Volume 2: Chapters 10 to 24 inclusive, covering electrical and magnetic properties of matter, including semiconductors and their applications in electronics, alternating current measurements, fluctuations and noise, magnetic resonance.

A number of minor errors have been corrected, and a section (20.8) has been added on Operational Amplifiers. We wish to thank Dr. F. N. H. Robinson for suggesting this, and Dr. J. F. Gregg, I. D. Morris, and J. C. Ward for help in its preparation. We are indebted to Dr. L. V. Morrison of the Royal Greenwich Observatory, Cambridge (Stellar Reference Frame Group), for the up-to-date plot of the variations in the length of the day, measured by the caesium clock, that now appears as Fig. 24.12. It is based on data published by the Bureau de l'Heure, Paris.

B. I. B.
B. B.

Contents

CONTENTS OF VOLUME 2

Volume 1

1. Electrostatics I

1.1. The electrical nature of matter

THE fundamental laws of electricity and magnetism were discovered by experimenters who had little or no knowledge of the modern theory of the atomic nature of matter. It should therefore be possible to present these laws in a textbook by dealing at first purely in macroscopic phenomena and then introducing gradually the details of atomic theory as required, developing the subject almost in the historical order of discovery. Instead, in this book a basic knowledge of atomic theory is assumed from the beginning, and the macroscopic phenomena are related to atomic properties throughout.

Conductors and insulators

For the purpose of electrostatic theory all substances can be divided into two fairly distinct classes: *conductors*, in which electrical charge can flow easily from one place to another; and *insulators*, in which it cannot. In the case of solids, all metals and a number of substances such as carbon are conductors, and their electrical properties can be explained by assuming that a number of electrons (roughly one per atom) are free to wander about the whole volume of the solid instead of being rigidly attached to one atom. Atoms which have lost one or more electrons in this way have a positive charge, and are called ions. They remain fixed in position in the solid lattice. In solid substances of the second class, insulators, each electron is firmly bound to the lattice of positive ions, and cannot move from point to point. Typical solid insulators are sulphur, polystyrene, and alumina.

When a substance has no net electrical charge, the total numbers of positive and negative charges within it must just be equal. Charge may be given to or removed from a substance, and a positively charged substance has an excess of positive ions, while a negatively charged substance has an excess of electrons. Since the electrons can move so much more easily in a conductor than the positive ions, a net positive charge is usually produced by the removal of electrons. In a charged conductor the electrons will move to positions of equilibrium under the influence of the forces of mutual repulsion between them, while in an insulator they are fixed in position and any initial distribution of charge will remain almost indefinitely. In a good conductor the movement of charge is almost instantaneous, while in a good insulator it is extremely slow. While there is no such thing as a perfect conductor or perfect insulator, such concepts are useful in developing electrostatic theory; metals form a good approximation to the former, and substances such as sulphur to the latter.

1.2. Coulomb's law and fundamental definitions

The force of attraction between charges of opposite sign, and of repulsion between charges of like sign, is found to be inversely proportional to the square of the distance between the charges (assuming them to be located at points), and proportional to the product of the magnitudes of the two charges. This law was discovered experimentally by Coulomb in 1785. In his apparatus the charges were carried on pith balls, and the force between them was measured with a torsion balance. The experiment was not very accurate, and a modern method of verifying the inverse square law with high precision will be given later (§ 1.3). From here on we shall assume it to be exact.

If the charges are q_1 and q_2, and r is the distance between them, then the force F on q_2 is along r. If the charges are of the same sign, the force is one of repulsion, whose magnitude is

$$F = C\frac{q_1 q_2}{r^2}.$$

The vector equation for the force is

$$\mathbf{F} = C\frac{q_1 q_2}{r^3}\mathbf{r}. \tag{1.1}$$

Here $\mathbf{F}$, $\mathbf{r}$ are counted as positive when directed from q_1 to q_2. Eqn (1.1) is the mathematical expression of Coulomb's law.

The units of $\mathbf{F}$ and $\mathbf{r}$ are those already familiar from mechanics; it remains to determine the units of C and q. Here there are two alternatives: either C is arbitrarily given some fixed numerical value, when eqn (1.1) may be used to determine the unit of charge, or the unit of charge may be taken as some arbitrary value, when the constant C is to be determined by experiment. The *Système International* (S.I.), which will be used throughout this book, makes use of the second method. The force F is in newtons, the distance r in metres, and an arbitrary unit, the coulomb, is used to measure the charges q_1 and q_2. The coulomb is directly related to the unit of current, the ampere, which is one coulomb per second; the ampere is defined by the forces acting between current-carrying conductors (see § 4.1). Eqn (1.1) for Coulomb's law is then analogous to that for gravitational attraction, except that it deals with electrical charges instead of masses; the constant of proportionality C must be determined by experiment. In the S.I., the constant C is written as $1/4\pi\epsilon_0$, the factor 4π being introduced here so that it occurs in formulae involving spherical rather than plane geometry. Eqn (1.1) therefore becomes

$$\mathbf{F}=\frac{1}{4\pi\epsilon_0}\frac{q_1q_2}{r^3}\mathbf{r}, \tag{1.2}$$

where **F** is in newtons (N), **r** in metres (m), and q in coulombs (C). The quantity ϵ_0 is known as the 'permittivity of a vacuum' (see § 1.5); its experimental value (see § 5.8) is $8{\cdot}85\times10^{-12}$ coulomb^2 newton^{-1} metre^{-2} ($\mathrm{C^2\,N^{-2}\,m^{-2}}$)—a more convenient name for this unit (see § 1.6) is farad metre^{-1} ($\mathrm{F\,m^{-1}}$).

Electric field and electric potential

The force which a charge q_2 experiences when in the neighbourhood of another charge q_1 may be ascribed to the presence of an 'electric field' of strength **E** produced by the charge q_1. Since the force on a charge q_2 is proportional to the magnitude of q_2, we define the field strength **E** by the equation

$$\mathbf{F}=\mathbf{E}q_2. \tag{1.3}$$

From this definition and Coulomb's law it follows that **E** does not depend on q_2, and is a vector quantity, like **F**. From eqn (1.2) we find that

$$\mathbf{E}=\frac{q_1}{4\pi\epsilon_0 r^3}\mathbf{r} \tag{1.4}$$

is the electric field due to the charge q_1.

If a unit positive charge is moved an infinitesimal distance d**s** in a field of strength **E**, then the work done by the field is $\mathbf{E}\cdot\mathrm{d}\mathbf{s}$, and the work done against the field is $-\mathbf{E}\cdot\mathrm{d}\mathbf{s}$. This follows from the fact that the force on unit charge is equal to the electric field strength **E**. The work done against the field in moving a unit positive charge from a point A to a point B will therefore be

$$V=-\int_{\mathrm{A}}^{\mathrm{B}}\mathbf{E}\cdot\mathrm{d}\mathbf{s}.$$

This is a scalar quantity known as the electric potential. If the field strength **E** is due to a single charge q at O, as in Fig. 1.1, then the force on unit charge at an arbitrary point P is along OP, and d**s** is the vector element $\mathrm{P_1P_2}$. Now $\mathbf{E}\cdot\mathrm{d}\mathbf{s}=E\cos\theta\,\mathrm{d}s=E\,\mathrm{d}r$, and hence

$$V_{\mathrm{B}}-V_{\mathrm{A}}=-\int_{\mathrm{A}}^{\mathrm{B}}E\,\mathrm{d}r=-\frac{q}{4\pi\epsilon_0}\int_{r_1}^{r_2}\frac{\mathrm{d}r}{r^2}=\frac{q}{4\pi\epsilon_0}\left(\frac{1}{r_2}-\frac{1}{r_1}\right).$$

Thus the difference of potential between A and B depends only on the positions of A and B, and is independent of the path taken between them.

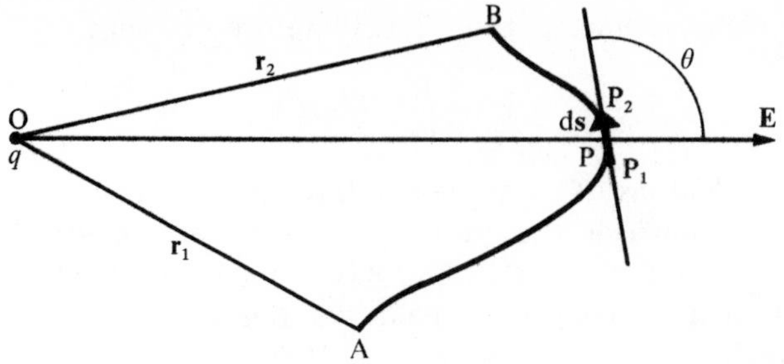

FIG. 1.1. Calculation of the potential difference between points A and B due to the field of a point charge q at O.

The potential at a point distance r from a charge q is the work done in bringing up unit charge to the point in question from a point at zero potential. By convention, the potential is taken as zero at an infinite distance from all charges, that is, $V=0$ for $r=\infty$. Therefore the potential at a point distance r from a charge q is

$$V = q/4\pi\epsilon_0 r. \tag{1.5}$$

The difference in potential dV between P_1 and P_2 (Fig. 1.1) distance $d\mathbf{s}$ apart is

$$dV = -\mathbf{E}.d\mathbf{s} = -(E_x\,dx + E_y\,dy + E_z\,dz).$$

Hence

$$\mathbf{E} = -\text{grad } V = -\nabla V, \tag{1.6}$$

where in Cartesian coordinates grad $V = \mathbf{i}\,\partial V/\partial x + \mathbf{j}\,\partial V/\partial y + \mathbf{k}\,\partial V/\partial z$ and $\mathbf{i}$, $\mathbf{j}$, $\mathbf{k}$ are unit vectors parallel to the x-, y-, and z-axes. The components of $\mathbf{E}$ along the three axes are

$$E_x = -\frac{\partial V}{\partial x}, \qquad E_y = -\frac{\partial V}{\partial y}, \qquad E_z = -\frac{\partial V}{\partial z}.$$

The negative sign shows that of itself a positive charge will move from a higher to a lower potential, and work must be done to move it in the opposite direction. (For vector relations, see Appendix A.)

The work done in taking a charge q round a closed path in an electrostatic field is zero. This can be seen from Fig. 1.2. The work done in taking the charge q round the path ABCA is

$$W = -q\oint \mathbf{E}.d\mathbf{s} = q(V_B - V_A) + q(V_C - V_B) + q(V_A - V_C) = 0,$$

and is independent of the path taken provided it begins and ends at the same point. Therefore the electric potential is a single-valued function of the space coordinates for any stationary distribution of electric charges; it has only one value at any point in the field.

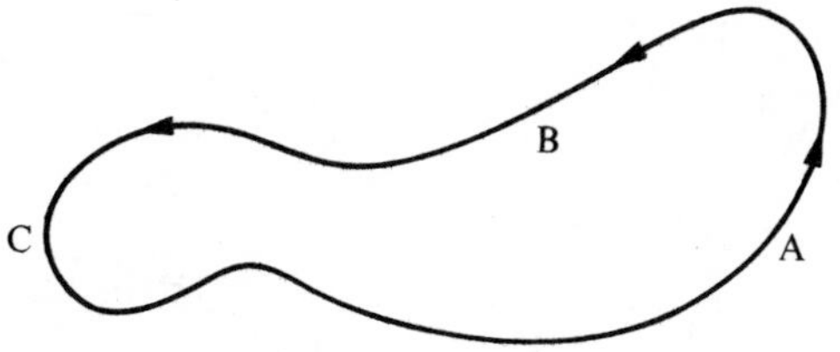

FIG. 1.2. The work done in taking an electric charge round a closed path in an electrostatic field is zero.

From the vector identity curl grad $V = 0$ or $\nabla \wedge (\nabla V) = 0$ (see Appendix § A.9, eqn (A.19)) it follows that curl $\mathbf{E} = 0$. Here curl $\mathbf{E}$ is a vector whose components are

$$\left(\frac{\partial E_z}{\partial y} - \frac{\partial E_y}{\partial z},\quad \frac{\partial E_x}{\partial z} - \frac{\partial E_z}{\partial x},\quad \frac{\partial E_y}{\partial x} - \frac{\partial E_x}{\partial y}\right).$$

These components can be shown to be zero by the use of elementary circuits (cf. Appendix § A.6), and the fact that no work is done in taking a charge round a closed path. The relation curl $\mathbf{E} = 0$ holds because $\mathbf{E}$ can be expressed as the gradient of a scalar potential: $\mathbf{E} = $ grad V. This is true in electrostatics, but does not hold when a changing magnetic flux threads the circuit (see § 5.1).

Since potential is a scalar quantity the potential at any point is simply the algebraic sum of the potentials due to each separate charge. On the other hand, $\mathbf{E}$ is a vector quantity, and the resultant field is the vector sum of the individual fields. Hence it is nearly always simpler to work in terms of potential rather than field; once the potential distribution is found, the field at any point is found by using eqn (1.6).

Units

From eqn (1.3) we obtain the unit of electric field strength. An electric field of 1 unit exerts a force of 1 newton on a charge of 1 coulomb. Electric field strengths can therefore be expressed in newton coulomb^{-1} (N C^{-1}).

The unit of potential is defined as follows: When 1 joule (J) of work is done in transferring a charge of 1 C from A to B, the potential difference between A and B is 1 volt (V). From eqn (1.6) E can be expressed in volt metre^{-1} (V m^{-1}), and this is the unit which is customarily used. It is easily verified that the two alternative units for E are equivalent.

Lines of force

A line drawn in such a way that it is parallel to the direction of the field at any point is called a line of force. Fig. 1.3 shows the lines of force for

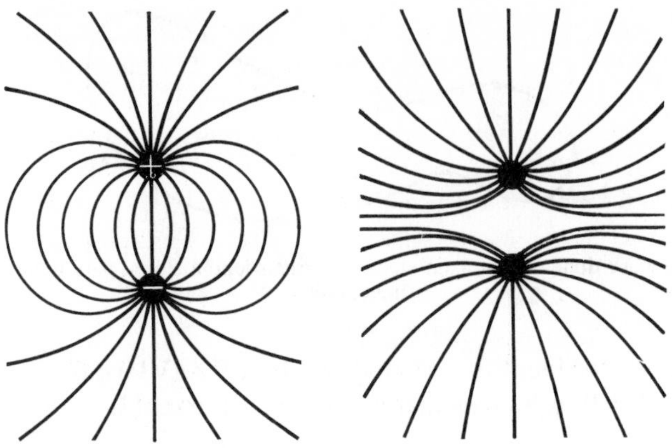

(a) (b)

FIG. 1.3. (a) Lines of force between equal charges of opposite sign. (b) Lines of force between equal charges of the same sign.

two equal charges. Lines of force do not intersect one another since the direction of the field cannot have two values at one point; they are continuous in a region containing no free charges, and they begin and end on free charges. The number of lines of force drawn through unit area normal to the direction of **E** is equal to the value of **E** at that point.

If a series of curves is drawn, each curve passing through points at a given potential, these equipotential curves cut the lines of force orthogonally. Equipotential curves are generally drawn for equal increments of potential; then **E** is greatest where the equipotentials are closest together.

1.3. Gauss's theorem

Let S be a closed surface surrounding a charge q, and let q be distant r from a small area d**S** on the surface S at A (Fig. 1.4(a)). The electric field strength **E** at A has the value

$$E = q/4\pi\epsilon_0 r^2.$$

The number of lines of force passing through an element of area dS is

$$\mathbf{E}.\mathrm{d}\mathbf{S} = E\cos\theta\,\mathrm{d}S = \frac{q\cos\theta\,\mathrm{d}S}{4\pi\epsilon_0 r^2},$$

where the outward normal to the surface element makes an angle θ with **E**. Now the solid angle subtended by dS at O is $\mathrm{d}\Omega = \cos\theta\,\mathrm{d}S/r^2$, and the value of $E\cos\theta\,\mathrm{d}S$ is therefore $q\,\mathrm{d}\Omega/4\pi\epsilon_0$. Hence the total number of lines of force passing through the whole surface is

$$\int E\cos\theta\,\mathrm{d}S = \frac{q}{4\pi\epsilon_0}\int \mathrm{d}\Omega = \frac{q}{\epsilon_0}, \qquad (1.7\text{a})$$

since a closed surface subtends a total solid angle of 4π at any point within the volume enclosed by the surface. If there are a number of charges $q_1, q_2, \ldots, q_n$ inside S, the resultant intensity of $\mathbf{E}$ at any point is the vector sum of the intensities due to each separate charge, and the integration of eqn (1.7a) may be carried out separately for each charge. In this way it is found that $\int E \cos\theta\, dS = \sum q/\epsilon_0$. On the other hand, the contribution of any charge outside S is zero, as may be seen from Fig. 1.4(b), since in this case

$$\int E \cos\theta\, dS = \frac{q}{4\pi\epsilon_0}\left(\int \frac{dS_1 \cos\theta_1}{r_1^2} - \int \frac{dS_2 \cos\theta_2}{r_2^2}\right) = 0.$$

We may summarize these results in the form

$$\int E \cos\theta\, dS = \int \mathbf{E}.d\mathbf{S} = \sum q/\epsilon_0, \tag{1.7b}$$

where the summation is to be taken only over the charges lying within the closed surface S. This is known as Gauss's theorem. We see that the integral of the normal component of $\mathbf{E}$ over the surface is equal to the total charge enclosed, divided by ϵ_0, irrespective of the way in which the charge is distributed.

If there exists throughout a volume enclosed by a surface S a charge distribution of varying density ρ_e, we have

$$\frac{1}{\epsilon_0}\int \rho_e\, d\tau = \int \mathbf{E}.d\mathbf{S} = \int \operatorname{div} \mathbf{E}\, d\tau, \tag{1.7c}$$

where $d\tau$ is an element of volume. The two volume integrals must be equal whatever the volume over which the integration takes place, and it therefore follows that the integrands themselves must be equal. Hence

$$\operatorname{div} \mathbf{E} = \frac{\partial E_x}{\partial x} + \frac{\partial E_y}{\partial y} + \frac{\partial E_z}{\partial z} = \frac{\rho_e}{\epsilon_0}, \tag{1.8}$$

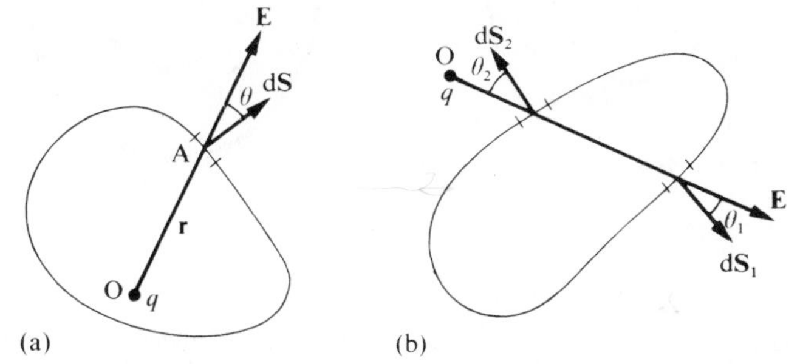

FIG. 1.4. Illustrating Gauss's theorem.

and this is the expression in differential form of Gauss's theorem. The transformation from a surface to a volume integral used above is due to Gauss (see Appendix § A.7).

One of the consequences of Gauss's theorem is that there can be no field within a conductor, nor can there be any volume distribution of charge within it. For, if there were such a charge distribution, a field would exist within the conductor, which would act on the charges. Since they are free to move in a conductor, they cannot then be in a state of equilibrium. Thus no electrostatic field can exist within the body of a conductor, and all parts of it must be at the same potential. If the conductor has a total charge different from zero, then this charge must reside entirely in a thin layer on the outer surface (see Fig. 1.5).

The fact that there can be no electric field within the body of a conductor has an important consequence in the case of a hollow closed conductor. If we apply Gauss's theorem to a surface S lying entirely within the conducting substance, as in Fig. 1.5(b), then $\int E \cos \theta \, dS = 0$, since $\mathbf{E} = 0$ everywhere over the surface. Hence the net charge inside the surface must be zero. This can be realized in two ways: (1) if there is a total charge q in the hollow space within the conductor, the lines of force from the charges comprising q must end on a distribution of charge on the inner surface of the conductor, and the total charge in this layer must be equal to $-q$; (2) if there is no charge in the hollow space, then there can be no field in this space. This last result is important, for many proofs

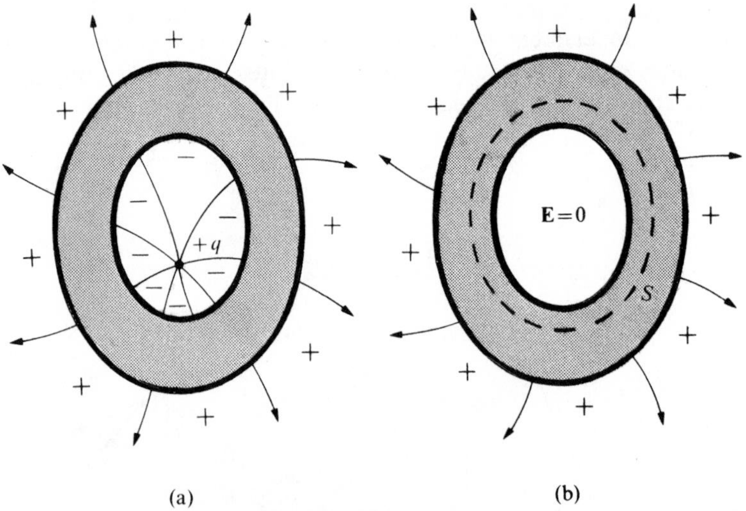

FIG. 1.5. Distribution of charge on a hollow conductor. (a) A hollow conductor with a point charge $+q$ inside, and induced charges $-q$ and $+q$ on the inside and outside surfaces. (b) The same conductor with no charge inside, and total charge $+q$ on the surface.

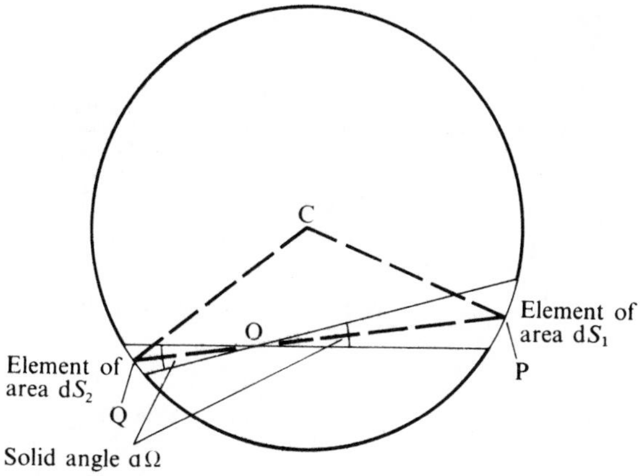

FIG. 1.6. The field inside a spherical conductor at a point O. Distances $OP = r_1$, $OQ = r_2$. From the geometry of the circle, $\angle OPC = \angle OQC = \theta$. Hence

$$d\Omega = \frac{dS_1 \cos\theta}{r_1^2} = \frac{dS_2 \cos\theta}{r_2^2}.$$

of the inverse square law (see below) depend on it. It means that if we put a closed conductor into a field, a charge distribution on the outer surface will be set up such that the field inside remains exactly zero.

Experimental proof of the inverse square law

Coulomb's attempts to check the inverse square law using a torsion balance were not capable of great accuracy, and most subsequent attempts have relied on the fact that the field inside a closed conductor is only zero if the inverse square law holds. We shall prove this for the special case of a spherical conductor.

In Fig. 1.6 let an elementary cone of solid angle $d\Omega$ be drawn with vertex at the point O within the sphere. This cone intersects the surface of the sphere in the elementary areas dS_1, dS_2 at distances r_1, r_2 from O. If the charge on the sphere has a uniform density σ_e per unit area, then the field at O due to the elements dS_1 and dS_2 will be

$$dE = \frac{\sigma_e}{4\pi\epsilon_0}\left(\frac{dS_1}{r_1^n} - \frac{dS_2}{r_2^n}\right)$$

assuming that the field of a point charge falls off as r^{-n}. But the solid angle $d\Omega = dS_1 \cos\theta / r_1^2 = dS_2 \cos\theta / r_2^2$, and we can therefore write

$$dE = \frac{\sigma_e \, d\Omega}{4\pi\epsilon_0 \cos\theta}\left(\frac{1}{r_1^{n-2}} - \frac{1}{r_2^{n-2}}\right).$$

This gives a resultant field towards the nearer element if $n < 2$, and towards the further element if $n > 2$. Clearly, the whole surface of the

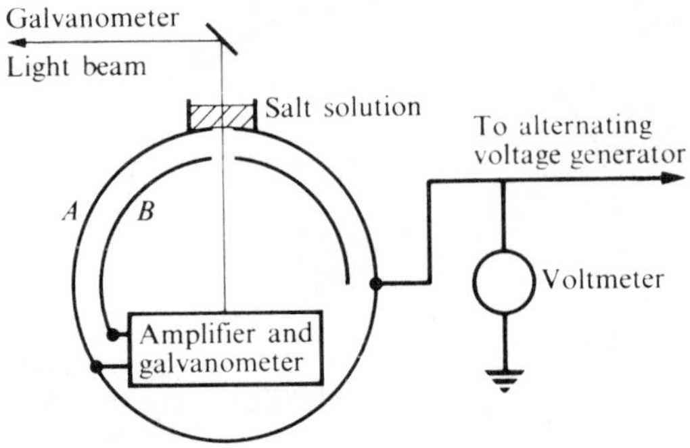

FIG. 1.7. Apparatus of Plimpton and Lawton for verifying the inverse square law.

sphere can be divided into elementary areas in this way, and the vector resultant of the fields at O will not be zero unless all the individual dE are zero, since there will be a resultant towards the nearer portion of the spherical surface if $n<2$, and vice versa. Thus, if it is shown experimentally that there is no field inside a charged sphere, it follows that the power of n in the inverse power law must be exactly 2.

This result was used to test the validity of the inverse square law by Cavendish and, later, by Maxwell. Maxwell had a spherical air condenser consisting of two concentric insulated spherical shells. The outer sphere had a small hole in it so that the inner one could be tested for charge by inserting through the hole an electrode connected to an electrometer. The two spheres were initially connected by a wire and charged to a high potential, and then insulated from one another. After earthing the outer sphere, the inner one was tested and found to have no charge. In this way Maxwell found that the value of n did not differ from 2 by more than one part in 21 600.

The experiment has been repeated by Plimpton and Lawton (1936) with a more sensitive detector, the electrometer being replaced by an amplifier and galvanometer. The detecting apparatus was placed inside the sphere A (see Fig. 1.7), and its conducting case, together with the hemisphere B, formed the inner conductor (it was shown that this does not necessarily have to be spherical in shape). The galvanometer deflection was observed through a small hole in the sphere A, covered by a wire grid immersed in salt solution so that A was effectively a closed conductor. It was found that this was essential for the field inside to be rigorously zero when $n=2$. The galvanometer was undamped so that it could swing at its natural period (0·5 s), and an alternating voltage of 3 kV, whose frequency was adjusted to synchronism with the galvanometer, was applied to the outer sphere. No potential difference between the inner and outer spheres was found, though a voltage of 10^{-6} V could have been detected. It was found that this was only true if the hole in A was covered with the salt solution. The galvanometer deflection observed if this was removed was used to check that the frequency of the alternating voltage was equal to the natural period of the galvanometer. From this experiment, Plimpton and Lawton concluded that n did not differ from 2 by more than one part in 10^9.

1.4. Electric dipoles

An electric dipole consists of two charges equal in magnitude but of opposite sign, separated by a small distance. Fig. 1.8 shows such a dipole with charges $+q$ and $-q$ separated by a distance a. Then the potential at a point P is

$$V=\frac{q}{4\pi\epsilon_0}\left(\frac{1}{r_2}-\frac{1}{r_1}\right). \tag{1.9}$$

If $a \ll r$, the quantities $1/r_1$ and $1/r_2$ may be expanded in a power series. Since $r_2^2=r^2-ra\cos\theta+\frac{1}{4}a^2$, we have

$$\frac{1}{r_2}=\frac{1}{r}+\frac{a}{2r^2}\cos\theta+\frac{a^2}{8r^3}(3\cos^2\theta-1)+\ldots,$$

and the expansion of $1/r_1$ differs only in that the signs are reversed of all terms with odd powers of $\cos\theta$. Hence

$$V=\frac{qa\cos\theta}{4\pi\epsilon_0 r^2}=\frac{p\cos\theta}{4\pi\epsilon_0 r^2} \tag{1.10a}$$

provided that the next term, which is smaller by a factor $(a/r)^2$, and all higher terms can be neglected. This approximation is justified if we let

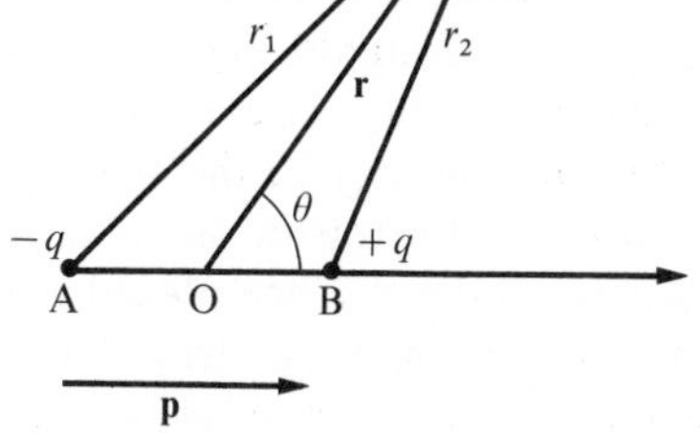

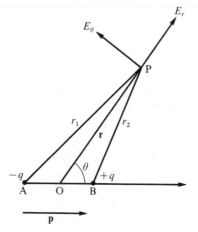

FIG. 1.8. The electric field due to an electric dipole.

$a/r \to 0$, keeping the product $qa = p$ constant; this approximation is that of a 'point dipole', and the electric dipole moment $\mathbf{p}$ is a vector whose direction is from the negative to the positive charge.

If $\mathbf{r}$ is a vector drawn from O as origin to P, then $p \cos\theta = \mathbf{p}.\mathbf{r}/r$, and

$$V = \frac{\mathbf{p}.\mathbf{r}}{4\pi\epsilon_0 r^3}. \tag{1.10b}$$

If O is a fixed point and P is regarded as a variable point, then

$$\mathbf{r}/r^3 = -\text{grad}(1/r),$$

so that the formula for the potential may also be written as

$$V = -\frac{1}{4\pi\epsilon_0}\{\mathbf{p}.\text{grad}_P(1/r)\}. \tag{1.11a}$$

Here the subscript P is added to the operator grad to denote that differentiation is with respect to P as the variable point. If we regard P as fixed, and move from A to B, then eqn (1.9) could have been written in the form

$$V = \frac{1}{4\pi\epsilon_0}\{\mathbf{p}.\text{grad}_O(1/r)\}. \tag{1.11b}$$

Here the subscript O denotes that O is now the variable point, and there is a change of sign from eqn (1.11a) because $\mathbf{r}$ is now measured in the opposite direction, from P to O instead of from O to P.

The components of the electric field strength at P can be calculated by differentiating the potential given by eqn (1.10a). The radial and azimuthal components are

$$E_r = -\left(\frac{\partial V}{\partial r}\right)_\theta = \frac{1}{4\pi\epsilon_0}\left(\frac{2p\cos\theta}{r^3}\right),$$

$$E_\theta = -\frac{1}{r}\left(\frac{\partial V}{\partial \theta}\right)_r = \frac{1}{4\pi\epsilon_0}\left(\frac{p\sin\theta}{r^3}\right). \tag{1.12}$$

These equations show that the electric field strength of a dipole falls off as $1/r^3$, and its potential as $1/r^2$, whereas the corresponding laws for a single pole are $1/r^2$ and $1/r$. The significance of this difference is that at large distances the fields of the two equal and opposite charges which comprise a dipole cancel one another in the first approximation (that is, terms varying as $1/r^2$ vanish), but terms in the next order ($1/r^3$ in the field) remain. Similarly, if two equal dipoles are placed end to end, giving a set of charges as in Fig. 1.9 (known as a quadrupole), their fields annul one another at large distances, and the potential of a quadrupole falls off as $1/r^3$ (see Problem 1.1), and its field as $1/r^4$.

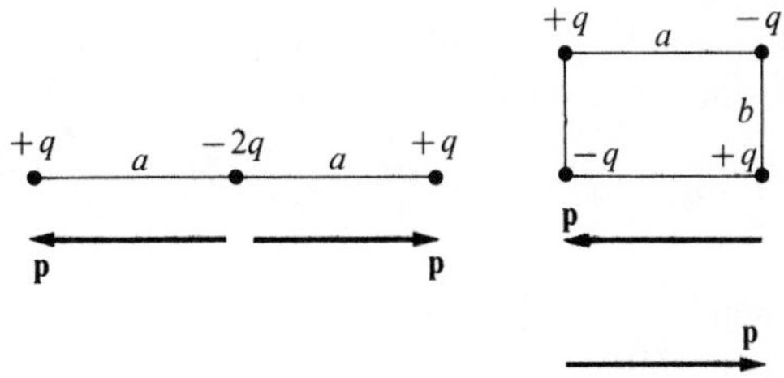

FIG. 1.9. Two quadrupole moments represented as an assembly of charges or a pair of dipoles; note that the net charge and net dipole moment are zero in each case.

If a dipole consisting of two charges $-q$ and $+q$ a distance $\mathbf{a}$ apart is placed in a uniform field, its potential energy U_P is (see Fig. 1.10)

$$U_P = q(V_B - V_A) = -qa \cos\theta\, E = -q\mathbf{a}.\mathbf{E} = -\mathbf{p}.\mathbf{E}, \tag{1.13}$$

where $\mathbf{p}$ is the dipole moment. This shows that the energy depends only on the angle which $\mathbf{p}$ makes with $\mathbf{E}$, and not on the position of the dipole. Hence there is no translational force acting on the dipole, but there is a couple

$$\mathbf{T} = -(\mathrm{d}U_P/\mathrm{d}\theta) = pE\sin\theta = \mathbf{p}\wedge\mathbf{E}, \tag{1.14}$$

which tends to turn the dipole into a position parallel to the field.

If the dipole is placed in a non-uniform field, a translational force is exerted on it, and we shall derive an expression for the x-component of this force. If E_x is the value of the x-component of the field at A, its value at B may be written as

$$E_x' = E_x + \left(\frac{\partial E_x}{\partial x}\right)a_x + \left(\frac{\partial E_x}{\partial y}\right)a_y + \left(\frac{\partial E_x}{\partial z}\right)a_z,$$

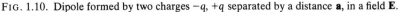

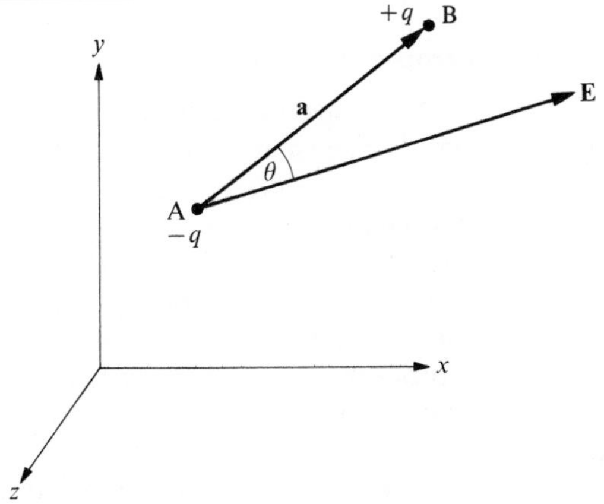

FIG. 1.10. Dipole formed by two charges $-q$, $+q$ separated by a distance $\mathbf{a}$, in a field $\mathbf{E}$.

where a_x, a_y, a_z are the components of **a** along the three axes. The x-component of the force on the dipole is therefore

$$F_x = -qE_x + qE'_x = qa_x\left(\frac{\partial E_x}{\partial x}\right) + qa_y\left(\frac{\partial E_x}{\partial y}\right) + qa_z\left(\frac{\partial E_x}{\partial z}\right)$$

$$= p_x\left(\frac{\partial E_x}{\partial x}\right) + p_y\left(\frac{\partial E_x}{\partial y}\right) + p_z\left(\frac{\partial E_x}{\partial z}\right). \qquad (1.15a)$$

Now

$$\frac{\partial E_x}{\partial y} = \frac{\partial}{\partial y}\left(-\frac{\partial V}{\partial x}\right) = \frac{\partial}{\partial x}\left(-\frac{\partial V}{\partial y}\right) = \frac{\partial E_y}{\partial x},$$

since the order of differentiation is immaterial, and similarly

$$\frac{\partial E_x}{\partial z} = \frac{\partial E_z}{\partial x}.$$

Hence the force component may also be written as

$$F_x = p_x\frac{\partial E_x}{\partial x} + p_y\frac{\partial E_y}{\partial x} + p_z\frac{\partial E_z}{\partial x}, \qquad (1.15b)$$

with similar expressions for the other components. This result may also be derived directly from the potential energy U_P of a dipole (eqn (1.13)) differentiation of which gives

$$F_x = -\left(\frac{\partial U_P}{\partial x}\right) = \frac{\partial}{\partial x}(p_xE_x + p_yE_y + p_zE_z),$$

where the components of **E** (but not those of **p**) are functions of position. The components of **F** are those of a vector which (see eqn (A.10) of Appendix A) may be written as

$$\mathbf{F} = (\mathbf{p}\cdot\text{grad})\mathbf{E}. \qquad (1.15c)$$

1.5. The theory of isotropic dielectrics

Faraday found that if a slab of insulating material was inserted between two metal plates across which a constant voltage was applied by means of a battery, the charge on the plates increased. If the insulator entirely filled the intervening space, the charge increased by a factor ϵ_r where ϵ_r is called the relative permittivity or dielectric constant of the insulator. It varies between 1 and 10 for most solid substances, being 1 for vacuum and 1·00057 for air at room temperature and pressure. To find how ϵ_r is related to the intrinsic properties of the material, or dielectric, it is necessary to consider what happens inside a dielectric when an electric field is applied to it.

Dielectric substances are insulators, and therefore do not contain free electrons. Each electron is bound to the ionic lattice by the electrostatic attraction between the negative electronic charge and the positive charges on the nuclei. In the absence of any external field, the electrons are distributed symmetrically with respect to the nuclei, but when a field is applied, the electrons are displaced in the direction opposite to that of the field, while the more massive nuclei are slightly displaced in the direction of the field. (The centre of gravity remains fixed, since there is no translational force on the system as a whole.) Each ion thus acquires an

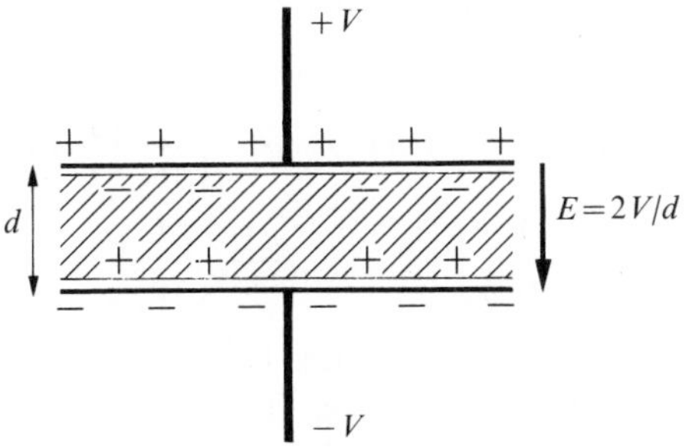

FIG. 1.11. Effect of a dielectric in increasing the capacitance between two parallel plates.

electric dipole moment which is parallel to and in the same direction as the applied field. If a slab of dielectric is placed between parallel metal plates as in Faraday's experiment and the voltage across the plates is constant, there will be an induced negative charge on the dielectric surface near the positive plate (Fig. 1.11) and a similar positive charge on the surface near the negative plate. There will be no resultant charge density within a simple uniform dielectric, as all the individual dipoles are aligned parallel to the field and hence each negative charge of one dipole is next to the positive charge of the next dipole. The surface charges on the dielectric will induce charges of opposite sign on the plates, and the charge on the plates is increased when the dielectric is inserted, if the voltage is kept constant—as Faraday found in his experiments.

The action of the electric field in giving each atom of a dielectric an induced dipole moment is termed 'polarization'. (We shall see in Chapter 10 that electric polarization can arise in other ways in substances containing molecules.) The macroscopic polarization is defined as the electric dipole moment per unit volume. It is found by taking the vector sum of the microscopic dipole moments in unit volume, and is therefore itself a vector, denoted by the symbol **P**.

The resultant moment for an element of volume $d\tau$ is $\mathbf{P}\,d\tau$, and the potential of such an element a distance r away is, by eqn (1.11b),

$$dV = \frac{1}{4\pi\epsilon_0}(\mathbf{P}\,d\tau)\cdot\text{grad}(1/r),$$

where the differentiation is with respect to the coordinates of the volume element containing the dipoles. The potential due to a finite volume of dielectric is then

$$V = \frac{1}{4\pi\epsilon_0}\int\{\mathbf{P}\cdot\mathrm{grad}(1/r)\}\,\mathrm{d}\tau.$$

But, from eqn (A.12) in Appendix A,

$$\mathrm{div}(\mathbf{P}/r) = \frac{1}{r}\,\mathrm{div}\,\mathbf{P} + \mathbf{P}\cdot\mathrm{grad}(1/r).$$

Hence

$$V = \frac{1}{4\pi\epsilon_0}\int\mathrm{div}(\mathbf{P}/r)\,\mathrm{d}\tau - \frac{1}{4\pi\epsilon_0}\int\frac{1}{r}\,\mathrm{div}\,\mathbf{P}\,\mathrm{d}\tau$$

$$= \frac{1}{4\pi\epsilon_0}\int\frac{1}{r}\,\mathbf{P}\cdot\mathrm{d}\mathbf{S} - \frac{1}{4\pi\epsilon_0}\int\frac{1}{r}\,\mathrm{div}\,\mathbf{P}\,\mathrm{d}\tau, \qquad (1.16)$$

where Gauss's theorem of divergence has been used in the transformation from a volume to a surface integral. These two terms in eqn (1.16) show that the resultant potential can be attributed to an apparent surface charge of density $P\cos\theta$, where θ is the angle which $\mathbf{P}$ makes with the normal to the surface of the dielectric, and an apparent volume distribution of charge whose density is

$$-\mathrm{div}\,\mathbf{P} = -\left(\frac{\partial P_x}{\partial x} + \frac{\partial P_y}{\partial y} + \frac{\partial P_z}{\partial z}\right). \qquad (1.17)$$

These apparent charge densities arise from the displacement, under the action of the electric field, of bound charges from their normal equilibrium positions in zero electric field. They are often referred to as the 'polarization charge' or the 'bound charge' to distinguish them from 'free charge' which is not bound to any particular point in the dielectric.

Gauss's theorem in dielectrics

In § 1.3 it was shown by Gauss's theorem that the integral $\int\mathbf{E}\cdot\mathrm{d}\mathbf{S}$ of the normal component of $\mathbf{E}$ over any closed surface is equal to the total charge within the surface, divided by ϵ_0. If the surface is within a dielectric medium, the total charge must include both the free charges and the polarization charges. The volume charge density is thus $\rho_e - (\mathrm{div}\,\mathbf{P})$, so that

$$\int\mathbf{E}\cdot\mathrm{d}\mathbf{S} = \int\mathrm{div}\,\mathbf{E}\,\mathrm{d}\tau = \frac{1}{\epsilon_0}\int(\rho_e - \mathrm{div}\,\mathbf{P})\,\mathrm{d}\tau$$

or

$$\int\mathrm{div}(\epsilon_0\mathbf{E}+\mathbf{P})\,\mathrm{d}\tau = \int\rho_e\,\mathrm{d}\tau.$$

Comparison of this with eqn (1.7c) shows that, in effect, $\epsilon_0\mathbf{E}$ has been replaced by $(\epsilon_0\mathbf{E}+\mathbf{P})$. We may define a new vector **D**, such that

$$\mathbf{D}=\epsilon_0\mathbf{E}+\mathbf{P}. \tag{1.18}$$

D is known as the 'electric displacement', and eqn (1.18) is valid even in an anisotropic medium, where **P** is not necessarily parallel to **E**. We may now write Gauss's theorem, in a dielectric medium, in the form

$$\int \mathbf{D}.\mathrm{d}\mathbf{S}=\int \operatorname{div}\mathbf{D}\,\mathrm{d}\tau=\int \rho_e\,\mathrm{d}\tau. \tag{1.19}$$

Since the volume integrals must be equal over any arbitrary volume, it follows that their integrands must be equal, that is,

$$\operatorname{div}\mathbf{D}=\rho_e, \tag{1.20}$$

which is the differential form of Gauss's theorem. It is easy to see that these equations reduce to those of § 1.3 *in vacuo*, where $\mathbf{P}=0$.

In a simple isotropic dielectric, **P** is parallel to **E**, and hence so also is **D**. We may write $\mathbf{P}=\chi_e\epsilon_0\mathbf{E}$, where χ_e is known as the 'electric susceptibility'. Then $\mathbf{D}=(1+\chi_e)\epsilon_0\mathbf{E}$, and writing

$$\epsilon=(1+\chi_e)\epsilon_0=\epsilon_r\epsilon_0, \tag{1.21}$$

we have

$$\mathbf{D}=\epsilon\mathbf{E}, \tag{1.22}$$

where ϵ is known as the 'permittivity'. In free space $\epsilon=\epsilon_0$, and thus ϵ_0 is the 'permittivity of a vacuum'. When a medium is present $\mathbf{D}/\mathbf{E}$ is increased by the ratio $\epsilon_r=\epsilon/\epsilon_0$, and the factor ϵ_r is known as the 'relative permittivity' of the medium.

In crystalline solids where the crystal structure has less than cubic symmetry, the polarization vector **P** is not necessarily parallel to the electric field strength vector **E**, and it follows from eqn (1.18) that the same is true for the displacement vector **D**. Under these conditions the relation between **P** and **E** is given by a set of equations of the type

$$P_x/\epsilon_0=\chi_{xx}E_x+\chi_{xy}E_y+\chi_{xz}E_z,$$
$$P_y/\epsilon_0=\chi_{yx}E_x+\chi_{yy}E_y+\chi_{yz}E_z,$$
$$P_z/\epsilon_0=\chi_{zx}E_x+\chi_{zy}E_y+\chi_{zz}E_z.$$

An analogous set of equations relates the components of **D** to those of **E**, and both χ_e and ϵ_r are 'tensor quantities'. We shall restrict our discussion to isotropic media, where $\chi_{xx}=\chi_{yy}=\chi_{zz}$ and the other elements, such as χ_{xy}, are zero, so that both χ_e and ϵ_r reduce to simple scalar quantities. Typical of such media are cubic crystals, non-crystalline solids, and fluids.

Our second restriction is to substances in which **P** is linearly proportional to **E**, so that the electric susceptibility χ_e is independent of field strength. In practice the largest static or slowly varying electric fields that can be applied are small compared with the electric fields that bind the electron to the atom, and the approximation that χ_e and ϵ_r are independent of field strength is a good one (as is implied in the older name of 'dielectric constant' for ϵ_r). At optical frequencies, in the intense electric fields that exist in light beams generated by high-power lasers, it is necessary to write the polarizibility as a power series

$$\chi_e = \chi_e^{(0)} + \chi_e^{(1)}E + \chi_e^{(2)}E^2 \ldots,$$

where the terms beyond $\chi_e^{(0)}$ represent non-linear effects which may be important in applications such as modulation or frequency-changing in optical beams, in the same way as non-linear effects in electric conduction are important in radio-electronics.

If we have a single point charge in a uniform dielectric of permittivity ϵ, we may apply Gauss's theorem over a sphere of radius r whose centre is at q. The surface integral reduces to $4\pi r^2 D = 4\pi r^2 \epsilon E = q$, so that

$$\epsilon \mathbf{E} = \mathbf{D} = q\mathbf{r}/4\pi r^3. \tag{1.23}$$

It follows that the force between two charges q_1, q_2 a distance $\mathbf{r}$ apart is

$$F = \frac{q_1 q_2}{4\pi\epsilon r^3}\mathbf{r}, \tag{1.24}$$

and the potential a distance r from a point charge q is

$$V = q/4\pi\epsilon r. \tag{1.25}$$

In a dielectric where χ_e is uniform, isotropic, and independent of field strength there is no volume distribution of charge, even when $\mathbf{E}$ is non-uniform, as in eqn (1.23). This follows from the fact that $\operatorname{div} \mathbf{E} = 0$, which leads to the volume distribution of charge in eqn (1.17) reducing to $-\operatorname{div} \mathbf{P} = -\chi_e \epsilon_0 \operatorname{div} \mathbf{E} = 0$. At a surface where the relative permittivity changes, there will be a surface distribution of apparent charge.

Some properties of **D** *and* **E**

It is important to distinguish clearly between the two vector quantities electric field strength **E** and electric displacement **D**. **E** is defined as the force acting on unit charge, irrespective of whether a dielectric medium is present or not. This definition is expressed in eqn (1.3). The displacement **D** is defined by eqn (1.18). The quantity $\mathbf{D}.\mathbf{dS}$ is sometimes known as the electric flux through the element of area $\mathbf{dS}$. From (1.19) the total flux is q through an area surrounding a charge q, and this flux is unaltered by the presence of a dielectric medium. The unit of flux is the coulomb, and the unit of D is the coulomb metre^{-2} (C m^{-2}).

Since **D** is a vector we may draw lines of displacement analogous to lines of force, such that the number passing through unit area is equal to

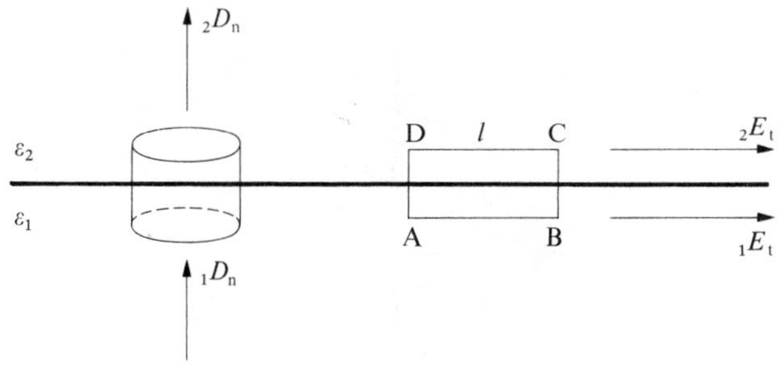

FIG. 1.12. Boundary conditions at the surface between two dielectrics.

the displacement. Also by Gauss's theorem lines of displacement are continuous in a space containing no free charges; they begin and end only on free charges. At the boundary of two dielectrics ϵ_1 and ϵ_2, if no free charge resides there, lines of **D** are continuous but lines of **E** are not, because lines of force end on both free and polarization charges, whereas lines of **D** end only on free charges. There is a polarization charge on the surface separating two dielectrics, since the induced moment per unit volume is different in the two media. Lines of **E** begin and end on this surface charge, but not lines of **D**. The rules governing the behaviour of **E** and **D** at the surface of a dielectric, or the boundary between two dielectrics, are embodied in two 'boundary conditions', which will now be derived.

To find the boundary condition for **D**, we apply Gauss's theorem to a small cylinder which intersects the boundary, as in Fig. 1.12, and whose axis is normal to the boundary. If the height of the cylinder is very small compared with its cross-sectional area, the only contribution to $\int \mathbf{D} \cdot \mathrm{d}\mathbf{S}$ over its surface will come from the components of **D** normal to the boundary. Since there is no free charge on the boundary, $\int \mathbf{D} \cdot \mathrm{d}\mathbf{S} = 0$, and hence

$$_2D_n = {_1D_n}, \tag{1.26}$$

where the symbols refer to the normal components of **D** on the two sides of the boundary.

The boundary condition for **E** is found by considering the work done in taking unit charge round a small rectangular circuit such as ABCDA in Fig. 1.12. If the sides BC, AD are very small compared with AB, CD, then the work done will be $l({_1E_t} - {_2E_t}) = 0$. Hence

$$_2E_t = {_1E_t}, \tag{1.27}$$

where the symbols refer to the tangential components of **E** on either side of the boundary. Thus eqns (1.26) and (1.27) are the two fundamental boundary conditions. It follows from them that lines of **D** will in general

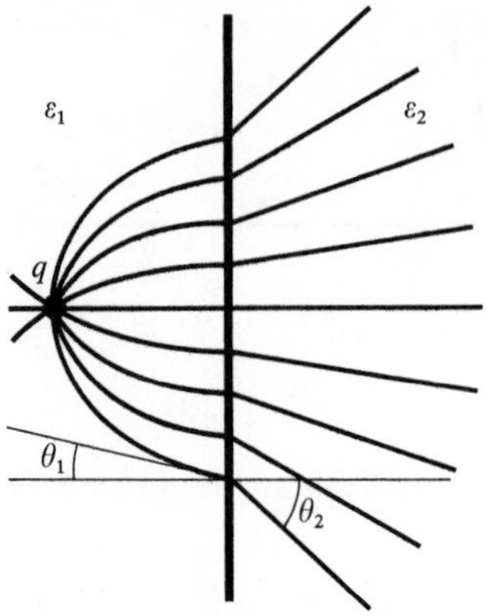

FIG. 1.13. Refraction of lines of displacement at the boundary between two dielectrics ($\epsilon_2 > \epsilon_1$). From eqns (1.26) and (1.27) we have

$$D_1 \cos\theta_1 = D_2 \cos\theta_2,$$

$$E_1 \sin\theta_1 = (D_1/\epsilon_1)\sin\theta_1 = E_2 \sin\theta_2 = (D_2/\epsilon_2)\sin\theta_2.$$

Hence
$$\epsilon_1 \cot\theta_1 = \epsilon_2 \cot\theta_2.$$

be refracted at the boundary between two dielectrics. A typical example is shown in Fig. 1.13.

Similar boundary conditions may be applied at the surface of a conductor. Since there can be no field inside the conductor, the tangential component of **E** just outside the conductor must also be zero, and any field at the surface must be normal to the surface. If we apply Gauss's theorem to an elementary cylinder intersecting the surface, as in Fig. 1.12, we have ${}_2D_n\, dS = \sigma_e\, dS$ (since ${}_1D_n = 0$ within the conductor), where σ_e is the charge density on the conducting surface. Since ${}_2\mathbf{D}$ must be normal to the surface, we have ${}_2\mathbf{D} = {}_2D_n$, and hence (dropping the subscripts)

$$D = \epsilon E = \sigma_e \tag{1.28}$$

at the surface of the conductor immersed in a medium of permittivity ϵ.

1.6. Capacitors and systems of conductors

If a charge Q is given to an isolated conductor its voltage is increased by an amount V. For a given conductor the ratio Q/V is independent of Q and depends only on the size and shape of the conductor. The ratio Q/V is called the capacitance of the conductor, and is denoted by C. If a second conductor which is earthed is brought close to the first one, a charge of opposite sign is induced on it. The potential difference falls, and since Q is constant on an isolated conductor, the capacitance has increased. The two conductors together form a capacitor, and the capacitance of the capacitor is defined as the ratio of the charge on either conductor to the potential difference between them. A capacitor is an instrument for storing charge, and a capacitor of large capacitance can store a correspondingly large quantity of charge for a given potential difference between the plates. The capacitance depends on the geometry of the conductors and the permittivity of the medium separating them.

In general, calculation of the capacitance of a conductor or a capacitor is difficult unless simple geometrical shapes are involved. The principle of the calculation may be illustrated by the case of an isolated sphere, of radius a, in an infinite dielectric. Suppose this carries a charge Q. Then by applying Gauss's theorem over a spherical surface of radius r, concentric with the sphere, we have

$$4\pi r^2 D = 4\pi r^2 \epsilon E = Q,$$

since, by symmetry, **D** and **E** are constant over the spherical surface and everywhere normal to it. Hence

$$E = \frac{Q}{4\pi\epsilon r^2},$$

which is the same as eqn (1.23) for a point charge Q. The potential of the sphere is

$$V = -\int_{\infty}^{a} E\,\mathrm{d}r = \frac{Q}{4\pi\epsilon a}.$$

Hence

$$C = Q/V = 4\pi\epsilon a. \tag{1.29}$$

A more practical example is provided by a pair of circular coaxial cylinders, as used in a 'coaxial cable'. In Fig. 1.14 the radii of the cylinders are a, b; they are assumed to be infinitely long, the space between them being completely filled by a uniform dielectric. The lines of

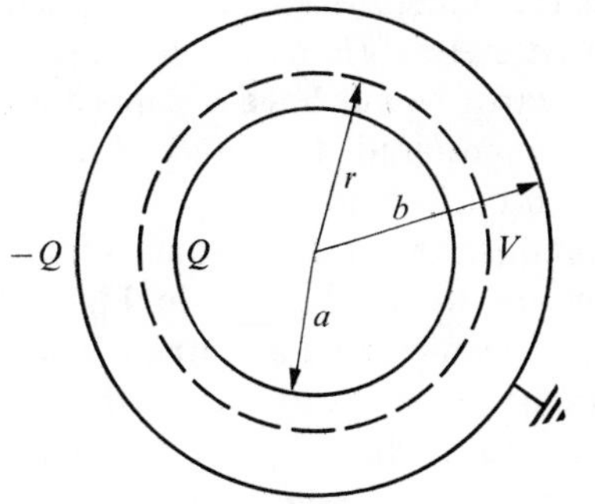

FIG. 1.14. Cross-section of two coaxial cylinders.

D and **E** are radial (normal to the equipotential surfaces), and we may apply Gauss's theorem to a coaxial cylindrical surface of radius r and length l. If $+Q$ is the charge per unit length on the inner cylinder,

$$2\pi r l D = 2\pi r l \epsilon E = +Ql,$$

from which $E = Q/2\pi r\epsilon$. The potential difference between the cylinders is

$$V_b - V_a = -\int_a^b E\,\mathrm{d}r = -(Q/2\pi\epsilon)\int_a^b r^{-1}\,\mathrm{d}r = -(Q/2\pi\epsilon)\ln(b/a),$$

and the capacitance of a length l is therefore

$$C = \frac{Ql}{V_a - V_b} = \frac{2\pi l\epsilon}{\ln(b/a)}. \qquad (1.30)$$

Another simple type of capacitor is formed by two plane parallel plates of area S and separation t. If the lateral dimensions of the plates are large compared with their separation (or if the plates are surrounded by 'guard rings' at the same potential), then the field between them is uniform and normal to the plates, being given by eqn (1.28) with $\sigma_e = Q/S$. Since the field is uniform, the potential difference between the plates is simply $V = Et$, and the capacitance is thus

$$C = Q/V = \epsilon S/t. \qquad (1.31)$$

This equation (and also eqns (1.29) and (1.30)) shows that the capacitance increases by a factor ϵ_r if the space between the plates is filled with a medium of relative permittivity $\epsilon_r = \epsilon/\epsilon_0$. This agrees with the original definition of dielectric constant by Faraday, mentioned at the beginning of § 1.5.

The unit of capacitance is called the farad (F); the plates of a capacitor of 1 F carry a charge of 1 C if their potential difference is 1 V. Reference to eqns (1.29)–(1.31) above, or more fundamentally, to eqn (1.25), shows that a capacitance has the dimensions of ϵ_0 multiplied by a length; hence the unit of ϵ_0 is the farad metre^{-1} (F m^{-1}). The farad is a very large unit (a sphere the size of the earth would have a capacitance of about 10^{-3} F), and the subdivisions microfarad (μF $= 10^{-6}$ F), and picofarad (pF $=$ 10^{-12} F) are commonly used instead.

A result often required is the net capacitance of a number of capacitors joined either in series, or in parallel, as in Fig. 1.15. If n capacitors $C_1, C_2, \ldots, C_n$ are joined in series, and a voltage V applied across them, a charge $+Q$ appears on the plate A and $-Q$ on the plate B. The plate 2 of the first capacitor will have charge $-Q$, and plate 1 of the second capacitor must therefore have charge $+Q$ since the two, though connected together, are otherwise isolated, and their total charge must

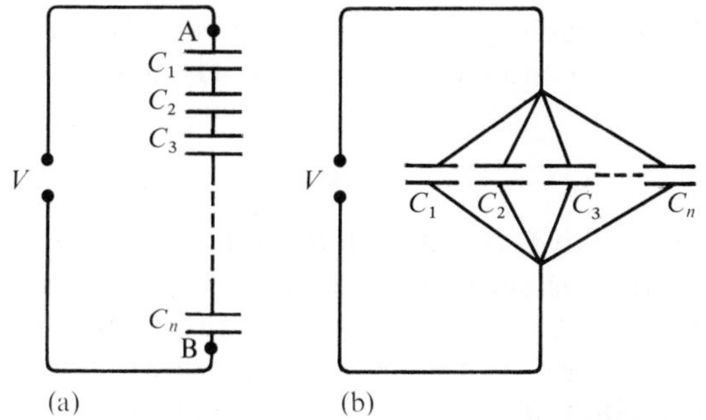

FIG. 1.15. (a) Capacitors in series. (b) Capacitors in parallel.

remain zero on connecting the battery. Thus each capacitor carries the same charge, irrespective of its size, and the potential across the kth capacitor is Q/C_k. Hence the total potential is $\sum_1^n (Q/C_k)$, and this equals Q/C, where C is the net capacitance. Equating these two results gives

$$\frac{1}{C}=\frac{1}{C_1}+\frac{1}{C_2}+\ldots+\frac{1}{C_k}+\ldots+\frac{1}{C_n}. \tag{1.32}$$

If the capacitors are joined in parallel, the voltage across each capacitor is equal to V. The total charge carried by all the capacitors is $Q = Q_1+Q_2+\ldots+Q_k+\ldots+Q_n = V(C_1+C_2+\ldots+C_k+\ldots+C_n) = VC$, where C is the net capacitance. Hence

$$C = C_1+C_2+\ldots+C_k+\ldots+C_n. \tag{1.33}$$

The potential energy of systems of charges and charged conductors

A system of electric charges possesses potential energy, since work must be done in bringing up any particular charge through the electrostatic field of the remaining charges. The energy depends only on the final state of the system and not on how the charges are established. We may suppose therefore that each charge is increased from zero to its final value in infinitesimal steps so that at any given instant each charge is αq_k, where q_k is its final value and α is a number less than unity which is the same for all charges. Then if V_k is the final value of the potential at the point occupied by q_k, the instantaneous value of the potential will be αV_k, and the work done in increasing the charge by $q_k\,d\alpha$ will be $(\alpha V_k)(q_k\,d\alpha)$. Thus the work done in increasing all the charges by a corresponding amount will be $\alpha\,d\alpha \sum_k q_k V_k$. The total work done is equal to the stored energy, which will therefore be

$$U=\sum_k q_k V_k \int_0^1 \alpha\,d\alpha = \tfrac{1}{2}\sum_k q_k V_k. \tag{1.34}$$

We may apply this result to a capacitor with two plates at potentials V_1, V_2 carrying charges $+Q$ and $-Q$ respectively. The energy of the capacitor will be

$$U=\tfrac{1}{2}QV_1-\tfrac{1}{2}QV_2=\tfrac{1}{2}QV=\tfrac{1}{2}CV^2=\tfrac{1}{2}Q^2/C, \tag{1.35}$$

where $V=V_1-V_2$ is the potential difference between the plates.

The energy of the system may be expressed in a different way which implies that it is distributed over the space between the charges occupied by their electrostatic field. Consider two nearby equipotential surfaces in this space which differ in potential by a small amount V, and are a distance $\mathrm{d}s$ apart. If two parallel conducting plates of area $\mathrm{d}S$ were inserted so as to coincide with these equipotentials, they would not alter the field distribution in any way. They would form a parallel-plate capacitor of capacitance C and energy $\frac{1}{2}CV^2$. But $C=\epsilon\,\mathrm{d}S/\mathrm{d}s$ and $V=-E\,\mathrm{d}s$, where E is the electric field at this point. The capacitor occupies a volume $\mathrm{d}\tau=\mathrm{d}S\,\mathrm{d}s$ and its energy is $\frac{1}{2}\epsilon E^2(\mathrm{d}S\,\mathrm{d}s)=\frac{1}{2}DE\,\mathrm{d}\tau$. We may therefore regard the energy as distributed throughout the field, the energy density at any point being $\frac{1}{2}DE$. This equation may be derived more rigorously by vector analysis, as follows.

In Fig. 1.16 suppose there exists a volume distribution of charge of density ρ_e per unit volume and a surface distribution of density σ_e per unit area. Then from eqn (1.34), if the summation is replaced by integrations, we have for the total energy

$$U=\tfrac{1}{2}\int\rho_e V\,\mathrm{d}\tau+\tfrac{1}{2}\int\sigma_e V\,\mathrm{d}S,$$

where the surface integral is taken over the surfaces of all the conductors present. By Gauss's theorem, $\rho_e=\operatorname{div}\mathbf{D}$, and hence, using a vector transformation (see Appendix A, eqn (A.12)),

$$\rho_e V=V\operatorname{div}\mathbf{D}=\operatorname{div}(V\mathbf{D})-\mathbf{D}\cdot\operatorname{grad}V.$$

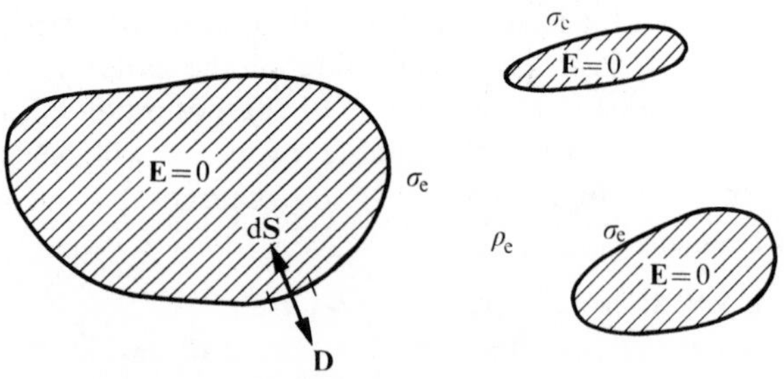

FIG. 1.16. Diagram to illustrate the calculation of the energy density of a system of surface and volume charges.

Therefore

$$\tfrac{1}{2}\int \rho_e V \,\mathrm{d}\tau = \tfrac{1}{2}\int \operatorname{div}(V\mathbf{D})\,\mathrm{d}\tau - \tfrac{1}{2}\int \mathbf{D}.\operatorname{grad} V \,\mathrm{d}\tau$$

$$= \tfrac{1}{2}\int V\mathbf{D}.\mathrm{d}\mathbf{S} + \tfrac{1}{2}\int \mathbf{D}.\mathbf{E}\,\mathrm{d}\tau.$$

The first integral must be taken over a closed surface bounding the whole volume, and also over the surface of each conductor. The closed surface may be taken at an infinitely large distance from all the charges, and its contribution to the surface integral then vanishes. For, at large distances, V varies as r^{-1} and D as r^{-2}, while $\mathrm{d}S$ increases with r^2; thus the integral is proportional to r^{-1} and tends to zero as r tends to infinity. The total energy may now be written

$$U = \tfrac{1}{2}\int \mathbf{D}.\mathbf{E}\,\mathrm{d}\tau + \tfrac{1}{2}\int V\mathbf{D}.\mathrm{d}\mathbf{S} + \tfrac{1}{2}\int \sigma_e V\,\mathrm{d}S,$$

where the surface integrals are taken over the surfaces of all the conductors. Since the integration is over the surface of the medium, $\mathrm{d}\mathbf{S}$ is a vector drawn outwards from the medium and hence into the conducting surface as in Fig. 1.16. But the normal component of $\mathbf{D}$ in this direction is $-\sigma_e$, from eqn (1.28), and hence $\mathbf{D}.\mathrm{d}\mathbf{S} = -\sigma_e\,\mathrm{d}S$, so that the two surface integrals in our expression for U cancel. Our final expression for U becomes

$$U = \tfrac{1}{2}\int \mathbf{D}.\mathbf{E}\,\mathrm{d}\tau. \tag{1.36}$$

Since $\mathbf{E}$ is zero within any conductor, we may regard the energy as distributed throughout the surrounding dielectric medium, with density $\tfrac{1}{2}\mathbf{D}.\mathbf{E}$. This expression is valid in anisotropic dielectrics, where $\mathbf{D}$ is not necessarily parallel to $\mathbf{E}$, but it assumes that $\mathbf{D}$ is always proportional to $\mathbf{E}$.

The results obtained above may easily be generalized to the case where $\mathbf{D}$ is not linearly proportional to $\mathbf{E}$ (which implies that in a capacitor containing a dielectric, Q is not linearly proportional to V). If a charge $\mathrm{d}Q$ is transferred from the plate at V_1 to that at V_2, the work done is

$$\mathrm{d}W = \mathrm{d}Q(V_1 - V_2) = V\,\mathrm{d}Q, \tag{1.37}$$

where V is the potential difference between the plates. In a parallel-plate capacitor with guard rings, with a separation t between the plates, the electric field strength $E = V/t$, while for an area S, $Q/S = D$. If an increase $\mathrm{d}Q$ in the charge occurs at constant V because of an increase $\mathrm{d}D$ in the dielectric, the work done per unit volume of dielectric is

$$V\,\mathrm{d}Q/tS = E\,\mathrm{d}D.$$

More generally, if $\mathbf{dD}$ is not parallel to $\mathbf{E}$, the increase in the stored energy per unit volume is

$$dU = \mathbf{E}.d\mathbf{D} = \epsilon_0\mathbf{E}.d\mathbf{E} + \mathbf{E}.d\mathbf{P}, \tag{1.38}$$

where $\epsilon_0\mathbf{E}.d\mathbf{E}$ is the change in the field energy in the absence of the dielectric, and $\mathbf{E}.d\mathbf{P}$ is the work done in polarizing the dielectric.

A similar modification can be applied to the more general treatment given above. Suppose the volume and surface charge distribution densities increase by $\delta\rho_e$ and $\delta\sigma_e$ respectively. Then the change in stored energy is

$$\delta U = \int (V\,\delta\rho_e)\,d\tau + \int (V\,\delta\sigma_e)\,dS,$$

where the integrations are over the volume of dielectric and the surfaces of the conductors respectively. Then the same transformations as used above, and consideration of the surface integrals, leads to

$$\delta U = \int (\mathbf{E}.d\mathbf{D})\,d\tau. \tag{1.39}$$

In calculating such changes in the electrical energy no account is taken of any heat flow into or out of the system; in thermodynamic terms, the change is adiabatic and, since it is reversible, it is a change at constant entropy. If a reversible change of the entropy also occurs, then the total change in the internal energy U' is

$$\delta U' = T\,\delta S + V\,\delta Q$$

or

$$\delta U' = T\,\delta S + \int (\mathbf{E}.d\mathbf{D})\,d\tau, \tag{1.40}$$

where the symbol U' is used for the internal energy other than that arising from electrical causes. Since the free energy is $F = U' - TS$, we have

$$dF = dU' - T\,dS - S\,dT$$

$$= -S\,dT + \int (\mathbf{E}.d\mathbf{D})\,d\tau, \tag{1.41}$$

showing that at constant temperature $(\mathbf{E}.d\mathbf{D})$ is the change in the free energy per unit volume. This relation is important in considering the thermodynamics of dielectrics, and of voltaic cells, where (see § 3.10) $V\,dQ$ is the change in the free energy when a charge dQ passes at constant temperature and V is the potential of the cell.

1.7. Stress in the electrostatic field

It has already been shown that the charge on a conductor resides in a thin surface layer. This is due to the mutual repulsion between charges of

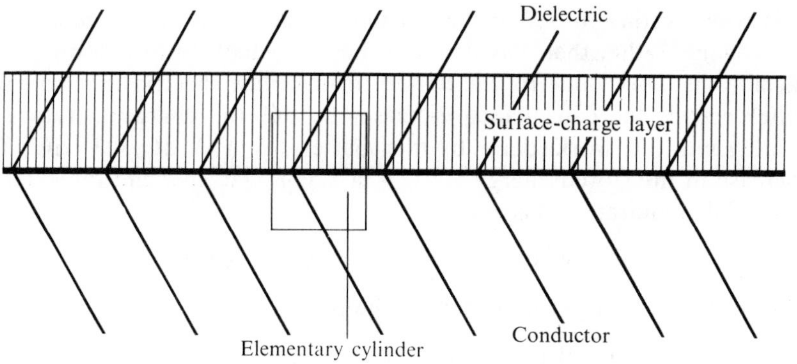

FIG. 1.17. Deduction of the tension on a charged conductor.

like sign, so that each portion of the charge on the conductor is trying to get as far away as possible from the remainder. This results in a tension acting on the surface of the conductor, whose magnitude will now be calculated. We shall assume that the charge of surface density σ_e is in a thin layer just outside the conducting surface, in a medium of permittivity ϵ, as in Fig. 1.17. By applying Gauss's theorem to a small cylinder with its axis normal to the surface, and with one end inside the conductor and the other within the surface layer, the field strength $\mathbf{E}'$ at this latter point is found to be $\mathbf{E}' = \sigma_e\alpha/\epsilon$, where $\sigma_e\alpha$ is the portion of the surface charge density lying between the conductor and the end of the cylinder. The force on the element of charge density $\sigma_e\,\mathrm{d}\alpha$ at the end of the cylinder is therefore $\mathbf{E}'(\sigma_e\,\mathrm{d}\alpha) = \sigma_e^2\alpha\,\mathrm{d}\alpha/\epsilon$, and the total force per unit area is

$$\frac{\sigma_e^2}{\epsilon}\int_0^1 \alpha\,\mathrm{d}\alpha = \frac{\sigma_e^2}{2\epsilon}. \tag{1.42}$$

At first sight the assumption that the charge layer resides in the dielectric may appear rather artificial, but the same result may be obtained by application of the principle of virtual work to special cases. For example, consider a parallel-plate capacitor with a medium of permittivity ϵ between the plates. If the separation between the plates is x, and their area S, the capacitance $C = \epsilon S/x$, and the stored energy is $\frac{1}{2}Q^2/C = \frac{1}{2}\sigma_e^2 S^2/C$, where σ_e is the charge density on the plates. If σ_e is kept constant, and the separation of the plates is increased, the rate of change of the stored energy is

$$\frac{\mathrm{d}U}{\mathrm{d}x} = \frac{\mathrm{d}}{\mathrm{d}x}\left(\frac{\sigma_e^2 Sx}{2\epsilon}\right) = \frac{\sigma_e^2 S}{2\epsilon}.$$

But $-(\mathrm{d}U/\mathrm{d}x)/S = -\sigma_e^2/2\epsilon$ is the tension per unit area on the plate, where the minus sign denotes that the tension acts in the opposite direction to the movement of the one plate, that is, in the direction of diminishing x.

It is interesting to derive the tension on the plates of this capacitor if the voltage, rather than the charge, is kept constant. In this case we must include the work done by the battery maintaining the constant potential difference. If the capacitance increases by dC, the charge increases by $V\,dC$, and the work done by the battery is $V(V\,dC) = V^2\,dC$. The increase in the stored energy is $dU = d(\frac{1}{2}CV^2) = \frac{1}{2}V^2\,dC$, and the external work dW required is therefore

$$\tfrac{1}{2}V^2\,dC - V^2\,dC = -\tfrac{1}{2}V^2\,dC = -dU.$$

Hence the tension per unit area on the plates is

$$-\frac{1}{S}\left(\frac{\partial W}{\partial x}\right) = -\frac{1}{S}\left(\frac{\partial W}{\partial C}\right)\left(\frac{\partial C}{\partial x}\right)$$

$$= -\frac{1}{S}\left(-\frac{V^2}{2}\right)\left(-\frac{\epsilon S}{x^2}\right) = -\frac{\epsilon E^2}{2} = -\frac{\sigma_e^2}{2\epsilon},$$

which is the same as before, as we should expect. The fact that the work done by the battery in maintaining the system at constant potential is just twice the increase in the stored energy is generally true, and our example is just a particular case. It is, however, probably more instructive to remember to put in the work done by the battery in working a particular problem at constant potential rather than to make use of a general theorem.

Stresses in dielectric media

Both Faraday and Maxwell used the concept of tubes of force. A tube of force contains an arbitrary but very large number of lines of force, and the number of tubes crossing unit area is equal to the electric field strength; similarly, the number of tubes of displacement per unit area is equal to **D**. They imagined these tubes to be in a state of tension, so that the force of attraction between two charges of opposite sign, for example, was transmitted along the tubes of force. It was also necessary to stipulate that there was a force of repulsion between tubes of force in a direction normal to their length, otherwise the tubes would all contract until they passed straight from one charge to another.

A macroscopic theory of such forces was developed by Maxwell using the concept of a 'stress tensor' (see Panofsky and Phillips 1955; Robinson 1973). The stress is regarded as being present in the field and acting on an element of area $d\mathbf{S}$ irrespective of whether $d\mathbf{S}$ is part of a real boundary or not. If it is not, there will be an equal and opposite stress on the other side, so that equilibrium is maintained. If $d\mathbf{S}$ is part of a real boundary of a dielectric, the field components are different on the two sides and there is a net force acting at the boundary. The value of the macroscopic theory lies in providing a quantitative treatment of effects which are not easy to interpret from the viewpoint of atomic theory.

References

PANOFSKY, W. K. H. and PHILLIPS, M. (1955). *Classical electricity and magnetism.* Addison-Wesley, New York.
PLIMPTON, S. J. and LAWTON, W. E. (1936). *Phys. Rev.* **50,** 1066.
ROBINSON, F. N. H. (1973). *Macrosopic electromagnetism.* Pergamon Press, Oxford.

Problems

1.1. An electric quadrupole (cf. Fig. 1.9) is formed by a charge $-2q$ at the origin and charges $+q$ at the points $(\pm a, 0, 0)$. Show that the potential V at a distance r large compared with a is approximately given by $V = +qa^2(3\cos^2\theta - 1)/4\pi\epsilon r^3$, where θ is the angle between r and the line through the charges.

1.2. Show that there is no translational force or couple on such an electric quadrupole in a uniform field. Prove that the couple on the quadrupole at a distance r from a point charge q' is $C = q(3q'a^2 \sin 2\theta)/4\pi\epsilon r^3$, where θ is the angle between r and the line through the charges (assume $r \gg a$).

1.3. Show that the force on an elementary dipole of moment **p**, distance **r** from a point charge q, has components

$$F_r = -\frac{qp\cos\theta}{2\pi\epsilon r^3}, \qquad F_\theta = -\frac{qp\sin\theta}{4\pi\epsilon r^3}$$

along and perpendicular to **r** in the plane of **p** and **r**, where θ is the angle which **p** makes with **r**.

1.4. Show that the potential energy of two coplanar dipoles p_1 and p_2 a distance r apart is

$$\frac{1}{4\pi\epsilon r^3}\left\{\mathbf{p}_1\cdot\mathbf{p}_2 - \frac{3(\mathbf{p}_1\cdot\mathbf{r})(\mathbf{p}_2\cdot\mathbf{r})}{r^2}\right\} = \frac{p_1 p_2}{4\pi\epsilon r^3}(\sin\theta_1\sin\theta_2 - 2\cos\theta_1\cos\theta_2),$$

where θ_1, θ_2 are the angles made by p_1 and p_2 respectively with the line joining their centres.

1.5. A charge q is placed at each of the four corners $(\pm a, 0, 0)$, $(0, \pm a, 0)$ of a square. Show that the potential at a point (x, y, z) near the origin is

$$V = \frac{q}{4\pi\epsilon a}\left\{4 + \frac{(x^2+y^2-2z^2)}{a^2} + \ldots\right\}$$

Verify that a charge of the same sign placed at the centre of the square is in stable equilibrium against a small displacement in the plane of the square, but is unstable for a displacement normal to this plane. This is an example of Earnshaw's theorem, which shows that a charge cannot rest in stable equilibrium in an electrostatic field.

1.6. The values of the vertical potential gradient of the earth at heights of 100 m and 1000 m above its surface are 110 V m^{-1} and 25 V m^{-1} respectively. What is the mean electrostatic charge per cubic metre of the atmosphere between these heights?

(*Answer:* $0{\cdot}835 \times 10^{-12}$ C m^{-3}.)

1.7. A parallel-plate capacitor with plates of area S and separation d has a block of dielectric, of relative permittivity ϵ_r, of cross-sectional area S, and thickness t ($t \ll d$), inserted in between the plates. Find the values of E and D in the space between the plates, in both air and dielectric. Show that the capacitance of the capacitor is $C = \dfrac{S\epsilon_r\epsilon_0}{\epsilon_r d-(\epsilon_r-1)t}$, and calculate the change in stored energy of the system when the dielectric is inserted (a) if the plates have a constant charge Q, and (b) if they are connected to a battery at a constant potential V.

$$\left(\textit{Answer: } (a)\ \Delta U = -\frac{Q^2(\epsilon_r-1)t}{2\epsilon_r\epsilon_0 S}; \qquad (b)\ \Delta U = +\frac{\epsilon_0 S V^2(\epsilon_r-1)t}{2d\{\epsilon_r d-(\epsilon_r-1)t\}}.\right)$$

1.8. For the capacitor consisting of two coaxial cylinders (Fig. 1.14), show by integration of eqn (1.36) that the stored energy per unit length is

$$\left(\frac{Q^2}{4\pi\epsilon}\right)\ln(b/a),$$

and that this is equal to $\frac{1}{2}Q^2/C$, where Q is the charge and C the capacitance per unit length.

If E_{max} is the greatest electric field strength that can exist anywhere in the dielectric without breakdown, show that the maximum voltage that can be applied between inner and outer cylinders is $aE_{max}\ln(b/a)$.

1.9. A capacitor is formed by two coaxial cylinders of radii a and b. The axes of the cylinders are vertical and the inner cylinder is suspended from a balance so that it hangs only partly within the outer cylinder. Find an expression for the mass which must be added to the other pan of the balance to maintain equilibrium when a voltage V is connected between the two cylinders.

$$\left(\textit{Answer: } mg = \frac{\pi\epsilon V^2}{\ln(b/a)}.\right)$$

1.10. The electrometer is an instrument for measuring voltages by means of the force on a charged conductor. An attracted-disc electrometer has a moving plate of area $0{\cdot}01\ \text{m}^2$, separated by a distance of 1 mm from the fixed plate. Calculate the force between the plates when the potential difference across them is 100 V. Calculate the sensitivity at this voltage in newtons per volt.

(*Answer:* $F = 4{\cdot}42\times10^{-4}$ N, $dF/dV = 8{\cdot}85\times10^{-6}\ \text{N V}^{-1}$.)

1.11. The upper disc of such an electrometer is suspended by a spring. In equilibrium, the separation between the two discs is x when a voltage V is applied and a when $V=0$. Show that the equilibrium in the former case is stable provided that $x > 2a/3$.

1.12. A sphere carrying a charge density σ_e per unit area is immersed in an infinite dielectric medium. Verify eqn (1.42) by using the principle of virtual work and allowing the radius to change infinitesimally.

1.13. If **n** is a unit vector normal to the surface, show that the boundary conditions (1.26)–(1.28) can be expressed in vectorial form as

$$\mathbf{n}\cdot({}_2\mathbf{D}-{}_1\mathbf{D}) = \sigma_e; \qquad \mathbf{n}\wedge({}_2\mathbf{E}-{}_1\mathbf{E}) = 0.$$

An infinitely thin surface charge density σ_e is a mathematical abstraction; we may assume instead that near the surface there is a volume charge density ρ_e

which falls rapidly to zero inside a conductor. If the surface is the plane $z = 0$, and the charge density is independent of x, y, show that the value of D_z just outside the surface is given by

$$D_z = \int_{-\infty}^{0} \rho_e \, dz = \sigma_e,$$

where σ_e is the total charge per unit area of the surface.

1.14. Assuming that the total charge Ze of an atomic nucleus is uniformly distributed within a sphere of radius a, show that the potential at a distance r from the centre ($r \leqslant a$) is

$$V = \left\{\frac{3}{2} - \frac{1}{2}\left(\frac{r}{a}\right)^2\right\} \frac{Ze}{4\pi\epsilon_0 a}.$$

Show that the electrostatic energy of such a nucleus is

$$U = \frac{3(Ze)^2}{20\pi\epsilon_0 a}.$$

This electrostatic energy reduces the binding energy, giving a small increase δm in the mass of the nucleus, such that $U = \delta mc^2$, where c = velocity of light, by the Einstein relation. δm reduces the 'mass defect'.

1.15. An atom with an electron in an s-state has a finite density $-\rho_e$ of electronic charge inside the nucleus. Assuming that ρ_e is very small compared with the nuclear charge density, use the formula for the potential inside the nucleus given in the previous problem to show that the potential energy associated with the electron density $-\rho_e$ inside the sphere of radius a is $-2Ze\rho_e a^2/5\epsilon_0$, and that this is greater by $Ze\rho_e a^2/10\epsilon_0$ than it would have been if the nucleus were a point charge.

Since isotopes of the same element have different nuclear radii, this energy forms part of the 'isotope shift', that is, the difference in frequency of spectrum lines from different isotopes.

2. Electrostatics II

2.1. The equations of Poisson and Laplace

IN a region where there exists a charge distribution of density ρ_e per unit volume, the differential form of Gauss's theorem is (eqn (1.20))

$$\operatorname{div} \mathbf{D} = \rho_e.$$

Now $\mathbf{E} = -\operatorname{grad} V$, and in an isotropic dielectric $\mathbf{D} = \epsilon\mathbf{E}$. Hence

$$\nabla^2 V = \operatorname{div}(\operatorname{grad} V) = -\operatorname{div} \mathbf{E} = -\rho_e/\epsilon. \tag{2.1}$$

This is known as Poisson's equation. If there is no free charge present, $\rho_e = 0$, and we have Laplace's equation

$$\nabla^2 V = 0. \tag{2.2}$$

The operator denoted by ∇^2 is a scalar operator, which has its simplest form in Cartesian coordinates, where Poisson's equation becomes

$$\frac{\partial^2 V}{\partial x^2} + \frac{\partial^2 V}{\partial y^2} + \frac{\partial^2 V}{\partial z^2} = -\frac{\rho_e}{\epsilon}. \tag{2.3}$$

Two other coordinate systems will be considered. These are spherical polar coordinates, where Poisson's equation becomes

$$\frac{1}{r^2}\frac{\partial}{\partial r}\left(r^2\frac{\partial V}{\partial r}\right) + \frac{1}{r^2 \sin\theta}\frac{\partial}{\partial\theta}\left(\sin\theta\frac{\partial V}{\partial\theta}\right) + \frac{1}{r^2\sin^2\theta}\frac{\partial^2 V}{\partial\phi^2} = -\frac{\rho_e}{\epsilon}, \tag{2.4}$$

and cylindrical polar coordinates, where we have

$$\frac{1}{r}\frac{\partial}{\partial r}\left(r\frac{\partial V}{\partial r}\right) + \frac{1}{r^2}\frac{\partial^2 V}{\partial\theta^2} + \frac{\partial^2 V}{\partial z^2} = -\frac{\rho_e}{\epsilon}. \tag{2.5}$$

In principle, eqn (2.1) enables us to calculate the potential distribution due to any given set of charges and conductors. A formal solution of Poisson's equation can be found,

$$V = \frac{1}{4\pi\epsilon}\int \frac{\rho_e \, d\tau}{r}, \tag{2.6}$$

but this holds only in a vacuum or an infinite dielectric medium. Eqn (2.6) can be regarded as a formal extension, in a dielectric medium, of eqn (1.5) for the potential of a point charge q, by regarding $\rho_e \, d\tau$ as the element of charge. It can be shown (see, for example, Stratton 1941

(§ 3.7)) that the integral in eqn (2.6) is convergent although the denominator vanishes for $r=0$, provided that ρ_e remains finite, since the contribution from a sphere of radius a to the potential at its centre vanishes as $a\rightarrow 0$. It can be shown also that V is continuous in a region where ρ_e is finite, continuous, but bounded (that is, the charge density does not stretch to infinity in all dimensions), and that it is continuous through a single surface layer of surface charge density σ_e. In the latter case this condition for V clearly makes the tangential components of $\mathbf{E}$ continuous (eqn (1.27)), though from eqn (1.28) the normal components of $\mathbf{D}$ are not (that is, the gradient of V changes as we cross the surface charge). In general we can take V to be a continuous function, even at a boundary, an exception being the case of a double charge layer (see Problem 2.19).

If conducting surfaces are present, additional terms are needed in (2.6) to allow for the charge distribution on their surfaces. In general we do not know what this distribution is, and we have to resort to a number of special methods, but we must be sure that any solution we obtain which satisfies the boundary conditions is the correct and only answer. That this is the case is shown by an important theorem, known as the 'uniqueness theorem'. This theorem (see Stratton 1941) shows that if two different potential distributions are assumed to satisfy Laplace's equation and the boundary conditions, their difference is zero. We now discuss a number of methods for the case of no free charges, where the solutions needed are of Laplace's rather than Poisson's equation.

The required solution may be a sum of a number of functions, each of which satisfies Laplace's equation; for, if the functions $V_1, V_2, \ldots, V_n$ are each individual solutions of Laplace's equation, then

$$V=a_1V_1+a_2V_2+\ldots+a_nV_n,$$

where $a_1, a_2, \ldots, a_n$ are a set of numerical coefficients, is also a solution. A series of functions, each of which is a solution of Laplace's equation, may sometimes be found by making use of the fact that if V_1 is a solution, so also are any differentials of V_1 with respect to the space coordinates. Thus in Cartesian coordinates the functions $\partial V_1/\partial x$, $\partial V_1/\partial y$, $\partial V_1/\partial z$, $\partial^2 V_1/\partial x^2$ $\partial^2 V_1/\partial x\,\partial y$, etc., all satisfy eqn (2.2) if V_1 does. The proof of this can be seen from a single example. On partial differentiation of eqn (2.3) with respect to x, we have (setting $\rho_e=0$)

$$0=\frac{\partial}{\partial x}\left\{\frac{\partial^2 V_1}{\partial x^2}+\frac{\partial^2 V_1}{\partial y^2}+\frac{\partial^2 V_1}{\partial z^2}\right\}$$

$$=\frac{\partial^2}{\partial x^2}\left(\frac{\partial V_1}{\partial x}\right)+\frac{\partial^2}{\partial y^2}\left(\frac{\partial V_1}{\partial x}\right)+\frac{\partial^2}{\partial z^2}\left(\frac{\partial V_1}{\partial x}\right)=\nabla^2\left(\frac{\partial V_1}{\partial x}\right),$$

since the order of differentiation is immaterial when x, y, z are independent coordinates. The value of this method lies in the fact that once a series of functions which satisfy eqn (2.2) is established, any linear combination of these functions may be taken, and if they can be chosen in such a way as to satisfy the boundary conditions by adjustment of the coefficients, they give the unique solution to the problem.

In theory, any problem involving electrostatic fields may be solved by finding a solution which satisfies eqn (2.2) and gives the right boundary conditions. In practice the problem is almost insoluble by ordinary mathematical methods except in cases where there is a high degree of symmetry. These may be handled by the use of a series of known functions, and some examples of this method are given below. We shall consider also another special method which can be applied to the case of one or two point charges near to a conducting surface of simple shape. Though a number of other problems may be handled by mathematical methods which are beyond the scope of this volume, most of the problems met with in practice, such as the design of electron guns to give a focused beam in a cathode-ray tube, are dealt with either by use of approximate solutions, or by numerical methods using digital computers.

2.2. Solutions of Laplace's equation in spherical coordinates

We shall consider first the case of spherical coordinates, and assume initially that we have symmetry about the polar axis so that V is independent of ϕ. Then Laplace's equation reduces to

$$\frac{\partial}{\partial r}\left(r^2\frac{\partial V}{\partial r}\right)+\frac{\partial}{\partial \mu}\left\{(1-\mu^2)\frac{\partial V}{\partial \mu}\right\}=0, \tag{2.7}$$

where μ has been written for $\cos\theta$. This has solutions of the form $V=r^lP_l$, where P_l is a function of $\mu=\cos\theta$ only, and l is an integer. If we substitute such a function in eqn (2.7), and divide through by r^l, we obtain Legendre's differential equation for P_l

$$\frac{\mathrm{d}}{\mathrm{d}\mu}\left\{(1-\mu^2)\frac{\partial P_l}{\partial \mu}\right\}+l(l+1)P_l=0.$$

It is readily seen that replacing l by $-(l+1)$ leaves this equation unaltered, so that $P_l=P_{-(l+1)}$, that is, $V=r^lP_l$ and $V=r^{-(l+1)}P_l$ are both solutions of eqn (2.7).

Solutions of Legendre's equation may be obtained by standard methods, but a quick alternative method is as follows. We know that $V=1/r$ is a solution, and hence so is any partial derivative of this such as $(\partial V/\partial z)$ under the conditions x, y constant. Since $r^2=x^2+y^2+z^2$, we have $2r(\partial r/\partial z)=2z$ when x, y are kept constant, so that

$$\left(\frac{\partial r}{\partial z}\right)_{x,y}=\frac{z}{r}.$$

Hence

$$-\left(\frac{\partial}{\partial z}\right)_{x,y}\left(\frac{1}{r}\right)=\frac{1}{r^2}\left(\frac{\partial r}{\partial z}\right)_{x,y}=\frac{z}{r^3}=r^{-2}\cos\theta$$

if we take z to lie along the polar axis, so that $z=r\cos\theta$. The two functions $V=r^{-1}$ and $r^{-2}\cos\theta$ are the first two types of solution in the inverse powers $r^{-(l+1)}$, and correspond to values of $l=0$ and 1 respectively. Thus $P_0=1$, and $P_1=\cos\theta$. Further functions may be generated by successive differentiation; thus

$$\left(\frac{\partial}{\partial z}\right)\left(\frac{z}{r^3}\right)=\left(1-\frac{3z^2}{r^2}\right)r^{-3}=(1-3\cos^2\theta)r^{-3}$$

gives the next function, which is proportional to P_2. A general formula for P_l is

$$P_l=\frac{1}{2^l l!}\left(\frac{\partial}{\partial\mu}\right)^l(\mu^2-1)^l, \tag{2.8}$$

where the numerical coefficients are such that $P_l=1$ at $\mu=1$, that is, at $\theta=0$. The first few functions are given in Table 2.1, together with the radial functions $r^{-(l+1)}$ and r^l with which they combine to give solutions of Laplace's equation.

TABLE 2.1

Some spherical harmonic functions

Legendre function	Function of r	
$P_0=1$	r^{-1}	1
$P_1=\cos\theta$	r^{-2}	r
$P_2=\frac{1}{2}(3\cos^2\theta-1)$	r^{-3}	r^2
$P_3=\frac{1}{2}(5\cos^3\theta-3\cos\theta)$	r^{-4}	r^3
$P_4=\frac{1}{8}(35\cos^4\theta-30\cos^2\theta+3)$	r^{-5}	r^4
$P_5=\frac{1}{8}(63\cos^5\theta-70\cos^3\theta+15\cos\theta)$	r^{-6}	r^5
$P_6=\frac{1}{16}(231\cos^6\theta-315\cos^4\theta+105\cos^2\theta-5)$	r^{-7}	r^6

Associated Legendre functions

Table 2.1 clearly does not contain all possible solutions of Laplace's equation, since we can find others by differentiation with respect to x, y. For example,

$$-\left(\frac{\partial}{\partial x}\right)_{y,z}\left(\frac{z}{r^3}\right)=\frac{3z}{r^4}\left(\frac{\partial r}{\partial x}\right)_{y,z}=\frac{3zx}{r^5}=3r^{-3}\cos\theta\sin\theta\cos\phi$$

must also be a solution. The fact that it contains ϕ shows that it is not a solution of eqn (2.7), but of the more general equation which includes the dependence on ϕ. This is

$$\frac{\partial}{\partial r}\left(r^2\frac{\partial V}{\partial r}\right)+\frac{\partial}{\partial\mu}\left\{(1-\mu^2)\frac{\partial V}{\partial\mu}\right\}+\frac{1}{(1-\mu^2)}\frac{\partial^2 V}{\partial\phi^2}=0, \tag{2.9}$$

where we have again written μ for $\cos\theta$. As before, we assume that there exists a solution of the form $V = r^l\Theta\Phi$, where Θ, Φ are functions only of θ, ϕ respectively. Then the differential equation for $\Theta\Phi$ is

$$\frac{\partial}{\partial\mu}\left\{(1-\mu^2)\frac{\partial(\Theta\Phi)}{\partial\mu}\right\}+l(l+1)(\Theta\Phi)+\frac{1}{1-\mu^2}\frac{\partial^2(\Theta\Phi)}{\partial\phi^2}=0, \qquad (2.10)$$

which can be written in the form

$$\frac{(1-\mu^2)}{\Theta}\left[\frac{\partial}{\partial\mu}\left\{(1-\mu^2)\frac{\partial\Theta}{\partial\mu}\right\}+l(l+1)\Theta\right]=m^2=-\frac{1}{\Phi}\frac{\partial^2\Phi}{\partial\phi^2}, \qquad (2.11)$$

where the variables are separated. The right-hand side has the solution

$$\Phi_m=(2\pi)^{-\frac{1}{2}}\mathrm{e}^{\mathrm{j}m\phi}, \qquad (2.12)$$

and the equation for Θ becomes

$$\frac{\partial}{\partial\mu}\left\{(1-\mu^2)\frac{\partial\Theta}{\partial\mu}\right\}+\left\{l(l+1)-\frac{m^2}{1-\mu^2}\right\}\Theta=0. \qquad (2.13)$$

It is apparent that the functions listed in Table 2.1 are solutions of this equation for the special case $m = 0$, where there is no dependence on ϕ. The solutions of eqn (2.13) are

$$\begin{aligned}P_{l,m}&=\sin^m\theta\left(\frac{\partial}{\partial\mu}\right)^m P_l=(1-\mu^2)^{\frac{1}{2}m}\left(\frac{\partial}{\partial\mu}\right)^m P_l\\&=\frac{(1-\mu^2)^{\frac{1}{2}m}}{2^l l!}\left(\frac{\partial}{\partial\mu}\right)^{m+l}(\mu^2-1)^l.\end{aligned} \qquad (2.14)$$

The functions Φ defined by eqn (2.12) are 'normalized', that is,

$$\int_0^{2\pi}\Phi_m^*\Phi_m\,\mathrm{d}\phi=\int_0^{2\pi}(2\pi)^{-1}\mathrm{e}^{-\mathrm{j}m\phi}\mathrm{e}^{\mathrm{j}m\phi}\,\mathrm{d}\phi=1, \qquad (2.15)$$

and they are also 'orthogonal', that is, $(m'\neq m)$

$$\int_0^{2\pi}\Phi_{m'}^*\Phi_m\,\mathrm{d}\phi=\int_0^{2\pi}(2\pi)^{-1}\mathrm{e}^{-\mathrm{j}m'\phi}\mathrm{e}^{\mathrm{j}m\phi}\,\mathrm{d}\phi=0. \qquad (2.16)$$

By using the Kronecker δ, whose properties are that $\delta(m', m)=1$ if $m'=m$, but $\delta(m', m)=0$ if $m'\neq m$, we can write these two equations in the short form

$$\int_0^{2\pi}\Phi_{m'}^*\Phi_m\,\mathrm{d}\phi=\delta(m', m). \qquad (2.17)$$

The functions $P_{l,m}$ are orthogonal but not normalized, and hence it is often convenient to work in terms of 'spherical harmonics' defined by

$$Y_{l,m} = \Theta_{l,m}\Phi_m, \tag{2.18}$$

where

$$\begin{aligned} \Theta_{l,m} &= (-1)^m\left\{\frac{(2l+1)(l-|m|)!}{2(l+|m|)!}\right\}^{\frac{1}{2}} P_{l,m} \quad \text{for } m \geqslant 0, \\ \Theta_{l,m} &= (-1)^m \Theta_{l,|m|} \qquad\qquad\qquad\qquad \text{for } m < 0. \end{aligned} \tag{2.19}$$

The functions $Y_{l,m}$ are both orthogonal and normalized, that is,

$$\int_0^{2\pi}\int_0^{\pi} Y^*_{l',m'}Y_{l,m} \sin\theta \, d\theta \, d\phi = \delta(l', l)\,\delta(m', m), \tag{2.20}$$

where the integration is over the solid angle 4π. Here the significance of the δ functions is that the integral is zero unless both $l' = l$ and $m' = m$, in which case it is unity.

Expressions for the first members of the series of spherical harmonics are listed in Table 2.2 (note that the signs of the functions used by different authors sometimes differ; we have followed the definition adopted by Ramsey (1956) and Brink and Satchler (1968)). A general proof of the orthogonality and normalization relations is tedious, but the reader may verify that they are correct by evaluating the integrals for some of the functions given in Table 2.2.

The spherical harmonics have many applications, some of which are discussed in the next section. From the atomic viewpoint, their particular interest is that $Y_{l,m}$ represents the angular variation of the wavefunction

TABLE 2.2

Some spherical harmonic functions

Y_{10}	$+(3/4\pi)^{\frac{1}{2}}\cos\theta$	C_{10}	$+\cos\theta$
Y_{1+1}	$-(3/8\pi)^{\frac{1}{2}}\sin\theta e^{+j\phi}$	C_{1+1}	$-2^{-\frac{1}{2}}\sin\theta e^{+j\phi}$
Y_{1-1}	$+(3/8\pi)^{\frac{1}{2}}\sin\theta e^{-j\phi}$	C_{1-1}	$+2^{-\frac{1}{2}}\sin\theta e^{-j\phi}$
Y_{20}	$+(5/16\pi)^{\frac{1}{2}}(3\cos^2\theta-1)$	C_{20}	$+\frac{1}{2}(3\cos^2\theta-1)$
Y_{2+1}	$-(15/8\pi)^{\frac{1}{2}}\cos\theta\sin\theta e^{+j\phi}$	C_{2+1}	$-(3/2)^{\frac{1}{2}}\cos\theta\sin\theta e^{+j\phi}$
Y_{2-1}	$+(15/8\pi)^{\frac{1}{2}}\cos\theta\sin\theta e^{-j\phi}$	C_{2-1}	$+(3/2)^{\frac{1}{2}}\cos\theta\sin\theta e^{-j\phi}$
Y_{2+2}	$+(15/32\pi)^{\frac{1}{2}}\sin^2\theta e^{+j2\phi}$	C_{2+2}	$+(3/8)^{\frac{1}{2}}\sin^2\theta e^{+j2\phi}$
Y_{2-2}	$+(15/32\pi)^{\frac{1}{2}}\sin^2\theta e^{-j2\phi}$	C_{2-2}	$+(3/8)^{\frac{1}{2}}\sin^2\theta e^{-j2\phi}$
Y_{30}	$+(7/16\pi)^{\frac{1}{2}}(5\cos^3\theta-3\cos\theta)$	C_{30}	$+\frac{1}{2}(5\cos^3\theta-3\cos\theta)$
Y_{3+1}	$-(21/64\pi)^{\frac{1}{2}}(5\cos^2\theta-1)\sin\theta e^{+j\phi}$	C_{3+1}	$-(3/16)^{\frac{1}{2}}(5\cos^2\theta-1)\sin\theta e^{+j\phi}$
Y_{3-1}	$+(21/64\pi)^{\frac{1}{2}}(5\cos^2\theta-1)\sin\theta e^{-j\phi}$	C_{3-1}	$+(3/16)^{\frac{1}{2}}(5\cos^2\theta-1)\sin\theta e^{-j\phi}$
Y_{3+2}	$+(105/32\pi)^{\frac{1}{2}}\cos\theta\sin^2\theta e^{+j2\phi}$	C_{3+2}	$+(15/8)^{\frac{1}{2}}\cos\theta\sin^2\theta e^{+j2\phi}$
Y_{3-2}	$+(105/32\pi)^{\frac{1}{2}}\cos\theta\sin^2\theta e^{-j2\phi}$	C_{3-2}	$+(15/8)^{\frac{1}{2}}\cos\theta\sin^2\theta e^{-j2\phi}$
Y_{3+3}	$-(35/64\pi)^{\frac{1}{2}}\sin^3\theta e^{+j3\phi}$	C_{3+3}	$-(5/16)^{\frac{1}{2}}\sin^3\theta e^{+j3\phi}$
Y_{3-3}	$+(35/64\pi)^{\frac{1}{2}}\sin^3\theta e^{-j3\phi}$	C_{3-3}	$+(5/16)^{\frac{1}{2}}\sin^2\theta e^{-j3\phi}$

The functions C_{lm} are related to Y_{lm} by $C_{lm} = \{4\pi/(2l+1)\}^{\frac{1}{2}}Y_{lm}$. Note that $C_{l0} = P_l$.

for an electron in an atom which has orbital angular momentum $\sqrt{\{l(l+1)\}}(h/2\pi)$, and a component $m(h/2\pi)$ of angular momentum along the z-axis (the polar axis), where h is Planck's constant.

2.3. The multipole expansion

For a charge distribution whose density is ρ_e in a volume element $d\tau$ at a point where the potential is V, the potential energy is

$$U_P = \int \rho_e V \, d\tau, \tag{2.21}$$

where both ρ_e and V are functions of position. If the charge distribution extends only over a small volume, we can expand V in a series

$$\begin{aligned} V = V_0 &+ x(\partial V/\partial x) + y(\partial V/\partial y) + z(\partial V/\partial z) + \\ &+ \tfrac{1}{2}x^2(\partial^2 V/\partial x^2) + \tfrac{1}{2}y^2(\partial^2 V/\partial y^2) + \tfrac{1}{2}z^2(\partial^2 V/\partial z^2) + \\ &+ \tfrac{1}{2}xy(\partial^2 V/\partial x\, \partial y) + \tfrac{1}{2}yx(\partial^2 V/\partial y\, \partial x) + \ldots . \end{aligned} \tag{2.22}$$

The potential energy then becomes

$$U_P = \int \rho_e V_0 \, d\tau + \int \rho_e x(\partial V/\partial x) \, d\tau + \int \rho_e y(\partial V/\partial y) \, d\tau + \int \rho_e z(\partial V/\partial z) \, d\tau + \ldots . \tag{2.23}$$

Here the first term is simply qV_0, where $q = \int \rho_e \, d\tau$ is the total charge. Since $\partial V/\partial x = -E_x$, etc., we can write the next three terms as

$$p_x(\partial V/\partial x) + p_y(\partial V/\partial y) + p_z(\partial V/\partial z) = -(p_x E_x + p_y E_y + p_z E_z) = -\mathbf{p}.\mathbf{E}, \tag{2.24}$$

and by comparison with eqn (1.13) we identify the quantities p_x, p_y, p_z as the components of the electric dipole moment of the charge distribution. This gives a general definition of the dipole moment of a charge distribution, the components being

$$p_x = \int \rho_e x \, d\tau, \qquad p_y = \int \rho_e y \, d\tau, \qquad p_z = \int \rho_e z \, d\tau, \qquad \text{or } \mathbf{p} = \int \rho_e \mathbf{r} \, d\tau. \tag{2.25}$$

If we have a number of point charges rather than a continuous distribution, the integrals can be replaced by a summation (a dipole consisting of two equal and opposite charges separated by a small distance, as defined in Chapter 1, is thus a special case). Note that in eqn (2.24) we have implicitly assumed that the first differentials of V are constant over the region occupied by charge, since only then can we take them outside the integration.

The definition of the dipole moment $\mathbf{p}$ is independent of the origin chosen for the coordinate system provided that the total charge is zero. If the origin is moved by a vector distance $\mathbf{r}_0$, so that $\mathbf{r}'=\mathbf{r}-\mathbf{r}_0$ is the position vector in the new coordinate system, then

$$\mathbf{p}=\int \rho_e \mathbf{r}\, d\tau=\int \rho_e(\mathbf{r}'+\mathbf{r}_0)\, d\tau=\int \rho_e \mathbf{r}'\, d\tau+\mathbf{r}_0\int \rho_e\, d\tau$$

$$=\int \rho_e \mathbf{r}'\, d\tau=\mathbf{p}',$$

provided that $\int \rho_e\, d\tau=0$. A similar transformation applies to a set of point charges q_i provided that $\sum q_i=0$. This means that the electric dipole moment is uniquely defined for a charge distribution which has a net charge of zero, such as a neutral molecule.

If the charge distribution has reflection symmetry in the plane $z=0$, that is, if the charge density ρ_e at the point (x, y, z) is the same as that at the point $(x, y, -z)$, the component p_z of the dipole moment will be zero, since in the integral $\int \rho_e z\, d\tau$ the contributions from the points (x, y, z) and $(x, y, -z)$ will be equal and opposite. Thus in a diatomic molecule, an electric dipole moment can exist parallel to the line joining the two nuclei if they are different (that is, if the molecule is heteronuclear, such as HCl), but not if they are identical, as in a homonuclear molecule (H_2, Cl_2).

Similar considerations apply to p_x, p_y of course, and show that a diatomic molecule can have no electric dipole moment perpendicular to the internuclear axis. The method can be extended to more complicated molecules, such as CH_3Cl, C_2H_6, C_6H_6.

In an atom or nucleus the charge distribution is expected to have reflection symmetry in three mutually perpendicular planes, so that there will be no permanent electric dipole moment in any direction. Note that three such reflections change the point (x, y, z) into $(-x, -y, -z)$ and are equivalent to inversion through the origin. The assumption we have made about the charge distribution is equivalent to assuming that the system is 'invariant under the parity operation', that is, that its properties are unaltered by inversion.

Expansion in spherical harmonics

Similar considerations can be applied to the higher terms in eqn (2.23), but it is obvious that the large number of terms makes the expansion in Cartesian coordinates clumsy to handle. We therefore turn to another method of expanding the potential of a point charge which makes use of spherical harmonics.

Assume that we have a charge at the point B and wish to know how its potential varies in the neighbourhood of the point O, for example, at the point A (Fig. 2.1(a)). This requires evaluation of the quantity $(R^2-2Rr\cos\theta+r^2)^{-\frac{1}{2}}$, which is the inverse of the distance AB. If $r<R$,

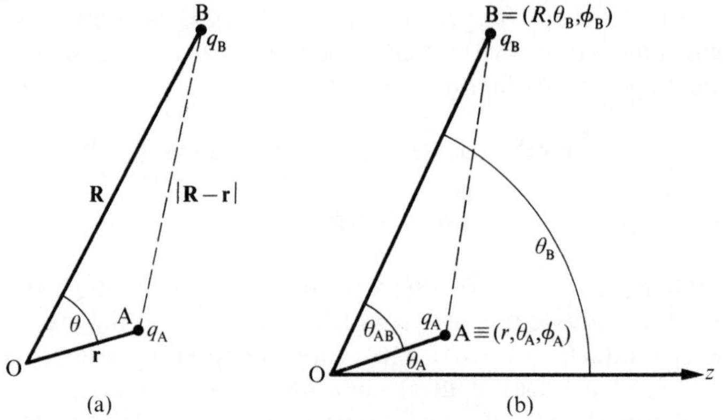

FIG. 2.1. Expansion of the potential at A due to a point charge q_B at B (or of the potential at B due to charge q_A at A) using spherical harmonics. In (b), the points A, B are not necessarily in the plane $\phi = 0$.

this function can be expanded in powers of (r/R):

$$\frac{1}{|\mathbf{R}-\mathbf{r}|} = (R^2 - 2Rr\cos\theta + r^2)^{-\frac{1}{2}}$$
$$= \frac{1}{R} + \frac{r}{R^2}P_1 + \frac{r^2}{R^3}P_2 + \ldots + \frac{r^l}{R^{l+1}}P_l + \ldots. \quad (2.26)$$

Here the functions P_l are the Legendre functions defined by eqn (2.8), as can readily be verified by direct expansion for the first terms.

A more general formula can be found where the points A, B in spherical coordinates are at (r, θ_A, ϕ_A) and (R, θ_B, ϕ_B) respectively. In Fig. 2.1(b), Oz is the polar axis, and the lines OA, OB (which are not necessarily coplanar) make angles θ_A, θ_B with it: the angle AOB is denoted by θ_{AB}. It can be shown that the Legendre function $P_l(\cos\theta_{AB})$ can be expressed in terms of θ_A, ϕ_A and θ_B, ϕ_B by the formula

$$P_l(\cos\theta_{AB}) = \frac{4\pi}{2l+1}\sum_{m=-l}^{+l}(-1)^{|m|}Y_{l,m}(\theta_A, \phi_A)Y_{l,-m}(\theta_B, \phi_B)$$
$$= \sum_{m=-l}^{+l}(-1)^{|m|}C_{l,m}(\theta_A, \phi_A)C_{l,-m}(\theta_B, \phi_B), \quad (2.27)$$

where the function

$$C_{l,m} = \left(\frac{4\pi}{2l+1}\right)^{\frac{1}{2}} Y_{l,m} \quad (2.28)$$

is also listed in Table 2.2. By the use of eqn (2.27) the potential at the point A due to a charge q_B at B may be written as

$$V = \frac{q_B}{4\pi\epsilon}\frac{1}{|\mathbf{R}-\mathbf{r}|}$$
$$= \frac{q_B}{4\pi\epsilon}\sum_{l=0}^{\infty}\sum_{m=-l}^{+l}(-1)^{|m|}\frac{r^l}{R^{l+1}}C_{l,m}(\theta_A, \phi_A)C_{l,-m}(\theta_B, \phi_B). \quad (2.29)$$

This shows that if we take A as a variable point, the potential at A due to the charge q_B at B has a series of components, and the magnitude of the component which varies as $r^l C_{l,m}(\theta_A, \phi_A)$ is determined by the value of $R^{-(l+1)} C_{l,-m}(\theta_B, \phi_B)$ at the point B. We may equally well take B to be a variable point at which we wish to find the potential due to a charge q_A at A; this is given by eqn (2.29) on replacing q_B by q_A. The potential at B then has a series of components, where the magnitude of the component varying as $R^{-(l+1)} C_{l,-m}(\theta_B, \phi_B)$ is determined by the value of $r^l C_{l,m}(\theta_A, \phi_A)$ at the point A.

If we have a number of point charges q_B at large distances, the magnitude of a given component in the potential at A near O can be found by the summation $\sum_B q_B R^{-(l+1)} C_{l,-m}(\theta_B, \phi_B)$, an example of which is given in Problem 2.18. Conversely, if we have a number of point charges q_A close to O, the magnitude of a given component in the potential at a distant point such as B can be found by the summation $\sum_A q_A r^l C_{l,m}(\theta_A, \phi_A)$. In either case the summation is replaced by an integration if we have a continuous distribution of charge. Note that the series expansion for the potential does not assume that $r \ll R$, but the terms will only converge rapidly if this is so.

The electrostatic energy of two charges q_A, q_B at the points A, B may be written by means of eqn (2.29) as

$$U_P = \frac{1}{4\pi\epsilon} \sum_{l=0}^{\infty} \sum_{m=-l}^{+l} (-1)^{|m|} \{q_A r^l C_{l,m}(\theta_A, \phi_A)\}\{q_B R^{-(l+1)} C_{l,-m}(\theta_B, \phi_B)\}, \quad (2.30)$$

which has the advantage that the terms involving the coordinates of the two charges have been separated. Thus if we have distributed charges with densities ρ_A, ρ_B at the points A, B the electrostatic energy is

$$\begin{aligned} U_P &= \frac{1}{4\pi\epsilon} \int\int \frac{\rho_A \rho_B \, d\tau_A \, d\tau_B}{|\mathbf{R} - \mathbf{r}|} \\ &= \frac{1}{4\pi\epsilon} \sum_{l=0}^{\infty} \sum_{m=-l}^{l} (-1)^{|m|} \int \rho_A r^l C_{l,m}(\theta_A, \phi_A) \, d\tau_A \times \\ &\quad \times \int \rho_B R^{-(l+1)} C_{l,-m}(\theta_B, \phi_B) \, d\tau_B \\ &= \frac{1}{4\pi\epsilon} \sum_l \sum_m A_{l,m} B_{l,-m}, \end{aligned} \quad (2.31)$$

where

$$A_{l,m} = \int \rho_A r^l C_{l,m}(\theta_A, \phi_A) \, d\tau_A, \quad (2.32)$$

$$B_{l,-m} = \int (-1)^{|m|} \rho_B R^{-(l+1)} C_{l,-m}(\theta_B, \phi_B) \, d\tau_B. \quad (2.33)$$

The quantities $A_{l,m}$ may be regarded as defining the components of the multipole moments of degree l, of the charge distribution near O, and it will be seen that they interact only with the conjugate components $B_{l,-m}$ which have the same value of l, m. The monopole component ($l=0$) contains only one term, while the dipole ($l=1$) components contain three terms which may easily be shown (Problem 2.15) to give the same interaction as eqn (2.24). In general the interaction energy involving r^l and $R^{-(l+1)}$ contains $(2l+1)$ terms, but the advantage of the quantities $A_{l,m}$, $B_{l,m}$ is that they can be expressed in terms of functions $C_{l,m}$ which are tabulated. We shall go no further than the quadrupole terms ($l=2$), which can be written out using Table 2.2. It can then readily be verified that, for the particular case where either charge distribution is symmetrical about the axis $\theta=0$, all the terms in eqn (2.31) vanish except that with $m=0$. If ρ_A has such symmetry (that is, it is independent of ϕ), its quadrupole interaction can be expressed in terms of a single component

$$A_{2,0}=\int \rho_A \tfrac{1}{2}r^2(3\cos^2\theta_A-1)\,d\tau_A=\tfrac{1}{2}qQ, \tag{2.34}$$

where the quantity (writing $r\cos\theta_A=z$)

$$Q=\frac{1}{|q|}\int \rho_A(3z^2-r^2)\,d\tau_A \tag{2.35}$$

is called the 'quadrupole moment' of the charge distribution, and has the dimensions of an area. It might be expected that one would take $q=\int \rho_A\,d\tau_A$, the total charge in the distribution, but for a nucleus by convention, q is taken as the charge on a single proton (not the total nuclear charge), and Q is expressed in terms of the unit of a barn (b) $=10^{-28}\,m^{-2}$. This unit is chosen because it is of the same order as the square of the nuclear radius (for an atom the quadrupole moment would be of order $10^{-20}\,m^{-2}$).

It has already been shown that invariance under the parity operation excludes the possibility of permanent electric dipole moments in atoms and nuclei. This may be expressed more generally using spherical harmonics. Inversion through the origin is equivalent to changing the point (r,θ,ϕ) into $(r,\pi-\theta,\pi+\phi)$. Since

$$P_{l,m}(\pi-\theta)=(-1)^{l-m}P_{l,m}(\theta), \qquad e^{jm(\pi+\phi)}=(-1)^m e^{jm\phi},$$

we have

$$C_{l,m}(\pi-\theta,\pi+\phi)=(-1)^l C_{l,m}(\theta,\phi), \tag{2.36}$$

and it follows that

$$\int \rho r^l C_{l,m}\,d\tau=0 \quad \text{if } l \text{ is odd.} \tag{2.37}$$

Thus invariance under the parity operation excludes the possibility of electric multipole moments of any odd degree.

The form in which the interaction energy is expressed in eqn (2.31) is very suitable to a case where one charge distribution (such as that of a nucleus) is confined to a small volume, but interacts with another charge distribution which is comparatively far away (such as the atomic electrons). The series then converges very rapidly, since $(r/R) \ll 1$; experimentally, nuclear electric multipole interactions higher than quadrupole are very difficult to detect. The convergence is much less rapid in atomic cases, such as the interaction between electrons within an atom, or between atomic electrons and the surrounding ions in a solid (which strongly affects their magnetic properties—see Chapter 14). However, integrals involving all but the first few values of l can be shown to vanish by means of orthogonality theorems, since the wavefunctions are themselves spherical harmonics.

2.4. Some electrostatic problems

Conducting sphere in a uniform field

Suppose an earthed conducting sphere of radius R is placed in a uniform field of strength $\mathbf{E}_0$. Then the field immediately around the sphere will become distorted owing to the induced charges on the surface of the sphere (Fig. 2.2), but the field at large distances will approach the value $\mathbf{E}_0$. In general the potential distribution can be expressed as a sum of terms of the type given in Table 2.1, with the condition that $V=0$ over the surface of the sphere. If we take a whole series of terms, it would turn out that the coefficients of most of them are zero. It is simpler to try a possible solution with a few terms whose nature is suggested by the symmetry of the problem; if it is then possible to satisfy the boundary condition $V=0$ at $r=R$ (taking the centre of the sphere as origin of coordinates), then this is the only correct solution.

If the polar axis is taken parallel to the uniform field strength $\mathbf{E}_0$, then

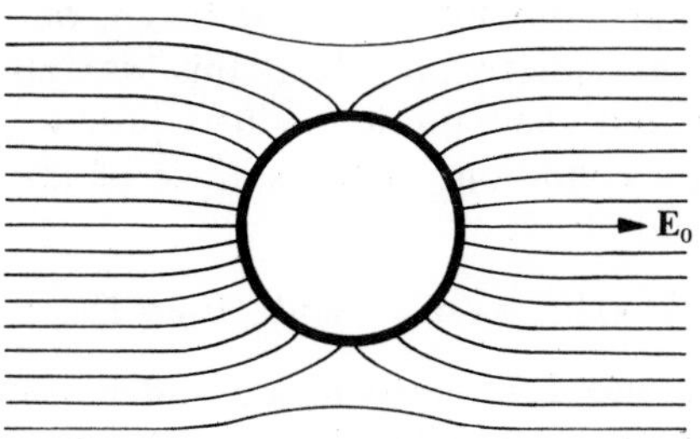

FIG. 2.2. The lines of force near a conducting sphere in a uniform electric field.

the potential at large distances is that of this field alone, so that

$$V = -E_0 r \cos\theta \quad \text{for } r \to \infty.$$

In order to make $V=0$ at $r=R$ for all values of θ, it seems likely that we can only add terms which vary with the same power of $\cos\theta$. Hence we try as a solution

$$V = -E_0 r \cos\theta + A r^{-2} \cos\theta. \tag{2.38}$$

It is clear that this satisfies our boundary condition if $E_0 R = A R^{-2}$; that is, $A = E_0 R^3$. Hence we have

$$V = -E_0 r \cos\theta (1 - R^3/r^3).$$

This shows that the potential outside the sphere is that due to the uniform field together with that of a dipole of moment $\mathbf{p} = 4\pi\epsilon \mathbf{E}_0 R^3$ situated at the centre of the sphere. Inside the sphere a solution of the type (2.38) is not acceptable, since it would give an infinite potential at the origin. Instead we must add a term $E_0 r \cos\theta$, which will just cancel the potential of the external field so that $V=0$ everywhere inside the sphere.

The magnitude of the induced charge at any point on the sphere can be found from the normal component of the field at the surface. This is

$$\begin{aligned} E_r = -\partial V/\partial r &= E_0 \cos\theta + 2E_0(R^3/r^3)\cos\theta \\ &= 3E_0 \cos\theta \quad \text{at } r = R. \end{aligned}$$

Hence the charge σ_e per unit area will be (from eqn (1.28))

$$\sigma_e = \epsilon E_r = 3\epsilon E_0 \cos\theta,$$

where ϵ is the permittivity of the medium surrounding the sphere.

Dielectric sphere in a uniform field

A slightly harder problem is that of a dielectric sphere of radius R and relative permittivity ϵ_1, surrounded by a medium of relative permittivity ϵ_2, and placed in a uniform field of strength $\mathbf{E}_0$, as in Fig. 2.3. Two separate potential functions must now be taken, one for inside and the other for outside the sphere (in effect, this was also required for the conducting sphere, but then inside we had just $V=0$). We must also satisfy the boundary conditions at the surface of the sphere, which are, from eqns (1.26) and (1.27),

$$\epsilon_1 E_{1r} = \epsilon_2 E_{2r} \quad \text{and} \quad E_{1t} = E_{2t},$$

where subscripts 1 and 2 refer to inside and outside the surface, and subscripts r and t refer to normal and tangential components respectively. Guided by the potential function for the case of the conducting sphere,

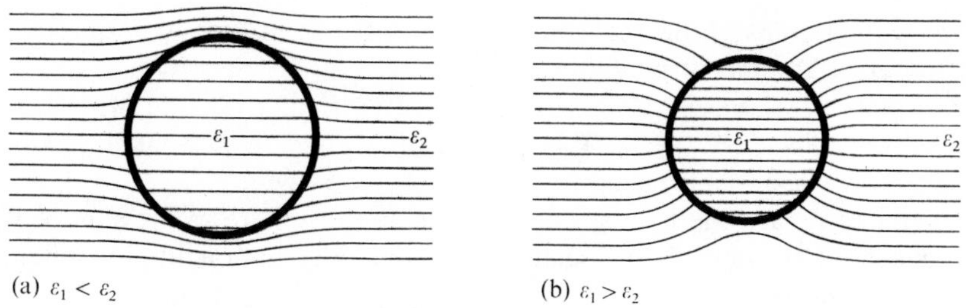

FIG. 2.3. The lines of electric displacement **D** due to a dielectric sphere of relative permittivity ϵ_1, in a uniform electric field in a medium of relative permittivity ϵ_2.

we shall assume

$$V_2 = -E_0 r \cos\theta + A_2 r^{-2}\cos\theta,$$
$$V_1 = B_1 r \cos\theta + B_2 r^{-2}\cos\theta$$

for outside and inside the sphere respectively. Clearly we cannot have $V_1 \to \infty$ for $r \to 0$, so that the coefficient B_2 must be zero. It is also obvious that V must be continuous at the boundary, since a discontinuity would give an infinite electric field there. Thus $V_1 = V_2$ at $r = R$; this automatically satisfies our second boundary condition since

$$E_t = E_\theta = -\frac{1}{r}\left(\frac{\partial V}{\partial \theta}\right),$$

and gives

$$B_1 R\cos\theta = -E_0 R\cos\theta + A_2 R^{-2}\cos\theta,$$

or

$$B_1 = A_2 R^{-3} - E_0. \tag{2.39}$$

The normal components of **E** at the surface are

$$E_{1r} = -(\partial V_1/\partial r)_{r=R} = -B_1\cos\theta$$

and

$$E_{2r} = -(\partial V_2/\partial r)_{r=R} = E_0\cos\theta + 2A_2R^{-3}\cos\theta.$$

Hence the first boundary condition gives

$$-B_1 = (\epsilon_2/\epsilon_1)(E_0 + 2A_2R^{-3}). \tag{2.40}$$

The solution of eqns (2.39) and (2.40) is

$$B_1 = -\left(\frac{3\epsilon_2}{\epsilon_1+2\epsilon_2}\right)E_0 \quad \text{and} \quad A_2 = \left(\frac{\epsilon_1-\epsilon_2}{\epsilon_1+2\epsilon_2}\right)R^3E_0,$$

so that the potential functions inside and outside the sphere are

$$V_1 = -\left(\frac{3\epsilon_2}{\epsilon_1+2\epsilon_2}\right)E_0 r\cos\theta, \tag{2.41}$$

$$V_2 = -\left(1 - \frac{R^3}{r^3}\,\frac{\epsilon_1-\epsilon_2}{\epsilon_1+2\epsilon_2}\right)E_0 r\cos\theta. \tag{2.42}$$

These equations show that the field strength $\mathbf{E}_1$ inside the sphere is parallel to $\mathbf{E}_0$, and of magnitude

$$\mathbf{E}_1 = \left(\frac{3\epsilon_2}{\epsilon_1+2\epsilon_2}\right)\mathbf{E}_0. \tag{2.43}$$

We can regard the field strength $\mathbf{E}_1$ inside the sphere as the sum of $\mathbf{E}_0$ and a field strength $\mathbf{E}_d$ associated with the polarization of the sphere, whose value is

$$\mathbf{E}_d = \mathbf{E}_1 - \mathbf{E}_0 = -\left(\frac{\epsilon_1-\epsilon_2}{\epsilon_1+2\epsilon_2}\right)\mathbf{E}_0. \tag{2.44}$$

When $\epsilon_1 > \epsilon_2$, $\mathbf{E}_d$ is in the reverse direction to $\mathbf{E}_0$, and it is known as the 'depolarization field'. It is proportional to $(\mathbf{P}_1-\mathbf{P}_2)$, where $\mathbf{P}_1 = (\epsilon_1-1)\epsilon_0\mathbf{E}_1$ is the actual polarization inside the sphere and $\mathbf{P}_2 = (\epsilon_2-1)\epsilon_0\mathbf{E}_0$ is the value it would have if the permittivity of the sphere were the same as that of the surrounding medium. The value of $(\mathbf{P}_1-\mathbf{P}_2)$ is readily found to be

$$\mathbf{P}_1-\mathbf{P}_2 = \epsilon_0\mathbf{E}_0(\epsilon_1-\epsilon_2)\left(\frac{1+2\epsilon_2}{\epsilon_1+2\epsilon_2}\right) \tag{2.45}$$

and

$$\mathbf{E}_d = -\frac{(\mathbf{P}_1-\mathbf{P}_2)}{\epsilon_0(1+2\epsilon_2)}. \tag{2.46}$$

Obviously both $(\mathbf{P}_1-\mathbf{P}_2)$ and $\mathbf{E}_d$ vanish when $\epsilon_1 = \epsilon_2$.

Outside the sphere, apart from the term $-\mathbf{E}_0 r \cos\theta$ arising directly from the field strength $\mathbf{E}_0$, the potential is just that of a point dipole $\mathbf{p}$ situated at the centre of the sphere and immersed in a uniform and infinite medium of relative permittivity ϵ_2. The value of $\mathbf{p}$ is

$$\mathbf{p} = V\frac{3\epsilon_2}{(1+2\epsilon_2)}(\mathbf{P}_1-\mathbf{P}_2), \tag{2.47}$$

where $V = 4\pi R^3/3$ is the volume of the sphere.

In the special case of a sphere in a vacuum, $\epsilon_2 = 1$ and $\mathbf{P}_2 = 0$; then $\mathbf{E}_d = -\frac{1}{3}(\mathbf{P}_1/\epsilon_0)$, and $\mathbf{p} = V\mathbf{P}_1$.

A depolarization field is present in all dielectric bodies but (1) only for a simple mathematical figure such as the ellipsoid is it a uniform field, and (2) it is only parallel to the polarization $\mathbf{P}$ when $\mathbf{P}$ is along a principal axis of the ellipsoid. Parallel to a principal axis such as the a-axis there exists a depolarization field strength $\mathbf{E}_d$ such that

$$\mathbf{E}_d = -\frac{\mathbf{P}_1-\mathbf{P}_2}{\epsilon_0\{1+(d_a^{-1}-1)\epsilon_2\}}, \tag{2.48}$$

where d_a is a pure number which depends only on the shape of the ellipsoid. The principal axes are mutually perpendicular, and any uniform polarization **P** can be resolved into components P_a, P_b, P_c; the factors d_a, d_b, d_c are unequal except for a sphere, so that the depolarizing field will be uniform but not in general parallel to **P**.

The ellipsoid is a useful example because it includes not only the sphere, but also the thin infinite slab and the long infinite cylinder as limiting cases. Mathematically it is rather tedious and the results are given in Appendix B. An important point is that for a finite slab and a finite cylinder the internal field and the polarization cannot be uniform, because of the boundary conditions at the corners.

Problems with cylindrical symmetry—conducting cylinder in a uniform field

In some three-dimensional problems the potential may be independent of one coordinate, and the problem then reduces to a two-dimensional one. It is often convenient to use cylindrical coordinates in such a case, taking the z direction as that in which the potential is invariant. Then putting $\rho_e = 0$ and $\partial^2 V/\partial z^2 = 0$ in eqn (2.5), we have for Laplace's equation

$$r\frac{\partial}{\partial r}\left(r\frac{\partial V}{\partial r}\right)+\frac{\partial^2 V}{\partial \theta^2}=0. \tag{2.49}$$

This is satisfied by a function of the form $V = r^n D_n$, where D_n is a function of θ alone (known as a cylindrical harmonic) which must satisfy the differential equation

$$\frac{\partial^2 D_n}{\partial \theta^2}+n^2 D_n = 0. \tag{2.50}$$

This equation is unchanged by the substitution of $-n$ for n, so that if $V = r^n D_n$ is a solution of Laplace's equation, so also is $V = r^{-n} D_n$. One solution is $V = \ln r$, and other solutions may be obtained either by partial differentiation with respect to $x = r\cos\theta$, or by direct solution of eqn (2.50).

A number of the simplest functions are given in Table 2.3; note that the general form of D_n will be $A_n \cos n\theta + B_n \sin n\theta$, where A_n and B_n are constants.

TABLE 2.3
Some cylindrical harmonic functions

Cylindrical harmonic	Corresponding solutions of Laplace's equation	
D_0	$\log r$	1
D_1	$r^{-1}(A\cos\theta+B\sin\theta)$	$r(A\cos\theta+B\sin\theta)$
D_2	$r^{-2}(A\cos 2\theta+B\sin 2\theta)$	$r^2(A\cos 2\theta+B\sin 2\theta)$
D_3	$r^{-3}(A\cos 3\theta+B\sin 3\theta)$	$r^3(A\cos 3\theta+B\sin 3\theta)$

The type of problem to which the solutions may be applied is illustrated by the case of a conducting circular cylinder, initially uncharged, lying with its axis at right-angles to a uniform field of strength $\mathbf{E}_0$. If the axis of the cylinder is taken as the z-axis, it is clear that the potential distribution will be independent of z. At large distances the potential will tend to $V=-E_0 r\cos\theta$, and we will assume that other terms required must also vary as $\cos\theta$. Then the potential outside the cylinder will be of the form

$$V=-E_0 r\cos\theta+Ar^{-1}\cos\theta.$$

To satisfy the boundary condition $V=0$ at $r=R$ for all values of θ, we must have $E_0 R=AR^{-1}$, so that the potential is

$$V=-E_0 r\cos\theta(1-R^2/r^2). \tag{2.51}$$

The first term is the potential of the external field, the second that of an extended dipole consisting of two parallel lines of positive and negative charge close to the z-axis.

In these problems, we have assumed a potential containing just the required number of terms. This is a matter of intelligent anticipation rather than guesswork or knowing the answer beforehand. If we had taken fewer terms, we could not have satisfied the boundary conditions. If we had taken more terms, the coefficients of the additional terms would have been found to be zero. In the case just considered, terms such as $\cos n\theta$ or $\sin n\theta$ could not satisfy the condition $V=0$ at $r=R$ for *all values of* θ, because the potential of the external field varies only as $\cos\theta$. We are justified in assuming that the solution we have found is the correct and only solution because of the uniqueness theorem. This theorem also justifies the use of another special method, which we shall now consider.

2.5. Electrical images

If we have two equal point charges of opposite sign separated by a certain distance $2a$, the plane passing through the midpoint of the line joining them and normal to this line is an equipotential surface at zero potential. Therefore if the negative charge (say) is replaced by a plane conducting sheet AB in Fig. 2.4, the field to the right of AB will remain unaltered. Conversely, if a point charge is placed in front of an infinite conducting plane, the resultant electric field to the right of AB will be the same as that produced by the original charge plus a negative charge an equal distance from the plane on the opposite side. The negative charge is the 'electrical image' of the original charge in the plane AB.

The method of images thus consists in replacing a conductor by a point charge such that the conducting surface is still an equipotential surface. Then Laplace's equation is still satisfied at all points outside the conductor, and by the principle of uniqueness, the problem of a point charge and its image is identical with that of a point charge and an infinite conducting surface as regards the region outside the conducting surface. Electrical

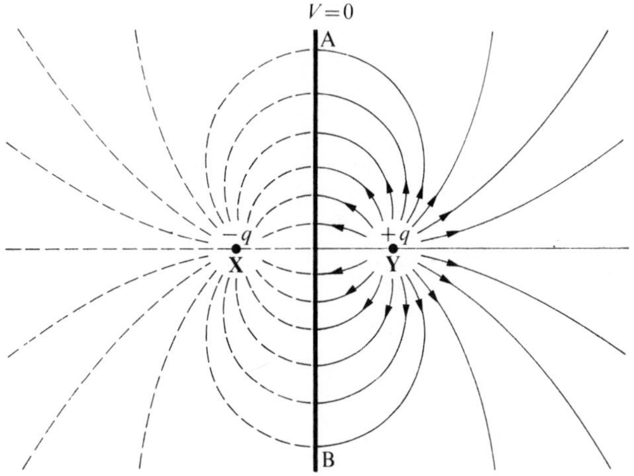

FIG. 2.4. A point charge $+q$ and its image charge $-q$ in an infinite conducting plane AB, showing the lines of force from $+q$ on the right of the plane which end on the surface charge on AB. The distance $XY = 2a$.

images are entirely virtual; a field on one side of a closed equipotential surface cannot be represented by an image on the same side of the surface, since this would give a singularity at the point occupied by the image charge. The field on the one side of the surface is identical with that which would be produced if the surface were replaced by an image charge on the other side of it.

Point charge and infinite conducting plane

The method of images will now be applied to a number of special cases, the simplest of which is that of a point charge q placed a distance a from an infinite conducting plane at zero potential. In this case it is obvious that the image must be a charge $-q$ at a distance a behind the plane, as in Fig. 2.5. The potential at an arbitrary point P is then

$$V = \frac{q}{4\pi\epsilon}\left\{\frac{1}{r} - \frac{1}{(r^2+4a^2+4ar\cos\theta)^{\frac{1}{2}}}\right\},$$

where ϵ is the permittivity of the medium outside the conductor. In order to calculate the charge density at any point on the plane, we must find the component of the electric field normal to the plane. This is (see Fig. 2.5)

$$E_x = E_r\cos\theta - E_\theta\sin\theta = -\frac{\partial V}{\partial r}\cos\theta + \frac{1}{r}\frac{\partial V}{\partial\theta}\sin\theta$$

$$= \frac{q}{4\pi\epsilon}\left\{\frac{\cos\theta}{r^2} - \frac{r\cos\theta+2a}{(r^2+4a^2+4ar\cos\theta)^{\frac{3}{2}}}\right\}$$

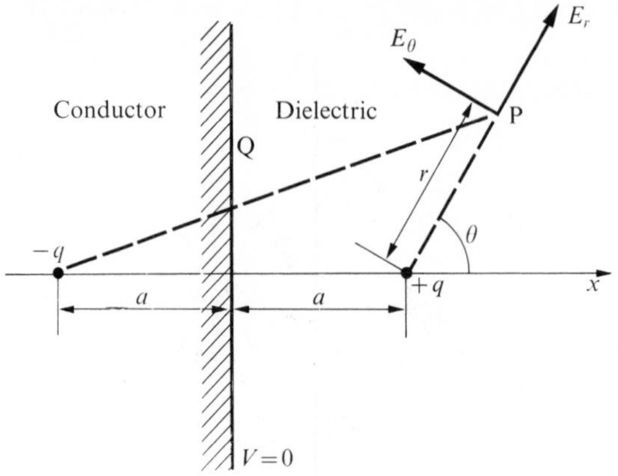

FIG. 2.5. Image of point charge in a conducting plane.

at the point P. At the point Q on the plane, $r\cos\theta = -a$, so that at Q

$$E_x = -qa/2\pi\epsilon r^3$$

(cf. Problem 2.6), where r is the distance of Q from the charge $+q$. The induced charge per unit area at Q is then

$$\sigma_e = \epsilon E_x = -qa/2\pi r^3. \tag{2.52}$$

The force exerted on the point charge by the induced charge on the plane is just equal to the force exerted on the charge by its image. That is, $F = -q^2/16\pi\epsilon a^2$, where the negative sign indicates that the charge is attracted towards the plane (see Problem 2.6).

Point charge and conducting sphere

A more difficult problem is that of a point charge q placed (*in vacuo*) a distance a from the centre of a conducting sphere of radius R (Fig. 2.6).

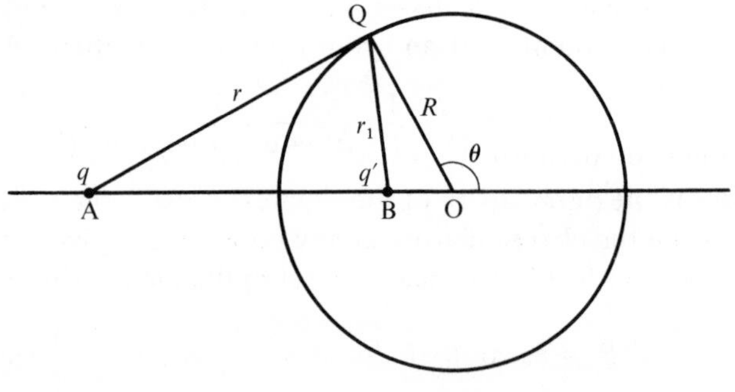

FIG. 2.6. A point charge q and its image q' in a conducting sphere of radius R.

OA $= a$, OB $= b$.

We shall consider first the case where the sphere is earthed and at zero potential. By symmetry, the image charge must be on the line through q to the centre of the sphere O, and we will assume that it consists of a single charge q' at a distance b from O. The potential at a point Q on the surface of the sphere is then

$$V=\frac{1}{4\pi\epsilon}\left(\frac{q}{r}+\frac{q'}{r_1}\right)$$

$$=\frac{1}{4\pi\epsilon}\left\{\frac{q}{(R^2+a^2+2aR\cos\theta)^{\frac{1}{2}}}+\frac{q'}{(R^2+b^2+2bR\cos\theta)^{\frac{1}{2}}}\right\}.$$

It is only possible to make $V=0$ over the whole surface of the sphere (that is, for all values of θ) if the functions in the denominators are similar functions of θ. This requires that we choose b so that $b/R=R/a$; that is, B is the inverse point to A in the sphere. Then the potential at Q is

$$V=\frac{q+(a/R)q'}{4\pi\epsilon(R^2+a^2+2Ra\cos\theta)^{\frac{1}{2}}},$$

and this will be zero if we make $q'=-q(R/a)$. Hence the image charge is of magnitude $-q(R/a)$ at the inverse point in the sphere, and the reader may verify, by integrating the charge density on the sphere, that the total charge on the sphere is equal to the image charge.

If the sphere is insulated and initially uncharged, the total charge on it must remain zero. It is therefore necessary to add a second image charge $-q'$ at such a point that the surface of the sphere remains an equipotential surface. This is accomplished by placing a charge $+q(R/a)$ at the centre of the sphere in addition to the charge $-q(R/a)$ at the inverse point. If the sphere is insulated but carries an initial charge Q the total charge at the centre would of course be $Q+q(R/a)$.

2.6. Line charges

Just as we have considered the mathematical abstraction of a point charge, so we may postulate a 'line charge' in which charge is uniformly distributed along an infinite straight line. Its strength is denoted by λ, the charge per unit length. To find the field of such a line charge, immersed in a medium of permittivity ϵ, we apply Gauss's theorem to a section of length t of a cylinder of radius r whose axis coincides with the line charge. This gives

$$\epsilon E(2\pi rt)=\lambda t,$$

since by symmetry the field strength **E** is everywhere normal to the axis. Hence $E=\lambda/2\pi\epsilon r$, and the potential at a distance r from the axis is

$$V=-\frac{\lambda}{2\pi\epsilon}\int\frac{\mathrm{d}r}{r}=-\frac{\lambda}{2\pi\epsilon}\ln r+V_0. \qquad (2.53)$$

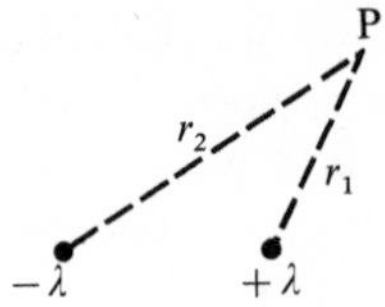

FIG. 2.7. Two parallel line charges normal to the plane of the paper, with charge $+\lambda$ and $-\lambda$ per unit length.

Here the constant V_0 cannot be defined by assuming $V=0$ at $r=\infty$ since the line charge itself extends to infinity.

If we have two parallel line charges of equal strength but opposite sign, as in Fig. 2.7, the potential at a point whose perpendicular distances from the line charges are r_1, r_2 respectively is

$$V=\frac{\lambda}{2\pi\epsilon}\ln(r_2/r_1)+V_0. \tag{2.54}$$

The equipotential surfaces given by this equation are shown in Fig. 2.8. They have the form of cylinders whose cross-sections form a set of coaxial circles with limiting points at the line charges. The surface whose potential is V_0 is the median plane (for which $r_1=r_2$) between the two line charges. From this it follows that the problem of a line charge parallel to a conducting plane can be solved by the method of images, using a line charge of opposite sign as image. We shall apply our results to a more realistic problem.

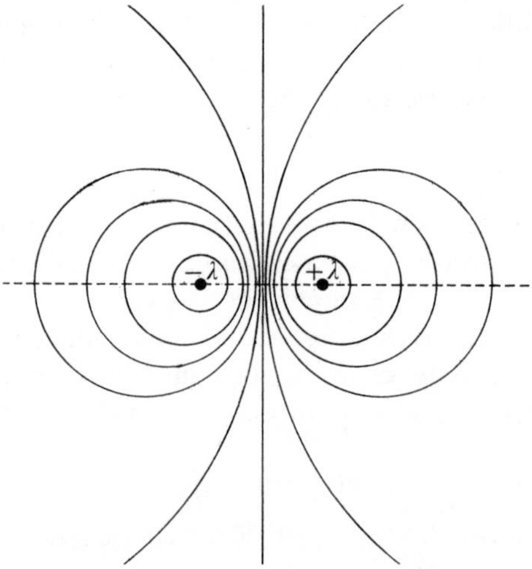

FIG. 2.8. The lines of constant potential for two parallel infinite line charges $+\lambda$ and $-\lambda$, normal to the plane of the paper. They form systems of coaxial circles with limiting points at the charges.

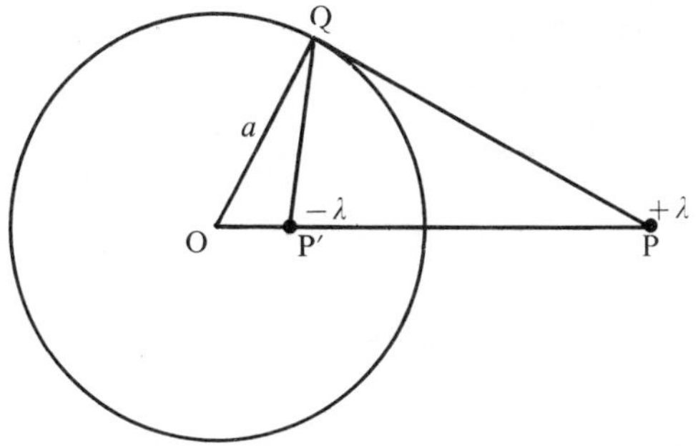

FIG. 2.9. The image of an infinite line charge in an infinite conducting cylinder.

Capacitance between two parallel infinite circular cylinders

Consider first an infinite line charge of strength λ which is parallel to an infinite cylinder of radius a. In the cross-section shown in Fig. 2.9, the line charge is at P and the axis of the cylinder at O. We imagine an image line charge of strength $-\lambda$ to be placed at P′, the position of P′ being chosen so that the circle formed by the cross-section of the conducting cylinder coincides with one of the family of coaxial circles which are the equipotentials of the line charge and its image. Then the potential at any point Q on the surface of the cylinder is

$$V = -\frac{\lambda}{2\pi\epsilon}\ln\left(\frac{QP}{QP'}\right) + V_0.$$

Since P and P′ are the limiting points of the family of coaxial circles, it follows that they are inverse points with regard to any one of these circles. Hence

$$QP/QP' = OP/OQ = OP/a,$$

and the potential at Q is

$$V = -\frac{\lambda}{2\pi\epsilon}\ln\left(\frac{OP}{a}\right) + V_0,$$

which is independent of the position of Q on the surface of the cylinder, showing that this is an equipotential.

We turn now to the case of two infinite parallel cylinders, each of radius a, whose axes are a distance $2d$ apart. Then in Fig. 2.10 the distance OO′ is $2d$, and P, P′ are the limiting points of a family of coaxial circles. P and P′ are chosen so that two of the circles coincide with the surfaces of the cylinders, making each of these an equipotential if line charges of strength λ and $-\lambda$ were placed at P and P′ respectively. Then the potential at an arbitrary point whose distance is r_1 from P and r_2 from P′ is

$$V = -\frac{\lambda}{2\pi\epsilon}\ln(r_1/r_2),$$

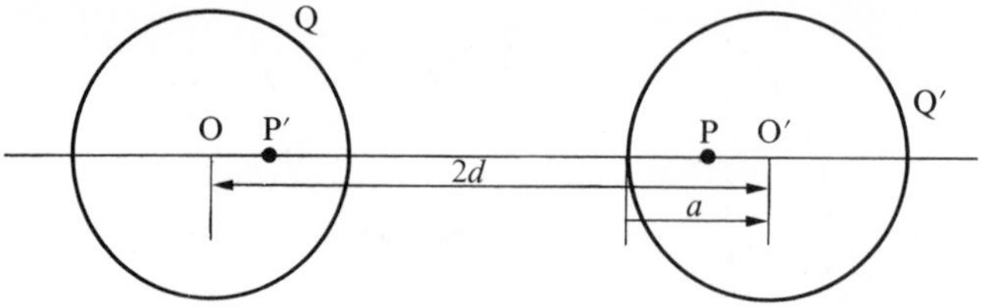

FIG. 2.10. Two infinitely long parallel conducting cylinders.

where the constant V_0 is zero if we take the median plane ($r_1 = r_2$) between the cylinders to be at zero potential. Then the potentials of the two cylinders are

$$V_Q = -\frac{\lambda}{2\pi\epsilon}\ln(OP/a) \quad \text{and} \quad V_{Q'} = +\frac{\lambda}{2\pi\epsilon}\ln(OP/a)$$

respectively. But $OP + OP' = 2d$, and $OP.OP' = a^2$, since P, P′ are inverse points in the circle of radius a centre O. Hence

$$OP = d + \sqrt{(d^2 - a^2)},$$

and the capacitance per unit length between the two cylinders is

$$C = \lambda/(V_{Q'} - V_Q)$$

$$= \frac{\pi\epsilon}{\ln[\{d + \sqrt{(d^2 - a^2)}\}/a]}. \qquad (2.55)$$

When $d \gg a$, this approaches the limiting value

$$C = \frac{\pi\epsilon}{\ln(2d/a)}. \qquad (2.56)$$

A similar problem is the capacitance of a horizontal telegraph wire with respect to the earth. This may be treated as an infinite cylinder of radius a a distance d above an infinite conducting plane. It is clear that the potential distribution will be the same as in the case of the two parallel cylinders if we assume that the conducting plane coincides with the median plane between the cylinders, which is the equipotential surface $V = 0$. Then the charge on the wire per unit length is λ, and the potential difference between it and the plane is just half that between the two cylinders in the previous problem. Hence the capacitance per unit length will be (assuming $d \gg a$)

$$C = \frac{2\pi\epsilon}{\ln(2d/a)}. \qquad (2.57)$$

Note that the approximation $d \gg a$ is tantamount to assuming that the wire behaves as if it had a line charge λ per unit length along its axis, since as (a/d) approaches zero the point P′ moves towards O in Fig. 2.10 and OP $\to 2d$.

2.7. Images in dielectrics

The potential distribution due to a point charge near a dielectric surface may sometimes be found by the method of images. We shall illustrate this type of problem by considering the case of a point charge q a distance a from a semi-infinite dielectric bounded by a plane surface. This problem is more complex than that of a point charge and conducting plane since a second image system is required to represent the field within the dielectric. It is not obvious that the field can be represented by that of a single point charge, but we shall assume that this is possible (if our assumption is wrong we shall not be able to satisfy the boundary conditions). We take therefore a single charge q_2 at a point B, as in Fig. 2.11, and the potential at point Q in the dielectric will then be

$$V_Q = \frac{q_2}{4\pi\epsilon_0 r_2}. \tag{2.58}$$

The field of the point charge q will polarize the dielectric and there will therefore be a surface charge on the dielectric which affects the field outside. We assume that this can be represented by an image charge q_1 at C in the dielectric, and the potential at point P outside the dielectric is then

$$V_P = \frac{1}{4\pi\epsilon_0}\left(\frac{q}{r} + \frac{q_1}{r_1}\right). \tag{2.59}$$

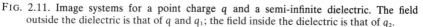

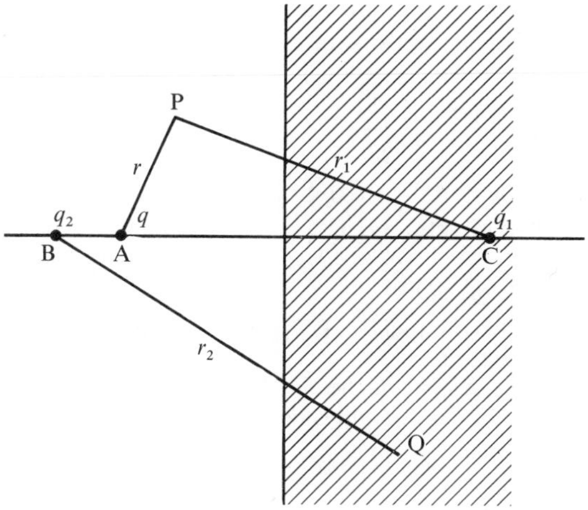

FIG. 2.11. Image systems for a point charge q and a semi-infinite dielectric. The field outside the dielectric is that of q and q_1; the field inside the dielectric is that of q_2.

By symmetry, q, q_1, and q_2 will all lie on a normal to the dielectric surface. Note that the relative permittivity ϵ_r does not appear in these equations since the effect of the dielectric is replaced by two image systems *in vacuo.*

To avoid an infinite electric field at the boundary we must assume that $V_P = V_Q$ at the boundary, and this automatically satisfies our first boundary condition, that the tangential components of **E** must be the same on either side of the boundary. It is clear that the condition $V_P = V_Q$ everywhere on the boundary can only be satisfied if r, r_1, and r_2 vary at the same rate as we move along the boundary, and we must therefore have $r = r_1 = r_2$, so that B coincides with A, and C is as far behind the surface as A is in front. In addition, $q + q_1 = q_2$. Our second boundary condition is that the normal components of **D** must be continuous at the boundary, that is, $(\partial V_P/\partial z) = \epsilon(\partial V_Q/\partial z)$ at the surface, which we take to be the plane $z = 0$. Now at an arbitrary point (x, y, z), $r_2 = \{x^2 + y^2 + (a+z)^2\}^{\frac{1}{2}}$, and $r_1 = \{x^2 + y^2 + (a-z)^2\}^{\frac{1}{2}}$, so that

$$\frac{\partial}{\partial z}\left(\frac{1}{r}\right) = \frac{\partial}{\partial z}\left(\frac{1}{r_2}\right) = -\frac{a+z}{r^3}, \qquad \frac{\partial}{\partial z}\left(\frac{1}{r_1}\right) = \frac{a-z}{r_1^3}.$$

Using these relations for the case $z = 0$, our second boundary condition becomes

$$\frac{a(q-q_1)}{4\pi\epsilon_0 r^3} = \frac{\epsilon_r a q_2}{4\pi\epsilon_0 r^3}.$$

Hence we have the relations

$$q + q_1 = q_2 \quad \text{and} \quad q - q_1 = \epsilon_r q_2,$$

which give $q_2 = 2q/(\epsilon_r + 1)$ and $q_1 = -q(\epsilon_r - 1)/(\epsilon_r + 1)$ for the image charges. The force of attraction on the charge q towards the dielectric is therefore

$$F = -\frac{qq_1}{4\pi\epsilon_0(2a)^2} = \frac{q^2(\epsilon_r - 1)}{16\pi\epsilon_0 a^2(\epsilon_r + 1)}. \tag{2.60}$$

The lines of displacement for the case of a point charge and an infinite dielectric are shown in Fig. 1.13.

References

BRINK, D. M. and SATCHLER, G. R. (1968). *Angular momentum* (2nd edn). Clarendon Press, Oxford.

RAMSEY, N. F. (1956). *Molecular beams.* Clarendon Press, Oxford.

STRATTON, J. A. (1941). *Electromagnetic theory.* McGraw-Hill, New York.

Problems

2.1. The polarization charge on the surface of a spherical cavity is $-\sigma_e \cos\theta$, at a point whose radius vector from the centre makes an angle θ with a given axis Oz. Prove that the field strength at the centre is $\sigma_e/3\epsilon_0$, parallel to Oz.

If the cavity is in a uniform dielectric subject to a field of strength $\mathbf{E}_0$ parallel to the direction $\theta = 0$, show that $\sigma_e = 3E_0\epsilon_o(\epsilon_r - 1)/(1+2\epsilon_r)$, where ϵ_r is the relative permittivity of the dielectric. Verify that this gives the correct value for the field strength at the centre of the cavity (eqn (2.43)) and note that σ_e is not simply $(\epsilon_r - 1)\epsilon_0 E_0$ because of the distortion of the field in the dielectric caused by the presence of the cavity.

2.2. A dipole $\mathbf{p}$ is situated at the centre of a spherical cavity of radius a in an infinite medium of relative permittivity ϵ_r. Show that the potential in the dielectric medium is the same as would be produced by a dipole $\mathbf{p}'$ immersed in a continuous dielectric, where

$$\mathbf{p}' = \mathbf{p}\frac{3\epsilon_r}{2\epsilon_e + 1},$$

and that the field strength inside the cavity is equal to that which the dipole would produce in the absence of the dielectric, plus a uniform field $\mathbf{E}_r$

$$\mathbf{E}_r = \frac{2(\epsilon_r - 1)}{2\epsilon_r + 1}\frac{\mathbf{p}}{4\pi\epsilon_0 a^3}.$$

$\mathbf{E}_r$ is known as the 'reaction field strength'. These formulae are used in Onsager's theory of dielectrics (see Chapter 10).

2.3. An infinite circular cylinder of relative permittivity ϵ_r is subjected in vacuum to a transverse uniform field of strength $\mathbf{E}_0$. Show that the field inside the cylinder is

$$\frac{2}{\epsilon_r + 1}\mathbf{E}_0,$$

and that the depolarizing field is $-\mathbf{P}_1/2\epsilon_0$.

2.4. Find an expression for the surface density of charge on an infinitely long conducting cylinder of radius a, placed with its axis at right-angles to a uniform electric field of strength $\mathbf{E}_0$, as a function of the polar angle θ.

(*Answer*: $\sigma_e = 2\epsilon E_o \cos\theta$.)

2.5. A uniform electric field of strength $\mathbf{E}$ is set up in an infinite dielectric. Show that (*a*) if a long needle-shaped cavity, whose lateral dimensions are very small compared with its length, is cut in the dielectric with its axis parallel to $\mathbf{E}$, then the field strength in this cavity is $\mathbf{E}$; (*b*) if a flat disc-shaped cavity, whose lateral dimensions are very large compared with its thickness, is cut with its plane normal to the direction of $\mathbf{E}$, then the field strength in the cavity is $\mathbf{D}/\epsilon_0$, where $\mathbf{D}$ is the displacement in the dielectric.

Verify that the field in an intermediate shape of cavity (such as in Problem 2.1) lies between these extreme values.

2.6. A point charge q is placed near an infinite conducting plane. Verify that the total charge on the plane is $-q$ by integrating eqn (2.52), and calculate the total force on the plane by integration of the tension per unit area (eqn (1.42)) over the area of the plane. Verify the expression given in the text for the field strength at Q (Fig. 2.5) by vector addition of the fields of the point charge and its image.

2.7. Show that the work done in bringing up a charge q from infinity to a distance a from a conducting plane at zero potential is $-q^2/16\pi\epsilon a$. Verify that the same result is obtained using eqn (1.34) (remember that the induced charge on the plane is at zero potential).

2.8. Show that the force on a charge q distance a *in vacuo* from the centre of an insulated and uncharged conducting sphere of radius R is $(a>R)$

$$F=\frac{q^2}{4\pi\epsilon_0}\left\{\frac{R}{a^3}-\frac{Ra}{(a^2-R^2)^2}\right\}.$$

2.9. A point charge is placed in a hollow metal sphere of radius R. If the charge is q and is a distance b from the centre of the sphere show that the force on it is

$$\frac{q^2}{4\pi\epsilon_0}\frac{Rb}{(R^2-b^2)^2}.$$

In what direction does the force act?

(*Hint.* Find the image of q outside the sphere such that the sphere is an equipotential surface.)

2.10. A point charge q is placed at a distance $3R$ from the centre of an isolated conducting sphere of radius R which already has a charge equal to q. Prove that the surface densities at points on the sphere nearest to and farthest from the point charge are in the ratio 8:29.

2.11. An elementary dipole of strength p is placed at a point P, outside and at a distance a from the centre C of an earthed conducting sphere of radius R. The axis of the dipole is in the direction CP. Prove that its image system consists of a point charge pR/a^2 and a dipole of strength pR^3/a^3, both situated at the point P′ which is inverse to P in the sphere.

2.12. Show that if an infinite line charge λ per unit length is at a distance d from an infinite conducting plane in a medium of permittivity ϵ, the surface density of charge in the plane is $\sigma_e=-d\lambda/\pi r^2$, where r is the shortest distance from the line of charge to the point in question.

2.13. Calculate the force per unit length on the infinite line charge of the last question.

(*Answer*: $F=-\lambda^2/4\pi\epsilon d$.)

2.14. Electric charge is distributed over a thin spherical shell with a density which varies in proportion to the value of a single function $P_l(\cos\theta)$ at any point on the shell. Show, by using the expansions (2.26) and (2.27) and the orthogonality relations for the Legendre functions, that the potential varies as $r^l P_l(\cos\theta)$ at a point (r,θ) inside the sphere and $r^{-(l+1)}P_l(\cos\theta)$ at a point (R,θ) outside.

2.15. Show that, for $l=1$, the quantities defined by eqn (2.32) are

$$A_{1,0}=p_z,\qquad A_{1,1}=-2^{-\frac{1}{2}}(p_x+jp_y),\qquad A_{1,-1}=2^{-\frac{1}{2}}(p_x-jp_y),$$

while for a point charge at B in Fig. 2.1 (b),

$$\frac{1}{4\pi\epsilon}B_{1,0}=-E_z,\qquad \frac{1}{4\pi\epsilon}B_{1,1}=-2^{-\frac{1}{2}}(E_x+jE_y),\qquad \frac{1}{4\pi\epsilon}B_{1,-1}=2^{-\frac{1}{2}}(E_x-jE_y),$$

and verify that this gives the same interaction energy as eqn (2.24).

2.16. For an atom in a p-state, the wavefunction is $\psi=f(R)Y_{1,0}$, and the charge density is $-e\psi^2$. Show that the atomic quadrupole moment (as defined by eqn

(2.35)) is $\frac{4}{5}\langle R^2\rangle$, where $\langle R^2\rangle$ is the mean-square distance of the electron from the nucleus.

2.17. The isotope of mass 35 of chlorine has a nuclear electric quadrupole moment Q. Show that for this nucleus in a chlorine atom whose wavefunction $\psi = f(R)Y_{1,0}$, the energy of interaction between the nuclear electric quadrupole moment and the electron is

$$U_P = -\frac{\frac{1}{5}e^2Q\langle R^{-3}\rangle}{4\pi\epsilon_0},$$

where $\langle R^{-3}\rangle$ is the mean inverse cube of the distance of the electron from the nucleus and e the electronic charge.

If $Q = -0{\cdot}079\,\text{b} = -7{\cdot}9\times10^{-30}\,\text{m}^2$, and $\langle R^{-3}\rangle = 5\times10^{31}\,\text{m}^{-3}$, show that the energy of interaction (U_P/h) expressed in frequency units (h is Planck's constant) is about 27 MHz.

2.18. Six equal charges q are placed at the points $(\pm R, 0, 0)$, $(0, \pm R, 0)$, $(0, 0, \pm R)$. Show that the terms of lowest degree in the potential at a point $(x, y, z) = (r, \theta, \phi)$ near the origin are

$$V = \frac{6q}{4\pi\epsilon R} + \frac{q}{4\pi\epsilon}\cdot\frac{r^4}{R^5}\cdot\frac{7}{2}\{C_{4,0} + (\tfrac{5}{14})^{\frac{1}{2}}(C_{4,4} + C_{4,-4})\},$$

where $C_{4,0} = P_4$ (which is given in Table 2.1), and

$$C_{4,\pm4} = (\tfrac{35}{128})^{\frac{1}{2}}\sin^4\theta\,e^{\pm j4\phi}.$$

(*Hints*. Since the system has inversion symmetry through the origin, terms in odd powers of r must vanish. Also V must have fourfold symmetry about the polar (z-)axis, so that rotations changing ϕ by $\frac{1}{2}\pi$ must leave V unchanged: only functions with $m = 0$ or 4 satisfy this condition.)

Note that in Cartesian coordinates

$$V = \frac{6q}{4\pi\epsilon R} + \frac{q}{4\pi\epsilon}\cdot\frac{1}{R^5}\cdot\frac{35}{4}(x^4 + y^4 + z^4 - \tfrac{3}{5}r^4),$$

showing that the x-, y-, z-axes are all equivalent (cubic symmetry).

2.19. An electric double layer is formed of two parallel planes a distance t apart, one with a surface charge density $+\sigma_e$ and the other with a surface charge density $-\sigma_e$. Show that the normal components of $\mathbf{D}$ are the same on either side of the double layer (but not in between), but that the values of the potential V differ by

$$V_2 - V_1 = P_s/\epsilon_0,$$

where $P_s = \sigma_e t$ is the dipole moment per unit area of the double layer, directed from side 1 to side 2.

3. Current and voltage

3.1. Introduction

In the previous chapters on electrostatics we have been concerned with stationary electric charges. If a free charge is placed in an electric field it will be acted on by a force, and will move in the direction of the lines of force. Thus, if an initial difference of potential exists in a conductor, the charges will move until they reach positions of equilibrium, and the whole of the conductor becomes an equipotential surface. But by connecting a battery between two points of a conductor, a permanent difference of potential may be maintained between these two points, and there will then be a continuing flow of charge. This constitutes an electric current, and the strength of the current I is defined by the rate at which charge passes any given point in the circuit. If we are dealing with a current extended in space, then we may define the current density $\mathbf{J}$ as the quantity of charge passing per second through unit area of a plane normal to the line of flow. The total current flowing through any surface is found by integrating the normal component of the current density: that is

$$I=\int \mathbf{J}.\mathrm{d}\mathbf{S}. \tag{3.1}$$

If the current is carried by particles of charge q with density n per unit volume and velocity $\mathbf{v}$, then the current density is

$$\mathbf{J}=nq\mathbf{v}. \tag{3.2}$$

Thus $\mathbf{J}$ is a vector whose direction is that of the velocity $\mathbf{v}$ of the carriers if q is positive.

In early experiments on electricity there was no evidence for the sign of the charges forming the current, since there was no means of distinguishing between a flow of positive charges in one direction and a flow of negative charges in the opposite direction. The positive direction of current flow was therefore taken as that in which a positive charge would move in an electric field. Thus in a circuit, the conventional direction for the flow of current is from the higher potential to the lower potential: for example, from the positive pole of a battery round the external circuit to the negative pole. It is customary to retain this convention, although the modern theory of metallic conduction shows that the positively charged ions are fixed, while a certain number of electrons are free to move about the body of the metal. Since the electrons are negatively charged, their direction of movement is opposite to that of the conventional current flow.

Measurement of q/m for carriers of electric current in a metal

The first direct experimental evidence that the carriers of electricity in a metal are electrons was supplied by the measurements of Tolman and Stewart (1917). The principle of this experiment depends on a comparison of the electric current with the momentum carried by the particles. If the current density is nqv, and the particles have mass m, then the momentum associated with the current crossing unit area of a plane normal to the direction of flow is nmv. Thus the ratio of the electric current density to the momentum 'current density' is simply equal to the ratio of charge to mass (q/m) of the carriers. The sign of q/m is obtained from comparison of the directions of flow of the electric current and momentum current. The method we shall now describe is that of a later experiment by Kettering and Scott (1944).

A circular coil is suspended by a thin fibre so that its plane is horizontal and it forms a torsional pendulum with very small damping. The coil, consisting of N turns of radius r, carries a current I. If the number of electrons per unit length of the wire is n, and they move with a mean velocity v, their angular momentum about the axis of the coil is

$$\Gamma = mrvn(2\pi rN),$$

since the total number of electrons in the coil is $n(2\pi rN)$. The current $I = nqv$, and hence we have

$$\Gamma = 2\pi r^2 N(m/q)I = 2AN(m/q)I,$$

where A is the area of the coil. In the experiment, a current I is maintained in the coil, and then suddenly reversed. This imparts an impulse 2Γ to the coil, whose angular momentum is thus altered by $4AN(m/q)I$. In practice the amplitude of swing of the coil is observed with the current flowing in one direction, and the current is reversed at the moment when the coil passes through its equilibrium position. This changes the amplitude of swing θ_0 by an amount

$$\Delta\theta_0 = 2\Gamma(T/2\pi\Im),$$

where 2Γ is the change of angular momentum due to the current reversal, T is the time of swing, and $\Im$ the moment of inertia of the coil. The value of q/m can thus be determined by measurement of the change in amplitude of oscillation for a given current I.

Although the theory of the experiment is simple there were many practical difficulties. Leads to the coil had to be brought in so that the free suspension by the torsion fibre was not affected. To eliminate vibration and disturbance due to changing magnetic fields, the apparatus was installed in an underground vault. The experiment was performed both with coils made of copper and aluminium. The values of m/q obtained were 5·64, 5·67, and $5{\cdot}79\times10^{-12}$ kg C^{-1} for three different copper coils, and $5{\cdot}66\times10^{-12}$ kg C^{-1} for an aluminium coil. The mean of these results, $5{\cdot}69\times10^{-12}$ kg C^{-1}, is in very close agreement with the reciprocal ($5{\cdot}686\times10^{-12}$ kg C^{-1}) of the most accurate determinations of $|q/m|$ for free electrons. The sign of m/q, obtained from the direction of the change of amplitude, corresponded to the carriers being negatively charged.

3.2. Flow of current in conductors

Since electric charge can neither be created nor destroyed, it follows that the rate of increase of the total charge inside any arbitrary volume must be equal to the net flow of charge into this volume. We have therefore

$$\frac{\partial}{\partial t}\left(\int \rho_e \, d\tau\right) = \int \frac{\partial \rho_e}{\partial t} \, d\tau = -\int \mathbf{J}.d\mathbf{S},$$

where the integrals are taken respectively over the volume and the surface bounding it. On transforming the surface integral into a volume integral, we have

$$\int \frac{\partial \rho_e}{\partial t} \, d\tau = -\int \operatorname{div} \mathbf{J} \, d\tau$$

or

$$\int \left(\operatorname{div} \mathbf{J} + \frac{\partial \rho_e}{\partial t}\right) d\tau = 0.$$

This integral must be zero whatever the volume over which we integrate, and this can only be true if the integrand is itself zero. We may therefore write

$$\operatorname{div} \mathbf{J} = -\frac{\partial \rho_e}{\partial t}, \qquad (3.3)$$

which is known as the equation of continuity. In the steady state $\partial \rho_e/\partial t = 0$, and therefore

$$\operatorname{div} \mathbf{J} = 0 \qquad (3.4)$$

in any region of current flow which does not contain a source or sink of current. Such a source or sink by which current may be injected into or withdrawn from a conducting region is known as an electrode. The total current flow to or from an electrode may be found by integrating the current density over any surface which totally encloses the electrode.

Ohm's law

It is found experimentally that in a metallic conductor at constant temperature the current density is linearly proportional to the electric field strength. This is expressed by the equation

$$\mathbf{J} = \sigma \mathbf{E}, \tag{3.5}$$

where the constant σ is known as the conductivity. Its reciprocal is known as the resistivity, and denoted by the symbol ρ. Clearly

$$\mathbf{E} = \rho \mathbf{J}, \tag{3.5a}$$

and it is obvious that the quantities denoted by σ and ρ have no connection with those denoted by σ_e and ρ_e in the previous chapters.

If a conducting wire of cross-section A carries a current I, then $I = JA$; if the current enters at a point where the potential is V_1 and leaves at a point a distance l away where the potential is V_2, then $E = -(V_2 - V_1)/l$. Hence

$$I = \sigma EA = \sigma A(V_1 - V_2)/l = (V_1 - V_2)/R, \tag{3.6}$$

where $R = l/A\sigma = \rho l/A$ is known as the resistance of the wire. Eqn (3.6), which expresses the fact that the voltage between the ends of a conductor is proportional to the current flowing, is known as Ohm's law. In S.I. units V is measured in volts (V), I in amperes (A) (1 A is that current which when flowing in two parallel straight conductors of infinite length and negligible cross-section, placed 1 m apart in a vacuum, would produce a force 2×10^{-7} N per m of length between the two conductors—see § 4.1), and the unit of resistance is the ohm (Ω). Its reciprocal, the unit of conductance, is the reciprocal ohm (or siemens = S). Since $\rho = R(A/l)$, the dimensions of resistivity are those of resistance × length; its unit is the ohm-metre (Ω m), while that of σ is (ohm-metre)$^{-1}$ ($\Omega^{-1}\,\text{m}^{-1}$).

Values of the resistivity at room temperature are given in Table 3.1 for

TABLE 3.1

Resistivity of some typical materials

Substance		Resistivity at 290 K (Ω m)	Temperature coefficient (K^{-1})
Pure metals	silver	$1{\cdot}6\times10^{-8}$	$3{\cdot}8\times10^{-3}$
	copper	$1{\cdot}72\times10^{-8}$	$3{\cdot}9\times10^{-3}$
	aluminium	$2{\cdot}83\times10^{-8}$	$3{\cdot}9\times10^{-3}$
	platinum	10×10^{-8}	$3{\cdot}9\times10^{-3}$
Alloys	constantan	$44{\cdot}2\times10^{-8}$	$\sim10^{-6}$
	manganin	44×10^{-8}	$\sim10^{-6}$
Semiconductors	pure silicon	$\sim2\times10^{3}$	negative
	pure germanium	$\sim0{\cdot}5$	negative
Insulators	glass plate	2×10^{11}	—
	sealing wax	$\sim10^{14}$	—
	sulphur	$\sim10^{15}$	—
	fused quartz	$>5\times10^{16}$	—

a number of substances, and it will be seen that they vary over an enormous range. For most metals the resistivity is of order $10^{-8}\,\Omega\,\mathrm{m}$, being generally higher for alloys than for pure metals. The monatomic semiconductors silicon and germanium have resistivities higher by several orders of magnitude, and it is a characteristic of all semiconductors that the resistivity is very sensitive to the presence of extremely small amounts of impurity. Finally there is again an enormous increase in resistivity to the substances which are classed as insulators.

By combining eqns (3.2) and (3.5) we find that

$$\sigma = nq(\mathbf{v}/\mathbf{E}) = nqu. \tag{3.7}$$

Here u is a quantity known as the mobility; it is equal to the mean drift velocity which the carriers acquire in unit electric field. For positive charges, $\mathbf{v}$ has the same direction as $\mathbf{E}$, and their ratio is positive. For negative charges, though $\mathbf{J} = nq\mathbf{v}$ has the same direction as $\mathbf{E}$, the drift velocity $\mathbf{v}$ has the opposite direction; however, it is not usual to give the mobility a sign, and this may be expressed by writing $u = |(\mathbf{v}/\mathbf{E})|$. In S.I. the mobility is expressed in units of $\mathrm{m\,s^{-1}}$ per $\mathrm{V\,m^{-1}}$ or $\mathrm{m^2\,V^{-1}\,s^{-1}}$.

If all current is carried by particles whose charge is numerically equal to that of the electron, variations in σ or ρ must come from changes in n, the number of carriers per unit volume, or in the mobility u. In a metal n is constant; at room temperature the resistivity of pure metals varies roughly as the absolute temperature, and this arises from the fact that the mobility u (and hence also the conductivity) varies as T^{-1}. For semiconductors the conductivity rises with temperature, and it is often possible to represent it by a relation of the type

$$\sigma = \sigma_0 \exp(-b/T). \tag{3.8}$$

The mobility is not drastically different from that in a metal, and both the low value of σ and its temperature variation can be explained by assuming that the number of free carriers varies as $\exp(-b/T)$. At low temperatures all charges are firmly bound, and the conductivity is very small. As the temperature rises, some bound charges are released to become free carriers, but this requires a certain amount of extra energy, so that the number released varies as $\exp(-b/T)$. The larger the value of b, the smaller the number of carriers; an 'insulator' may be regarded as an extreme case in which b is so large that at room temperature the number of free carriers is negligible. This view is supported by the fact that at very high temperatures all 'insulators' acquire a certain amount of conductivity.

Theoretical basis of Ohm's law

When an electric field of strength $\mathbf{E}$ acts on a free particle of charge q, the particle is accelerated under the action of the force and its velocity increases at the rate

$$d\mathbf{v}/dt = (q/m)\mathbf{E}. \tag{3.9}$$

If there are n particles per unit volume, the current density $\mathbf{J} = nq\mathbf{v}$, and (3.9) therefore leads to a current increasing at the rate

$$d\mathbf{J}/dt = n(q^2/m)\mathbf{E}, \tag{3.9a}$$

an equation which is clearly at variance with the experimental observation, expressed in Ohm's law, that $\mathbf{J}$ is proportional to $\mathbf{E}$. The latter is, however, a steady-state result, and it can be reconciled with eqn (3.9**a**) if we assume that this expresses only a transient effect, and that there are other forces which prevent the current from increasing indefinitely. That such other forces must be present follows also from the observation that a current ceases to flow when the electric field driving it is removed, since the particles carrying the current have a net forward momentum $nm\mathbf{v} = \mathbf{J}(m/q)$ which must be destroyed in some way. A natural assumption is that this occurs through collisions between the charged particles and other constituents of the conducting medium.

If the particle last made a collision at time $t = 0$, after which its velocity was $\mathbf{v}_0$, then its displacement $\mathbf{s}$ at time t is

$$\mathbf{s} = \mathbf{v}_0 t + \tfrac{1}{2}(q/m)\mathbf{E}t^2.$$

On averaging over a distribution of times, we have

$$\bar{\mathbf{s}} = \tfrac{1}{2}(q/m)\mathbf{E}\overline{t^2},$$

where we have assumed that the velocity after a collision is completely random, and averages to zero, so that $\overline{\mathbf{v}_0} = 0$ and $\overline{\mathbf{v}_0 t} = 0$. This gives an average drift velocity under the influence of the field strength $\mathbf{E}$ of

$$\frac{\bar{\mathbf{s}}}{t} = \tfrac{1}{2}(q/m)\mathbf{E}\left(\frac{\overline{t^2}}{t}\right).$$

If we now assume that the chance of a free time t between collisions is proportional to $\exp(-t/\tau)$, simple integrations show that $\overline{t^2} = 2\tau^2$ and $\bar{t} = \tau$. Thus τ is the mean free time between collisions, and the drift velocity is

$$\bar{\mathbf{s}}/\bar{t} = (q/m)\mathbf{E}\tau.$$

The mobility is therefore

$$u = |(q/m)|\,\tau, \tag{3.10}$$

while the current density and conductivity become

$$\mathbf{J} = n(q^2/m)\mathbf{E}\tau \tag{3.11}$$

and

$$\sigma = n(q^2/m)\tau. \tag{3.12}$$

These equations show that both $\mathbf{J}$ and σ are independent of the sign of q.

The effects of collisions are formally similar to the presence of a viscous medium in which the particle is subjected to a retarding force proportional to its velocity. If such a force is added to eqn (3.9) we obtain

$$\mathrm{d}\mathbf{v}/\mathrm{d}t+\mathbf{v}/\tau=(q/m)\mathbf{E}, \tag{3.13}$$

where the constant τ clearly has the dimensions of time. If this equation is multiplied by nq, we have a differential equation for the current density $\mathbf{J}=nq\mathbf{v}$,

$$\mathrm{d}\mathbf{J}/\mathrm{d}t+\mathbf{J}/\tau=n(q^2/m)\mathbf{E}, \tag{3.13a}$$

whose solution is

$$\mathbf{J}=\sigma\mathbf{E}\{1-\exp(-t/\tau)\}. \tag{3.14}$$

This result shows that if the electric field of strength $\mathbf{E}$ is switched on at $t=0$, when the current is zero, then the new equilibrium current $\mathbf{J}=\sigma\mathbf{E}$, corresponding to Ohm's law, is established in a time of order τ, the mean time between collisions. For a metal such as copper, assuming one charge carrier per atom, we obtain from the conductivity at room temperature a value for τ of order 10^{-14} s. This shows that for most practical purposes the steady state is reached almost instantaneously after the electric field is switched on or off, or changed in any way. For this reason the constant τ is known as the relaxation time for current flow. It should not be confused with the time constant derived below for a different process, the establishment of a uniform *charge* distribution within a conductor.

Charge distribution and current flow

Although a large current may be flowing in a conductor the net charge density is everywhere the same, and if, as is generally the case, the conductor as a whole carries no charge, the net charge density is zero. If the charge density is not uniform, local electric fields are set up and charges move in such a way as to equalize the charge distribution. The time required for such a process is extremely short for all substances except the best insulators. From eqns (1.20), (3.3), and (3.5) we have

$$-\partial\rho_e/\partial t=\operatorname{div}\mathbf{J}=\operatorname{div}(\sigma\mathbf{E})=(\sigma/\epsilon)\rho_e, \tag{3.15}$$

a differential equation whose solution is

$$\rho_e=\rho_{e0}\exp(-t/\tau'), \tag{3.16}$$

where $\tau'=\epsilon/\sigma$. For a metal (taking $\epsilon=\epsilon_0$) this time is of order $(10^{-11}/10^8)$ s $=10^{-19}$ s, and only in an insulator whose conductivity is less than about $10^{-12}\,\Omega^{-1}\,\mathrm{m}^{-1}$ can a non-uniform charge distribution exist for more than a few seconds. This is the reason why in experiments on electrostatics extreme precautions are needed to achieve high insulation.

In the steady state, for a conducting medium where σ is uniform and isotropic, div $\mathbf{J} = \sigma$ div $\mathbf{E} = 0$, and since $\mathbf{E} = -\text{grad}\,\mathbf{V}$ we have

$$\nabla^2 V = 0,$$

so that Laplace's equation holds, as in electrostatics. If two perfectly conducting electrodes are immersed in an infinite medium of finite conductivity, the potential distribution in the medium is the same as in a condenser whose plates are formed by the two conductors; for the potential must satisfy Laplace's equation in each case with the boundary conditions V = constant on the surface of the conductors. In the medium, the lines of current flow are orthogonal to the lines of constant V, and coincide with the lines of $\mathbf{E}$. Since the resistance R of the medium between the conductors, and the capacity C of the capacitor formed when the medium is replaced by a dielectric, depend essentially on the distribution of the lines of $\mathbf{E}$, there is a simple relation between them.

The analogous equations for the two cases are

$$\left.\begin{array}{cc} \mathbf{D} = \epsilon \mathbf{E} & \mathbf{J} = \sigma \mathbf{E} \\ \epsilon\,\text{div}(\text{grad}\,V) = 0 & \sigma\,\text{div}(\text{grad}\,V) = 0 \\ \int \mathbf{D}.\text{d}\mathbf{S} = Q & \int \mathbf{J}.\text{d}\mathbf{S} = I. \end{array}\right\} \qquad (3.17)$$

Since $C = Q/V$, and $1/R = I/V$, C/ϵ is equivalent to $(1/R)/\sigma$; that is,

$$RC = \epsilon/\sigma. \qquad (3.18)$$

The product RC, like the quotient ϵ/σ above, has the dimensions of a time constant (see § 5.3).

3.3. The voltaic circuit

The mechanism by which a battery acts as the source of a constant potential will be discussed in § 3.10. For the present purpose the battery will merely be regarded as maintaining a potential difference between its two terminals. Fig. 3.1 illustrates the usual notation for the case of a battery B which causes a current I to flow through the resistance R connected across the terminals P, Q. Since the battery drives the current round any circuit attached to it, the potential difference it produces is

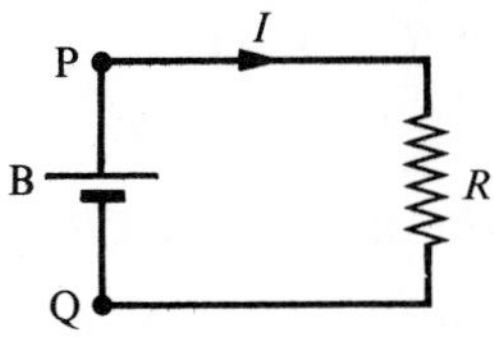

FIG. 3.1. A battery B sending a current I through a resistance R.

often called the 'electromotive force' or e.m.f. The total e.m.f. is equal to the line integral $\oint \mathbf{E}.\mathbf{ds}$ taken round the circuit; since no work is done by any external agency we must have

$$0 = V - \oint \mathbf{E}.\mathbf{ds}.$$

Hence, using eqn (3.6),

$$V = \oint \mathbf{E}.\mathbf{ds} = El = I(l/\sigma A) = IR \qquad (3.19)$$

if the battery of e.m.f. V is connected to a single conductor of conductivity σ, cross-section A, and length l. Eqn (3.19) is the basis of all calculations on resistance networks.

When a current flows through a conductor of finite resistance, charge is being transferred from a point at one potential to a point at a different potential. The direction of positive current flow is to a place at a lower potential so that there is a loss of electrical energy which appears as heat in the conductor. If a charge dQ flows between two points differing in potential by V in a time dt, the energy lost per second (power) is

$$-dW/dt = V(dQ/dt) = VI = V^2/R = I^2R, \qquad (3.20)$$

where R is the resistance between the two points. The unit of power is the volt-ampere, known as the watt (W). In an extended medium of conductivity σ, the power dissipated in an element of cross-section dS and length ds is $VI = (E\,ds)(J\,dS) = (E\,ds)(\sigma E\,dS) = \sigma E^2\,d\tau$, where $d\tau$ is the volume of the element. This result can also be obtained directly from eqn (3.2), since the power expended by a force $q\mathbf{E}$ moving n particles per unit volume with velocity $\mathbf{v}$ is

$$n(q\mathbf{E}).\mathbf{v} = \mathbf{J}.\mathbf{E} = \sigma E^2 = \rho J^2. \qquad (3.21)$$

The total power dissipated is found by integration over the whole volume of the conductor.

In practice it is found that the e.m.f. produced by a battery is not quite constant, but drops slightly when a current is drawn from it. The variation is the same as would be produced by an ideal source of e.m.f. V_0 equal to that produced by the battery on open-circuit, less the potential drop in a resistance R_i. The equivalent circuit is shown in Fig. 3.2, and R_i is known as the 'internal resistance' of the battery. If the battery is used to supply power to a load of resistance R, as in Fig. 3.2, the current I which flows is given by the equation $V_0 = I(R_i + R)$. The power dissipated in the load is therefore

$$-dW/dt = I^2R = V_0^2R/(R_i + R)^2. \qquad (3.22)$$

If the load can be varied, so that R is adjustable, then differentiation of this expression shows that it has a maximum value when $R = R_i$. This

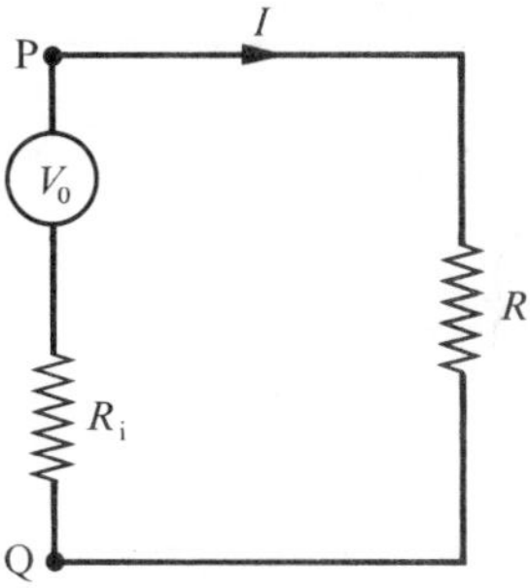

FIG. 3.2. Representation of a battery by an open-circuit e.m.f. V_0 and an internal resistance R_i.

is an example of the 'maximum power theorem', which states that, if a variable load is to be matched to a source of power so that the maximum power is to be dissipated in the load, its resistance must be adjusted to be equal to the internal resistance of the source. The greatest power which can be obtained is thus $\frac{1}{4}V_0^2/R_i$, and this is known as the 'available power' of the source. It should be noted that with a given load, and a range of batteries of the same voltage but of different internal resistance, maximum power is obtained with the battery of lowest internal resistance, so that the maximum power theorem does not apply to the converse problem.

It is often convenient to work in terms of conductance G rather than resistance R, where $G = 1/R$. Instead of (3.19), we have then

$$I = GV, \tag{3.23}$$

and the power dissipation in G is

$$-dW/dt = IV = I^2/G = GV^2. \tag{3.24}$$

In dealing with sources which produce an almost constant current rather than an almost constant e.m.f. it may be convenient to use the concept of a 'current generator', which produces a constant current I_0, shunted by an internal conductance G_i, as in Fig. 3.3. When connected to an external conductance G, the potentials across G and G_i must be the same, so that $I_i/G_i = I/G$ and the total current $I_0 = I_i + I$. Hence the current through G is $I = I_0G/(G+G_i)$, and the power dissipated in G is

$$-dW/dt = I_0^2G/(G+G_i)^2, \tag{3.25}$$

whose maximum value, if G is variable, is $\frac{1}{4}I_0^2/G_i$. This is another example of the maximum power theorem.

The current generator shunted by an internal conductance is completely equivalent to the voltage generator with an internal resistance. In Fig. 3.2 the current through $R = 1/G$ is $V_0/(R+R_i)$, and this is identical

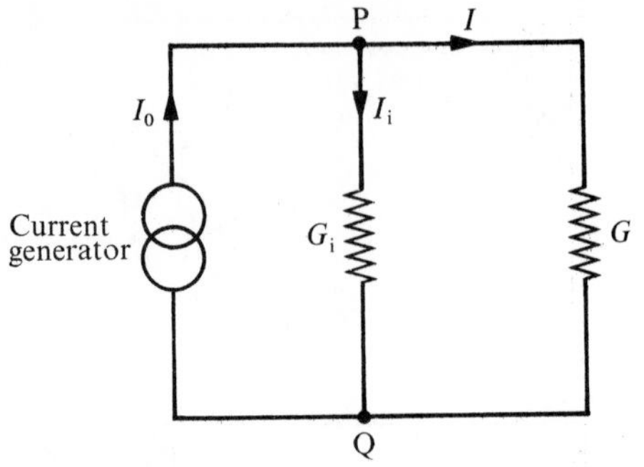

FIG. 3.3. The 'current-generator' circuit. The source provides a constant current I_0, shunted by a conductance G_i. It is equivalent to the 'voltage–generator' circuit of Fig. 3.2 provided $I_0 = G_i V_0$ and $R_i = 1/G_i$.

with $I_0G/(G+G_i)$ for all values of R provided that $I_0 = G_i V_0$ and $R_i = 1/G_i$.

3.4. Resistance networks

In a complicated network of resistances containing many branches, the calculation of the currents in the various branches is based on two laws due to Kirchhoff. They are extensions of the principles used above:

(1) the algebraic sum of all the currents meeting at a point is zero;
(2) the algebraic sum of the potential differences across the resistances in any closed circuit is equal to the total e.m.f. in that circuit.

The first law follows from the equation of continuity, since there can be no accumulation of charge at any point. It can be written

$$\sum I_k = 0. \tag{3.26}$$

The second law is an extension of eqn (3.19), and can be written

$$\sum_j V_j = \sum_k I_k R_k = \sum_k I_k/G_k, \tag{3.27}$$

where I_k is the current in the resistance R_k or conductance G_k.

These laws can immediately be applied to find the equivalent resistance of a number of resistances $R_1, R_2, \ldots, R_n$ in series or in parallel, as in Fig. 3.4. In the former case the voltage across all the resistances is

$$V = IR_1 + IR_2 + \ldots + IR_n = I(R_1 + R_2 + \ldots + R_n) = IR,$$

where R is the equivalent resistance. Hence

$$R = R_1 + R_2 + \ldots + R_n = \sum_k R_k. \tag{3.28a}$$

When the resistances are in parallel, the voltage across each is the same.

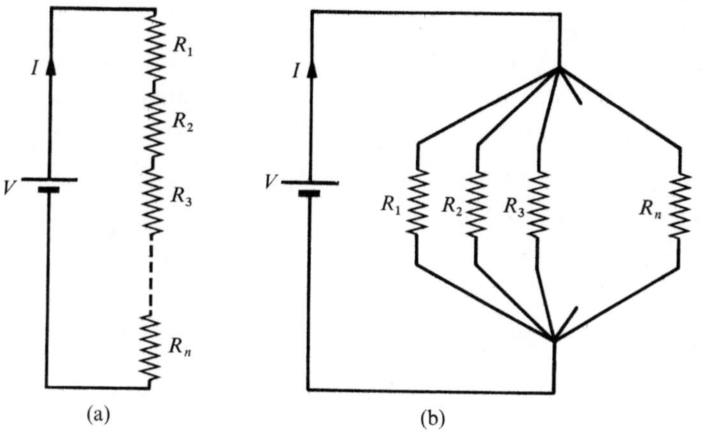

FIG. 3.4. Resistance arranged (a) in series, (b) in parallel.

The total current I is

$$I = \frac{V}{R_1} + \frac{V}{R_2} + \ldots + \frac{V}{R_n} = V\left(\frac{1}{R_1} + \frac{1}{R_2} + \ldots + \frac{1}{R_n}\right) = \frac{V}{R},$$

where R is the equivalent resistance. Hence in this case

$$\frac{1}{R} = \frac{1}{R_1} + \frac{1}{R_2} + \ldots + \frac{1}{R_n} = \sum_k \frac{1}{R_k}. \tag{3.28b}$$

In terms of conductances, it is clear that for a set of conductances in series

$$\frac{1}{G} = \sum_k \frac{1}{G_k}, \tag{3.29a}$$

while for conductances in parallel

$$G = \sum_k G_k. \tag{3.29b}$$

If a network has many branches, the problem of finding the current in each branch is best solved by the method of cyclic currents. Fig. 3.5 is

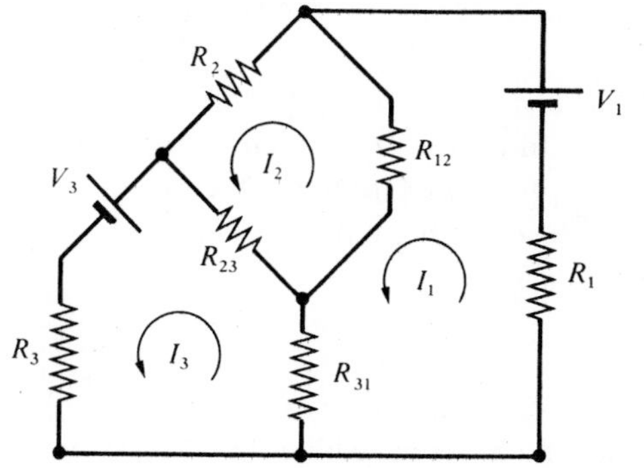

FIG. 3.5. A general network with cyclic currents.

part of a network in which there are n cyclic currents all flowing in an anticlockwise sense. Such a system of currents satisfies the first of Kirchhoff's laws automatically. Then the current through R_1 is I_1, but the current through R_{12} is (I_1-I_2). For circuit (1), for example, we have

$$V_1 = I_1R_1+(I_1-I_3)R_{13}+(I_1-I_2)R_{12} = I_1R_{11}-I_2R_{12}-I_3R_{13},$$

where $R_{11} = R_1+R_{12}+R_{13}$. In general, if $R_{qq} = R_q+\sum_p R_{qp}$,

$$\begin{aligned}
V_1 &= +R_{11}I_1 \pm R_{12}I_2 \pm R_{13}I_3 \pm \ldots \pm R_{1q}I_q \pm \ldots \pm R_{1n}I_n,\\
V_2 &= \pm R_{21}I_1 + R_{22}I_2 \pm R_{23}I_3 \pm \ldots \pm R_{2q}I_q \pm \ldots \pm R_{2n}I_n,\\
&\cdots\\
V_q &= \pm R_{q1}I_1 \pm R_{q2}I_2 \pm R_{q3}I_3 \pm \ldots + R_{qq}I_q \pm \ldots \pm R_{qn}I_n,\\
&\cdots\\
V_n &= \pm R_{n1}I_1 \pm R_{n2}I_2 \pm R_{n3}I_3 \pm \ldots \pm R_{nq}I_q \pm \ldots + R_{nn}I_n.
\end{aligned}$$

V_q is the total e.m.f. in the qth circuit, and is positive if it acts in the direction of I_q. The subscripts to the resistances denote the currents which flow through them. Then if Δ_{pq} is the cofactor of R_{pq} in the determinant

$$\Delta = \begin{vmatrix} +R_{11} & \cdot & \cdot & \pm R_{1q} & \cdot & \cdot & \pm R_{1n} \\ \cdot & \cdot & \cdot & \cdot & \cdot & \cdot & \cdot \\ \pm R_{p1} & \cdot & \cdot & \pm R_{pq} & \cdot & \cdot & \pm R_{pn} \\ \cdot & \cdot & \cdot & \cdot & \cdot & \cdot & \cdot \\ \pm R_{n1} & \cdot & \cdot & \pm R_{nq} & \cdot & \cdot & +R_{nn} \end{vmatrix},$$

$\Delta_{pq} = \Delta_{qp}$, since R_{pq} is identical with R_{qp}.

The reciprocity theorem

If we solve these equations for the current I_q when there is only one source of e.m.f. V_p in the circuit, then we find

$$I_q = V_p\Delta_{pq}/\Delta.$$

Similarly, the current I_p, when there is only the e.m.f. V_q in circuit, is

$$I_p = V_q\Delta_{qp}/\Delta.$$

Since $\Delta_{pq} = \Delta_{qp}$, we have the 'reciprocity theorem', which states that a given e.m.f. in the pth branch will produce the same current in the qth branch of a circuit as the same e.m.f. in the qth branch would produce in the pth branch.

Thévenin's theorem

The analysis of resistance networks may often be simplified by the use of a theorem due to Thévenin, which may be stated as follows: Suppose two terminals A, B emerge from a network for connection to an external circuit. Let V_{AB} be the voltage measured across these terminals when the external circuit is an open-circuit, and let I_{AB} be the current flowing through the external circuit when it is a short-circuit of zero resistance. Then the resistance of the network measured across the terminals A, B when all internal sources of e.m.f. have been short-circuited is given by $R_{AB} = V_{AB}/I_{AB}$. Thus, as far as the external circuit is concerned, the network behaves as a generator of e.m.f. equal to V_{AB} with an internal resistance R_{AB}.

A general proof of this theorem, which depends on the components of the network obeying Ohm's law, can be found in Duffin (1965). There is an equivalent theorem, due to Norton, which states that the network can also be represented by a current generator shunted by an internal conductance.

3.5. The potential divider and resistance bridge

If a pair of resistances R_1, R_2 are connected in series to a source of potential V_0, as in Fig. 3.6(a), the potential across the resistance R_2 is $V_0R_2/(R_1+R_2)$, and this potential may be varied by adjusting R_1 or R_2. Such a device is called a potential divider and is the basis of methods of comparing two potentials. The equivalent circuit is shown in Fig. 3.6(b). The open-circuit potential across the terminals A, B is clearly $V_{AB} = V_0R_2/(R_1+R_2)$, while if V_0 is short-circuited the resistance measured between terminals A, B corresponds to R_1 and R_2 in parallel. By Thévenin's theorem, this gives the value of R_{AB} as

$$R_{AB} = \frac{R_1R_2}{R_1+R_2}, \tag{3.30}$$

and it is easy to verify that the current through a resistance R connected across A, B is the same as would result from the equivalent circuit in Fig. 3.6(b). The equivalent current circuit is given in Fig. 3.6(c), where $I_{AB} = V_0/R_2 = G_{AB}V_{AB}$ with $G_{AB} = 1/R_{AB}$.

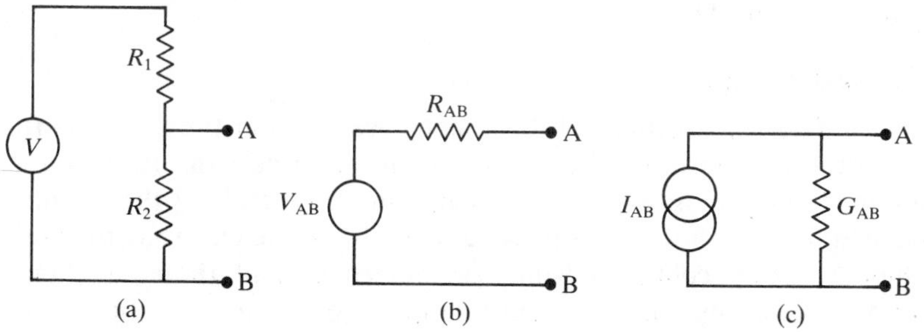

FIG. 3.6. The potential divider (a) and its equivalent circuits in terms of a voltage generator (b) or a current generator (c).

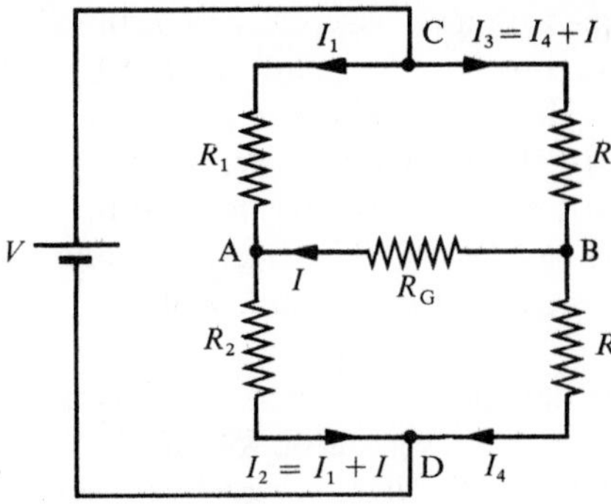

FIG. 3.7. Wheatstone's bridge. The current I through the detector (resistance R_G) is zero if $R_1/R_2 = R_3/R_4$. In the nomenclature of Callendar (1910) used in Problem 3.3, $R_4 = R$ (the unknown resistance), $R_2 = mR'$, $R_3 = nR'$, and $R_1 = nmR'$, so that the bridge is balanced when $R' = R$.

The value of a resistance R can be determined by measuring the voltage across it and the current through it, but this requires instruments for the accurate measurement of both voltage and current. In Wheatstone's bridge an unknown resistance can be determined in terms of other resistances without an accurate measurement of current or voltage. The circuit is shown in Fig. 3.7; R_4 is the unknown resistance, while R_1, R_2, and R_3 are known resistances, one of which must be variable. A current detector of resistance R_G is connected between A and B, but this is required only to determine when no current flows through R_G, that is, when the voltage at terminals A and B is the same. At this null point R_1, R_2 form a simple potential divider and the voltage across R_2 is $V_{AD} = VR_2/(R_1+R_2)$; similarly that across R_4 is $V_{BD} = VR_4/(R_3+R_4)$. These two voltages are equal provided that

$$R_1/R_2 = R_3/R_4. \tag{3.31}$$

This result is independent of any internal resistance of the battery and of the battery voltage itself, since V is just the voltage applied between terminals C and D.

3.6. Electron optics

In the following sections of this chapter we consider the flow of current due to the motion of charged particles, first in a vacuum, then in a gas, and finally in liquids and solids (metals). In §§ 3.6 and 3.7 the vacuum is assumed to be such that ions travel directly from one electrode to another without making collisions with gas molecules, and their equation of motion is given by eqn (3.9). On multiplication by $m\mathbf{v}$ we obtain

$$m\mathbf{v}.\left(\frac{\mathrm{d}\mathbf{v}}{\mathrm{d}t}\right) = \frac{\mathrm{d}}{\mathrm{d}t}\left(\tfrac{1}{2}mv^2\right) = q\mathbf{E}.\mathbf{v},$$

which on integration gives

$$\int_1^2 \mathrm{d}(\tfrac{1}{2}mv^2) = q\int_1^2 (\mathbf{E}.\mathbf{v})\,\mathrm{d}t = q\int_1^2 \mathbf{E}.\mathrm{d}\mathbf{s}. \tag{3.32}$$

If the particle starts from rest at a point where the potential $V=0$, we have

$$\tfrac{1}{2}mv^2+qV=0. \tag{3.33}$$

This result, which represents the conservation of energy, shows that the particle velocity v is proportional to $|V|^{\frac{1}{2}}$ irrespective of the direction of the velocity.

In an evacuated field-free space electrons travel in straight lines, and a beam of electrons leaving an electrode will eventually diverge. In an electronic vacuum device this must be counteracted in order to direct the electrons to a particular electrode or to a spot of small dimensions, as in a cathode-ray tube. The required convergence of the beam can be achieved by means of electric (or magnetic) fields, and is known as 'focusing'. The technology used is known as 'electron optics', though it is equally applicable to any charged particles; only the principles are outlined below.

As the name implies, there is a close analogy between the behaviour of electrons in an electrostatic field and that of light in a refracting medium. The basis of geometrical optics is Snell's law: when a light ray passes from a medium of refractive index n_1 through a plane boundary to another medium n_2, the angles of incidence and refraction obey the relation

$$n_1 \sin \alpha_1 = n_2 \sin \alpha_2.$$

In electron optics the corresponding case is that of an electron in a field-free space (that is, a region of constant potential V_1) crossing into a region at another potential V_2, as in Fig. 3.8. At the boundary there exists an electric field which accelerates the electron in the direction normal to the boundary, while the component of velocity parallel to the boundary remains unchanged. If the initial and final velocities are v_1 and v_2, the components parallel to the boundary are $v_1 \sin \alpha_1$ and $v_2 \sin \alpha_2$, so that we have

$$v_1 \sin \alpha_1 = v_2 \sin \alpha_2.$$

If the electron started from rest at a point where the potential is zero, this may be written

$$V_1^{\frac{1}{2}} \sin \alpha_1 = V_2^{\frac{1}{2}} \sin \alpha_2, \tag{3.34}$$

since v is proportional to $V^{\frac{1}{2}}$. Note that the electron velocity plays the same role as the refractive index and does not correspond to the velocity of light in the medium.

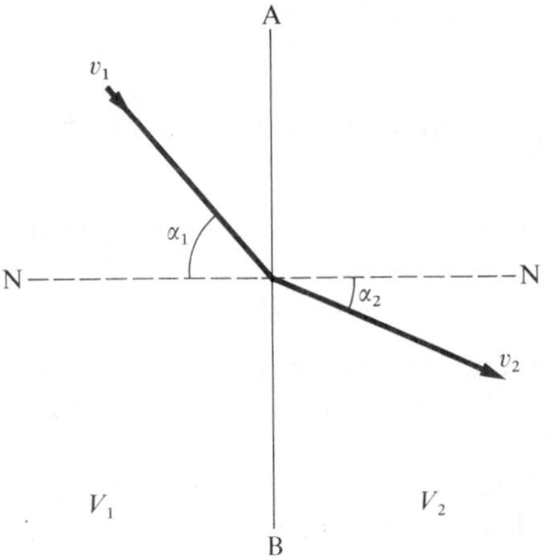

FIG. 3.8. 'Refraction' of an electron on crossing a boundary AB between two regions of potential V_1 and V_2. NN is the normal to this boundary.

The example just given is a special case of a more general correspondence based on Fermat's principle of least time in optics and Hamilton's principle of least action in mechanics. The former states that the path of a light ray is such that the time taken between any two points of the path is an extremum (generally a minimum). Thus we have

$$t=\int \mathrm{d}t=\int \frac{\mathrm{d}s}{v}=\frac{1}{c}\int n\,\mathrm{d}s=\text{minimum}. \tag{3.35}$$

Hamilton's principle states that the path of a particle is such that the line integral of its momentum is a minimum, that is,

$$\int mv\,\mathrm{d}s=\text{mimimum}. \tag{3.36}$$

So long as the mass of the particle is constant (that is, so long as relativity corrections are negligible) the analogy between electron velocity and refractive index is complete.

Although the formula for the focal length of a thin lens can be calculated quite simply in optics, the equivalent calculation for electron optics is generally very difficult. We shall content ourselves by showing how a focusing action can be obtained in a simple case. Fig. 3.9 shows a pair of parallel conducting planes each with a small circular aperture. The planes are at potentials V_1, V_2, and the potentials outside the planes away from the aperture are constant and equal to V_1 and V_2. Near the aperture the equipotentials are curved and bulge out as shown. If an electron travelling parallel to the z-axis (that is, normal to the planes) enters the aperture, it finds itself in a region where the lines of electric field, which are normal to the equipotentials, have a radial component. This gives the electron an acceleration normal to the axis, and it emerges into the field-free region with a component of velocity to or away from the axis. If the electron is in the xz plane as shown in the diagram, the x component of velocity given to it is

$$v_x = -\int_0^t \frac{e}{m} E_x \, dt = +\frac{e}{m}\int_0^t \left(\frac{\partial V}{\partial x}\right) dt. \qquad (3.37)$$

If the electron enters the aperture at a small distance h from the axis, then we may make the approximation

$$\left(\frac{\partial V}{\partial x}\right)_{x=h} = \left(\frac{\partial V}{\partial x}\right)_{x=0} + h\left(\frac{\partial^2 V}{\partial x^2}\right).$$

Since the potential satisfies Laplace's equation

$$\frac{\partial^2 V}{\partial x^2} + \frac{\partial^2 V}{\partial y^2} + \frac{\partial^2 V}{\partial z^2} = 0,$$

where, by symmetry, $\partial^2 V/\partial x^2 = \partial^2 V/\partial y^2$, and $(\partial V/\partial x) = (\partial V/\partial y) = 0$ on the axis, we have

$$\left(\frac{\partial V}{\partial x}\right)_{x=h} = -\frac{h}{2}\left(\frac{\partial^2 V}{\partial z^2}\right),$$

from which

$$v_x = -\int \frac{1}{2}\frac{e}{m} h\left(\frac{\partial^2 V}{\partial z^2}\right) dt = -\frac{1}{2}\frac{e}{m} h \int \left(\frac{\partial^2 V}{\partial z^2}\right)\frac{dz}{v_z},$$

where $v_z = dz/dt$ is the z component of the instantaneous velocity at any point. On emerging from the aperture the electron has a velocity $v_2 = (2eV_2/m)^{\frac{1}{2}}$ which is independent of h, and it moves at an angle θ with the

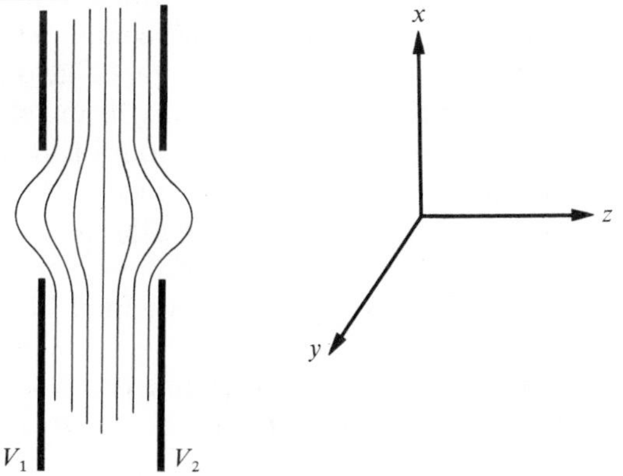

FIG. 3.9. A simple electron lens, consisting of two parallel conducting planes at potentials V_1, V_2 with apertures. The bulging of the equipotential surfaces near the apertures is shown.

axis where $\sin\theta = v_x/v_2$. Since v_x is proportional to h, for small values of θ (where the difference between $\sin\theta$ and $\tan\theta$ is negligible) all electrons will move towards (or appear to diverge from) a particular point on the axis, whose distance from the aperture is

$$-h/\tan\theta = -hv_2/v_x.$$

As the electrons were assumed to enter parallel to the axis, this is one focal point of the lens. If v_x is negative when h is positive we have a converging lens whose focal length is given by

$$\frac{1}{f_2} = -\frac{v_x}{hv_2} = \frac{e/2m}{(2eV_2/m)^{\frac{1}{2}}}\int\left(\frac{\partial^2 V}{\partial z^2}\right)\frac{dz}{v}$$

$$= \frac{e/2m}{(2eV_2/m)^{\frac{1}{2}}}\int\left(\frac{\partial^2 V}{\partial z^2}\right)\frac{dz}{(2eV/m)^{\frac{1}{2}}} = \frac{1}{4V_2^{\frac{1}{2}}}\int\frac{1}{V^{\frac{1}{2}}}\left(\frac{\partial^2 V}{\partial z^2}\right)dz, \qquad (3.38)$$

where we have assumed that $v_z = v$, the actual velocity at any point, and substituted $(2eV/m)^{\frac{1}{2}}$ for it inside the integral.

If the electrons had entered the lens from the right, they would have been brought to a focus at a point f_1, where

$$\frac{1}{f_1} = -\frac{1}{4V_1^{\frac{1}{2}}}\int\frac{1}{V^{\frac{1}{2}}}\left(\frac{\partial^2 V}{\partial z^2}\right)dz. \qquad (3.39)$$

Hence $f_1/f_2 = -(V_1/V_2)^{\frac{1}{2}}$, a formula which is exactly analogous to the optical case of a thin lens with initial and final media of different refractive indices n_1 and n_2.

Inspection of eqns (3.38) and (3.39) for the focal lengths shows that they require a knowledge of the variation of the potential on the axis of the lens. It is only possible in very simple cases to derive an analytical expression for V in the aperture, and in general the variation of V is calculated by numerical methods, by which an approximate solution of Laplace's equation with the required boundary conditions may be found. The variation of V always occupies a finite distance, and if it occupies a distance comparable with either of the focal lengths the electron lens is a 'thick lens' rather than a 'thin lens'. The behaviour of the system is again similar to that of an optical system, and is defined if the cardinal points are determined. These can be found by tracing the paths of a number of electrons through the system. One method of doing this is to divide the potential field into thin slices along the equipotential surfaces, and treat each slice as a thin lens.

The approximations made above in expanding $\partial V/\partial x$ near the axis and retaining only the first term are equivalent to the approximations made in 'Gaussian optics' in treating only rays near the axis. It is to be expected, therefore, that electron-lens systems will suffer from defects similar to those of optical systems, such as spherical aberration, etc. The equivalent of 'chromatic aberration' arises when not all electrons enter the system with the same velocity, since this corresponds to a variable refractive index. If the spread in velocity is due only to the Maxwellian distribution of velocity on emission from the cathode, chromatic aberration is small.

3.7. Space charge and Child's law

In the previous section the mutual repulsion of the electrons was neglected; this is justifiable only when the charge density is small. One result of the mutual repulsion is that a beam of electrons, initially moving parallel to one another, will gradually spread sideways, but this can be overcome by 'focusing' with a suitable electron-optical system. When the charge density is large, a much more important effect arises from the repulsive forces in the direction of motion, which can change the forward velocity and hence the total current. Such high charge densities occur in thermionic vacuum tubes, where a copious stream of electrons is emitted from a hot cathode (see Chapter 21). Under normal conditions the current flow in such devices is controlled by the charge density or 'space charge' near the cathode.

To examine the effect of this space charge on the flow of current, we shall consider the case of a diode where the cathode and anode form parts of parallel planes denoted respectively by the equations $x = 0$ and $x = d$. We shall further assume that the potential of the cathode is zero, while that of the anode is V_a. The potential between the electrodes can be determined by solving Poisson's equation

$$\frac{d^2V}{dx^2} = -\frac{\rho_e}{\epsilon_0} = \frac{ne}{\epsilon_0}, \tag{3.40}$$

where $-e$ is the electronic charge, and n the number of electrons per cubic metre. If the mass of an electron is m, and its velocity u at the point x where the potential is V, then the energy equation gives

$$\tfrac{1}{2}mu^2 = eV, \tag{3.41}$$

while the current density is

$$|J| = neu. \tag{3.42}$$

(Here the flow is unidirectional and it is not necessary to treat J as a vector quantity; we have avoided the negative sign which denotes that the direction of positive current flow is from anode to cathode.) The velocity can be eliminated between these equations, giving

$$ne = |J|\left(\frac{m}{2eV}\right)^{\frac{1}{2}}.$$

Substitution of this in Poisson's equation gives

$$d^2V/dx^2 = aV^{-\frac{1}{2}},$$

where $a = (|J|/\epsilon_0)(m/2e)^{\frac{1}{2}}$. This equation may be integrated if both sides are multiplied by $2(\mathrm{d}V/\mathrm{d}x)$, giving

$$\left(\frac{\mathrm{d}V}{\mathrm{d}x}\right)^2 - \left(\frac{\mathrm{d}V}{\mathrm{d}x}\right)_0^2 = 4aV^{\frac{1}{2}}.$$

$(\mathrm{d}V/\mathrm{d}x)_0$ is the electric field at the cathode, where V and x are zero. Since the constant a is proportional to $|J|$, it is evident that the maximum current density is attained when $(\mathrm{d}V/\mathrm{d}x)_0 = 0$. Then we may write

$$\mathrm{d}V/\mathrm{d}x = 2a^{\frac{1}{2}}V^{\frac{1}{4}},$$

integration of which gives

$$V^{\frac{3}{4}} = \tfrac{3}{2}a^{\frac{1}{2}}x, \tag{3.43}$$

where the constant of integration is zero because $V = 0$ at $x = 0$. Since at the anode $x = d$ and $V = V_a$, we have

$$V_a^{\frac{3}{2}} = 9ad^2/4 = (9/4\epsilon_0)(m/2e)^{\frac{1}{2}}d^2\,|J|, \tag{3.44}$$

showing that the current density $|J|$ is proportional to the three-halves power of the anode voltage, and inversely proportional to the square of the separation between cathode and anode. This relation was first derived by Child, and is sometimes known as Child's law.

Since the current density is independent of x, it follows from eqn (3.42) that the density of electrons is greatest where their velocity is smallest, that is, near the cathode. It is this concentration of electrons which reduces the electric field at the cathode, since their electric field is oppositely directed at this point to that due to the positive potential on the anode. The electron concentration cannot rise to a greater value than that required to make $\mathrm{d}V/\mathrm{d}x$ zero at the cathode, since no electrons could then leave the cathode, and the space charge would fall as electrons move away to the anode, without their being replenished from the cathode. Near the anode the electric field is greater than that due to the anode potential alone, because the field is here increased by the repulsive force due to the negative space charge near the cathode. The potential variation is shown in Fig. 3.10. Curve A shows the linear potential gradient which would exist in the absence of space charge, while curve B is that calculated above, on the assumption that the electrons are emitted from the cathode with zero velocity. It is easily seen from eqn (3.43) that the equation of curve B may be written in the form $V/V_a = (x/d)^{\frac{4}{3}}$. Owing to the finite velocity of emission, electrons can leave the cathode even when there is a small reverse electric field, and the space charge can then increase to the extent of setting up a potential minimum, as shown by curve C. The depth of this minimum is of the order W/e, where W is the average energy of the emitted electrons, since only those electrons with

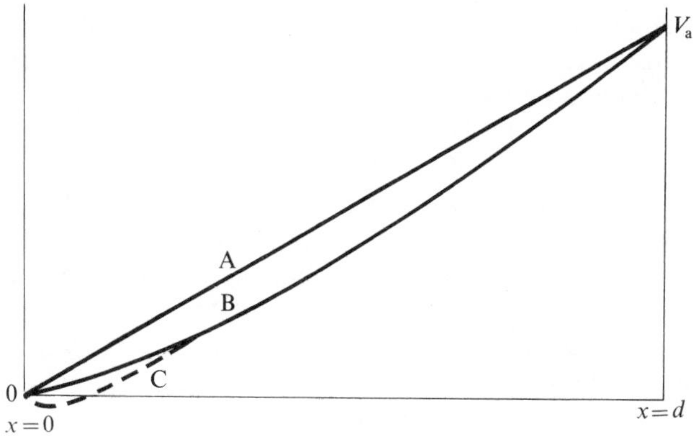

FIG. 3.10. Potential distribution in a diode with plane-parallel electrodes. Curve A, no space charge; curve B, space charge limited, electrons emitted with zero velocity; curve C, space charge limited, electrons emitted with finite velocity.

sufficient energy to penetrate the potential minimum will eventually reach the anode.

If numerical values are inserted the equation for the current density becomes

$$|J| = 2{\cdot}34\times10^{-6}\, V_a^{\frac{3}{2}}/d^2, \tag{3.45}$$

where $|J|$ is in amperes per square metre, V_a in volts, and d in metres. For example, if $V_a = 100$ V and $d = 10^{-2}$ m, then $|J| = 23{\cdot}4$ A m^{-2}. The three-halves power law has been shown to hold also for electrodes in the shape of coaxial circular cylinders and of concentric spheres. By a dimensional argument it may be shown to hold for any electrode geometry, assuming always that the electrons are emitted with zero velocity, though the numerical constant will obviously change. It does not hold indefinitely as V_a is increased, and the current will depart from the three-halves power law as soon as the space charge is no longer sufficiently dense to nullify the electric field at the cathode. Ultimately the current will be determined solely by the rate of emission of electrons from the cathode and will become independent of the anode voltage.

3.8. Conduction of electricity through gases

Under perfect conditions a gas consists of uncharged molecules, and therefore behaves as an insulator since there are no charged particles present to carry a current. In practice, owing to cosmic rays and radioactive background (especially in the walls of the containing vessel), there are always a few ions present, which are sufficient to initiate a spark discharge at sufficiently high electric fields (of the order of 3×10^6 V m^{-1} in air at atmospheric pressure), but at low fields the current passing is negligibly small unless ions are deliberately produced in the gas, or electrons are liberated at one of the electrodes (the cathode). The essential distinction between the two cases is that at low fields the current is limited by the supply of ions through external action (X-rays or ultraviolet light releasing electrons from the electrodes or from the gas molecules) while at high fields new ions are created by collisions between charged particles (accelerated by the applied electric field) and neutral molecules. A typical current–voltage characteristic is shown in Fig. 3.11. At very low voltages the current is proportional to the voltage, but at higher voltages it rises less rapidly and reaches a constant value, independent of the voltage over a wide range.

These results can be understood with the aid of a simple model. Suppose the electrodes consist of two infinite parallel planes; in Fig. 3.12 an electrode at $x = +d/2$ is maintained at potential $+V_0/2$, while that at $x = -d/2$ is at $-V_0/2$. We consider the numbers of positive and negative charges in a volume of unit cross-section, bounded by the planes x and $x + \mathrm{d}x$. Through the action of X-rays or a similar source of ionization, an equal number ($\beta\,\mathrm{d}x$) of positive and of negative ions is formed per unit time. Some of these are lost by recombination, the rate of which (the same for positive and for negative ions) is $-\alpha np\,\mathrm{d}x$, where n and p are the numbers of negative and positive ions per unit volume. The rate at which negative ions move in across the plane x is $n\dot{x}_n$, and the rate at

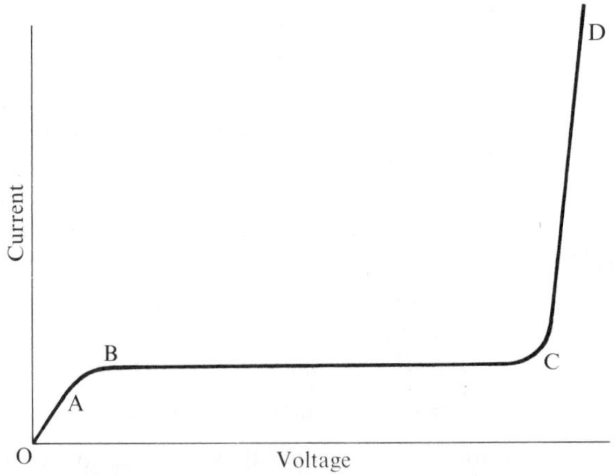

FIG. 3.11. Current–voltage characteristic of a gas.

OA Ohm's law is obeyed; most ions formed are lost by recombination.

BC All ions formed are swept to electrodes before recombination can take place.

CD Fresh ions are formed by collision when electrons can reach the ionization potential of the gas molecules between collisions (Townsend discharge).

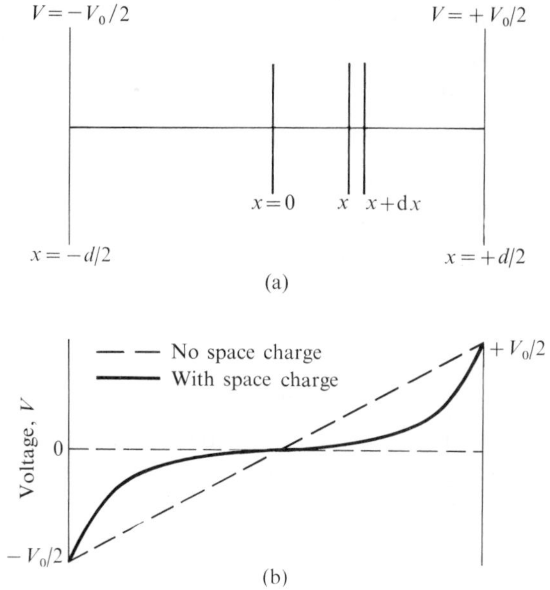

FIG. 3.12. Flow of current in an ionized gas. (a) Coordinate system for two parallel plane electrodes with potentials $\pm V_0/2$ at $x = \pm d/2$. (b) Potential distribution between the electrodes.

which they move out across the plane $(x+\mathrm{d}x)$ is $-n\dot{x}_n - \mathrm{d}(n\dot{x}_n)$, so that the net rate of gain is $-\mathrm{d}(n\dot{x}_n)$. Similarly the net rate of gain of positive ions is $-\mathrm{d}(p\dot{x}_p)$. In the steady state we have

$$\begin{aligned} \beta - \alpha np - \frac{\mathrm{d}}{\mathrm{d}x}(n\dot{x}_n) &= 0. \\ \beta - \alpha np - \frac{\mathrm{d}}{\mathrm{d}x}(p\dot{x}_p) &= 0. \end{aligned} \tag{3.46}$$

Subtraction of these two equations gives

$$\frac{\mathrm{d}}{\mathrm{d}x}(p\dot{x}_p - n\dot{x}_n) = 0.$$

If the ions carry charges q and $-q$ this means that the current density J, given by

$$J = q(p\dot{x}_p - n\dot{x}_n),$$

is a constant independent of x. This result is true only in so far as the loss of ions by recombination at the side walls, which we have not included, can justifiably be neglected.

An exact solution of our equations in the general case is difficult, and we shall consider only limiting cases. When the applied field is sufficiently high (about 2×10^3 V m^{-1} at atmospheric pressure), the ions move quickly to the electrodes and we may neglect the effect of recombination. Eqns (3.46), with $\alpha = 0$, may then be integrated to give

$$\begin{aligned} p\dot{x}_p &= \beta(x-\tfrac{1}{2}d), \\ n\dot{x}_n &= \beta(x+\tfrac{1}{2}d). \end{aligned} \tag{3.47}$$

Here the constants of integration are obtained from the fact that no ions are emitted by the electrodes, and they are formed only in the body of the gas; thus the positive ion density will be zero at the positive electrode ($p = 0$ at $x = +d/2$), and similarly $n = 0$ at $x = -d/2$. Then the current density is

$$J = q(p\dot{x}_p - n\dot{x}_n) = -q\beta d. \tag{3.48}$$

This result shows that the current is independent of the voltage, corresponding to the portion BC of the current–voltage characteristic in Fig. 3.11. The current density is proportional to βd, the total number of ions formed in unit cross-section between the electrodes per second. (The negative sign in (3.48) reflects the fact that the direction of positive current is towards the negative electrode.)

The effect of space charge may be deduced qualitatively from Poisson's equation, which gives

$$\mathrm{d}^2 V/\mathrm{d}x^2 = -\epsilon_0 \rho_e = \epsilon_0 q(n-p). \tag{3.49}$$

For small densities of charge $\mathrm{d}V/\mathrm{d}x$ will be constant as indicated by the broken line in Fig. 3.12(b), corresponding to a uniform electric field, but this will not be true at higher degrees of ionization. Since $p = 0$ at the positive electrode, $\mathrm{d}^2V/\mathrm{d}x^2$ is positive near $x = +d/2$ and, since $n = 0$ at the negative electrode, $\mathrm{d}^2V/\mathrm{d}x^2$ will be negative near $x = -d/2$. This gives a potential distribution of the type indicated by the solid line in Fig. 3.12(b), with the electric fields strongest near the electrodes.

When the voltage applied between the electrodes is small, the ionic velocities will also be small and we can neglect the last term in each of eqns (3.46). Then $p = n = (\beta/\alpha)^{\frac{1}{2}}$ and the space-charge density is zero everywhere, because the densities of positive and negative ions are equal. The field is uniform and the strength is $E = -(V_0/d)$. If the mobilities of positive and negative ions are u_p, u_n then, allowing for their opposite directions of motion,

$$\begin{aligned} \dot{x}_p &= u_p E = -u_p(V_0/d), \\ \dot{x}_n &= -u_n E = +u_n(V_0/d). \end{aligned}$$

The current density is

$$J = -q(\beta/\alpha)^{\frac{1}{2}}(u_p + u_n)(V_0/d). \tag{3.50}$$

Since J is proportional to V_0, Ohm's law is obeyed and we have the straight line corresponding to the initial part OA of the current–voltage characteristic in Fig. 3.11.

The mobility of an ion may be estimated in the following way. After collision with a gas molecule the ion is initially moving in a random direction with random velocity, and is then accelerated by the electric field strength E. The ions crossing the plane at which we wish to calculate the current density have come, on average, a distance L since the last collision, and the average time since the last collision is $\tau = L/c$, where c is the random velocity. The average velocity acquired under the action of the field of strength E is therefore $(q/M)E\tau$, and the mobility is

$$u = (q/M)\tau = (q/M)(L/c), \tag{3.51}$$

where M is the mass of the ion. Here L is equal to the mean free path, and not half the mean free path as is sometimes assumed; for a discussion of this point see, for example, Kennard (1938) (section 82) or Morse (1964) (Chapter 14). For ions of molecular size, L and c have the usual values given by kinetic theory: $L = 1/\sqrt{2}n\pi\sigma^2$, where n is the number of molecules per unit volume of diameter σ; and $c = (8kT/\pi M)^{\frac{1}{2}}$, where k is Boltzmann's constant and T is the absolute temperature.

At atmospheric pressure, both positive and negative ions have mobilities of the order of $10^{-4}\,\mathrm{m^2\,V^{-1}\,s^{-1}}$, in fair agreement with values calculated using eqn (3.51). As the pressure is lowered, the mobility increases inversely with the density for positive ions, corresponding to the expected increase in mean free path, but for negative ions it increases much more rapidly. This is due to the fact that at low pressures most of the negative ions are electrons rather than heavy charged molecules. At a given pressure, the ratio of electrons to heavy negative ions varies markedly from gas to gas; some molecules, such as Cl_2, readily attach electrons to form negative ions, while others such as H_2 do not. The mobility of electrons is much greater than that of heavy ions, mainly owing to their small mass, but also partly owing to their longer mean free paths. Since the diameter of an electron is negligible, its collision diameter with a gas molecule is only $\frac{1}{2}\sigma$, and since its velocity is much greater than that of the gas molecules the factor $\sqrt{2}$ introduced by Maxwell to allow for the relative velocities is absent, so that the electron mean free path is $4/n\pi\sigma^2$, or $4\sqrt{2}$ times that of a heavy ion. In addition, the average loss of energy by an electron in an elastic collision with a molecule is very small (see Problem 3.12), and the average energy of the electrons when a field is applied is much higher than that of the gas molecules or heavy ions. We may express this by saying that the 'mean temperature' of the electrons is higher than that of the gas. As the pressure is reduced, and the mean free path increases, the energy gained by an electron from the applied field increases and the effective electron temperature rises. The energy gained is proportional to the product of the mean free path and the applied field strength E, and since the mean free path is inversely proportional to the pressure p it follows that the conditions are a function of E/p. At low values of E/p the energy gained by an electron between collisions is small, and it makes only elastic collisions with the gas molecules, but at high values of E/p the mean electron temperature rises and the number of electrons in the high energy tail of the energy distribution increases rapidly. Those which have a few electron volts of energy can make inelastic collisions in which most of the energy is transferred to the colliding molecule. The effect on the molecule will now be discussed.

On quantum theory the total energy, kinetic plus potential, of an electron bound in an atom can only have certain allowed values, and in the normal state the electrons in an atom are in the lowest allowed levels, this is the 'ground state' of the atom. The different energy levels can be plotted on an 'energy-level diagram' (such as Fig. 14.2). The atom cannot exist with intermediate values of the energy, and if it is in an excited state (one of the higher energy levels) it may return to the ground state by emitting its excess energy as a quantum of light whose frequency ν is defined by the equation

$$W_1 - W_2 = h\nu. \tag{3.52}$$

For a molecule the energy-level diagram is similar to but rather more complicated than that of an atom.

If an electron with sufficient energy collides with an atom or molecule in its ground state, it may transfer some of its kinetic energy to the molecule and raise it to an excited state. For this to be possible the electron must have at least as much energy as the difference between the ground state and the first excited state of the molecule, and the potential through which the electron must be accelerated to obtain this energy is called the 'resonance potential' of the molecule. In general the molecule will get rid of this extra energy by emitting a photon (light quantum) within about 10^{-8} s, and the gas thus becomes luminous when the electrons gain sufficient energy from the applied electric field to raise the molecules into these excited states. As the energy of the electrons increases, the molecules are raised into higher excited states, corresponding to a bound electron being in an orbit of larger radius, and finally the molecule may be ionized, that is, an electron is completely removed, leaving the molecule as a positively charged ion. The energy required to do this (expressed in electronvolts) is called the ionization potential of the molecule. This process of ionization through electron impact increases the number of charged ions and electrons, and when the value of E/p is large enough for it to occur, the current through the gas is greatly increased. The steep rise in current CD with applied voltage shown after the saturation plateau BC in Fig. 3.11 is due to the formation of ions by collision; it was extensively investigated by Townsend and is known as the Townsend discharge. With specially designed electrodes the voltage in this region becomes almost independent of current, and small gas-filled tubes have been used as voltage stabilizers.

Ionization by electron collision is the primary process in producing fresh ions in the body of the gas. Experiments have shown that collisions with positive ions are much less effective in causing ionization (owing to their shorter mean free paths, heavy ions pick up less energy from the applied field than electrons), and this process can be neglected in comparison. The most important secondary processes for producing further charged particles occur at the cathode, from which electrons are emitted under the action of (1) bombardment by positive ions, (2) the photoelectric effect caused by photons emitted by excited molecules, and (3) bombardment by excited molecules. The relative importance of these three processes varies with the conditions; in general (1) is more important with cathode surfaces of high work function, and (2) with surfaces of low work function. The main types of excited molecules reaching the cathode are those in 'metastable states', that is, molecules in certain excited states which cannot return to the ground state by emitting a photon, and so have much longer lives than the 10^{-8} s mentioned above.

Secondary ionization processes occurring within the body of the gas, which (except at high pressure) appear to be less important than those at the cathode, are (4) photo-ionization, in which high-energy photons emitted by one molecule are absorbed by another, and may have sufficient energy to ionize it. This occurs mostly with the high-frequency ultraviolet radiation which is found in high-voltage discharge tubes; (5) as the temperature of the gas rises owing to the conversion of electrical energy to heat energy through collisions between molecules and ions, neutral molecules may have sufficient kinetic energy of random motion to ionize other molecules by collision. This process is sometimes called thermal ionization.

The Townsend discharge

Suppose we have two plane parallel electrodes a distance d apart and an electric field is applied between them. Let the negative electrode (the cathode) be illuminated with ultraviolet light which causes n_0 electrons to be emitted per second. These electrons are accelerated, and if the value of E/p is sufficiently high, they will produce further ions by collision. If n electrons cross a plane at a distance x from the cathode per second, then the number formed by ionization in the next element of distance $\mathrm{d}x$ will be proportional both to n and dx, so that we write

$$dn = C_1 n\, dx,$$

which on integration gives $n = n_0 \exp(C_1 x)$, and the current at the anode is therefore

$$I = n_0 q \exp(C_1 d) = I_0 \exp(C_1 d), \tag{3.53}$$

where I_0 is the current due to the original n_0 electrons alone. For low currents this equation is in good agreement with experiment, but at high currents the current shoots up rapidly towards infinity. This is due to the secondary processes, which increase the supply of ions, principally by causing the emission of more electrons from the cathode. If now the total emission of electrons from the cathode is n_0', the number of extra electrons produced in the gas by primary ionization must be $n_0'\{\exp(C_1 d)-1\}$, and this will also be the number of positive ions. The number of excited molecules emitting photons, and the number of molecules in excited states, will also be proportional to this number, and hence so also will be the number of secondary electrons emitted from the cathode, whatever the mechanism. Hence

$$n_0' = n_0 + C_2 n_0'\{\exp(C_1 d)-1\}, \tag{3.54}$$

from which

$$n_0'/n_0 = [1 - C_2\{\exp(C_1 d)-1\}]^{-1},$$

and the total current will be

$$I = n_0' q \exp(C_1 d) = I_0 \frac{\exp(C_1 d)}{1 - C_2\{\exp(C_1 d)-1\}}. \tag{3.55}$$

The total current will become infinite when the denominator is zero, that is, when

$$\exp(C_1 d) = (C_2 + 1)/C_2. \tag{3.56}$$

This implies that a finite current will pass when this condition is satisfied, even if I_0 is zero. The voltage at which this occurs is known as the sparking or breakdown potential V_s. Experimentally it was discovered by Paschen that for a given gas V_s depends only on the product pd of the gas pressure p and the electrode separation d. This is known as Paschen's law, and it holds up to very high pressures; it follows from the Townsend theory (above), for the constant C_1 is the number of ions produced by an electron in going unit distance. This number must be proportional to the number of molecules per unit volume, and hence to the pressure, and it also depends on the average energy gained by an electron between collisions. This energy varies as EL, where L is the mean free path, and since $E = V/d$ and L is inversely proportional to p, we have

$$C_1 d = (pd)\, F(V/pd), \tag{3.57}$$

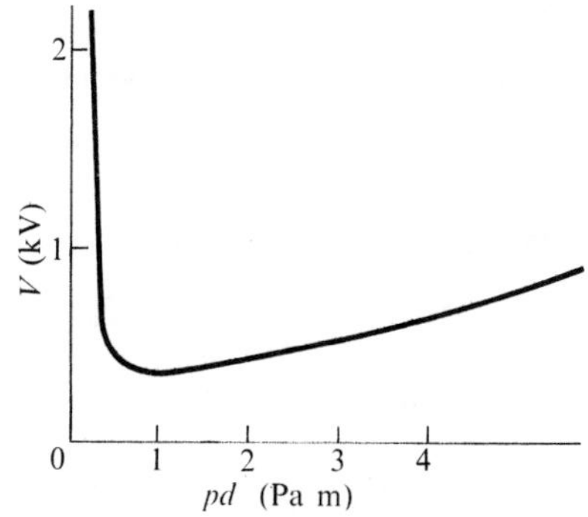

FIG. 3.13. Breakdown voltage V for air plotted against pd. V in volts, p in pascals, and d in metres.

where $F(V/pd)$ is some single-valued function of (V/pd). Since C_2 is a constant it follows from eqn (3.56) that the sparking potential V_s is a function only of pd for a given gas.

Inspection of eqn (3.54) shows that when the condition of eqn (3.56) is fulfilled, $C_2\{\exp(C_1 d)-1\}=1$, that is, the secondary processes produce all the electrons leaving the cathode. These electrons increase at an exponential rate, and the discharge current rises very rapidly (in a time of the order 10^{-7} s at atmospheric pressure) and a spark passes. A typical curve for the variation of V_s with the product (pressure × electrode separation) is shown in Fig. 3.13. The sharp rise at the low-pressure end is due to the low density, when the chance of an electron encountering a molecule is small and few ions are formed by collision. In the high-pressure region collisions are frequent, but the mean free path is small so that few electrons gain sufficient energy from the field between collisions to cause ionization. Thus for any given electrode separation d, there is always a certain pressure at which the sparking potential is a minimum.

Later work on the Townsend discharge (see Llewellyn Jones 1953) shows that the theory given above holds over a very wide range of values of pd. At the higher pressures (pd somewhat greater than 10^2 Pa×metres and sparking potentials of 10–100 kV) (Pa = pascal) positive ions cannot reach the cathode in the duration $\approx 10^{-7}$ s found experimentally for a spark, and most of the secondary emission from the cathode is due to photons; however, cathode emission is then probably less important than ionization in the body of the gas. At pressures of the order of 10^7 Pa, and with gaps of the order of centimetres, Paschen's law breaks down. This is due to the high fields at the cathode ($\sim 10^8$ V m^{-1}), which cause appreciable field emission (see §11.5), a process which does not depend on the number of ions formed in the body of the gas, as assumed in eqn (3.54). Since both field emission and photoelectric emission depend on the work function (see §§ 11.4, 11.5), the nature of the cathode surface becomes increasingly important at high pressures.

3.9. Conduction of electricity through liquids

Certain liquids, such as hydrocarbons, are extremely good insulators, while others, such as water, have an appreciable conductivity. Solutions of some salts in water have a conductivity of the order of 10^{-5} times that of metals, and such salts are known as ionic compounds. An example is sodium chloride, which in a simple picture is formed by the transfer of one electron from the sodium atom to the chlorine atom, so that the molecule consists of a positively charged sodium ion and a negatively charged chlorine ion. Such a molecule has a permanent electric dipole moment, and is called a 'polar molecule'; in a non-polar molecule, such as hydrogen, the electrons are shared equally between the two atoms, and there is no permanent dipole moment. The water molecule is itself strongly polar, and when a substance such as sodium chloride is dissolved in it, the electric fields of the water molecules are strong enough to dissociate the solute molecules into separate Na^+ ions and Cl^- ions. There are also composite ions formed by groups of molecules which have gained or lost an electron. The solution is called an 'electrolytic solution', the solute being known as the 'electrolyte'. The degree of dissociation of the electrolyte in solution is determined by the dynamic equilibrium between recombination and dissociation. With 'strong electrolytes', such as sodium chloride, the dissociation is practically complete at all ordinary concentrations; for 'weak electrolytes', such as acetic acid, the degree of dissociation is greatest at high dilution and falls steadily with increasing concentration according to the law

$$\alpha^2/(1-\alpha) = K/c,$$

where α is the fraction of solute molecules dissociated, c the concentration, and K is a constant which depends on the temperature.

If a potential difference is applied between two electrodes in an electrolytic solution, a current will flow through the solution. The current is carried by both positive and negative ions, which are produced by the decomposition of the electrolyte. If n_1, n_2 are the numbers of ions per unit volume with charges q_1, q_2 and mobilities u_1, u_2, the current density is

$$\mathbf{J} = (n_1\,|q_1|\,u_1 + n_2\,|q_2|\,u_2)\mathbf{E} = \sigma\mathbf{E}. \tag{3.58}$$

The specific conductivity σ is found to be proportional to the concentration for very weak solutions, but increases less rapidly at higher concentrations. For weak electrolytes, the conductivity is determined by the degree of dissociation—that is, the numbers of ions present—indicating that the mobility is independent of concentration. For strong electrolytes, where dissociation is practically complete at all concentrations, the mobility falls at high concentrations, because each ion attracts round itself an 'atmosphere' of ions of opposite sign, effectively making it bigger and retarding its progress through the solution. The mobility of the ions in a liquid, as we should expect, is always very small compared with that in a gas.

Electrolysis

Metallic ions and hydrogen (H^+) always travel to the cathode, or negative electrode, and negative ions to the anode. This transfer of charged ions by an electric current is called electrolysis. For example, if two copper electrodes are immersed in copper-sulphate solution, copper is dissolved from the anode, and deposited on the cathode. With carbon electrodes in a brine solution, hydrogen appears at the cathode and chlorine at the anode. There are two fundamental laws, discovered by Faraday, which are obeyed by all electrolytes.

(1) The mass of a given substance liberated at one electrode is proportional to the total charge which has passed.

(2) The mass of a given substance liberated at an electrode by unit charge is proportional to the chemical equivalent of that substance.

These two laws can be condensed into the following form: If the mass liberated at an electrode is m when a current I passes for a time t, then $m = ZIt$, where Z is a constant for a given element, called the *electrochemical equivalent*. The chemical equivalent of an ion is the atomic weight divided by the valency, the valency being equal to the number of electronic charges carried by the ion. Then if A is the atomic weight and v the valency,

$$m = (A/v)\frac{It}{F}, \tag{3.59}$$

where F is a universal constant known as the Faraday. The Faraday is the charge of electricity which liberates one gram equivalent (A/v) of an ion in electrolysis. If N_A (Avogadro's number) is the number of atoms in a mole, the total charge carried by a gram equivalent is $(N_A/v)(ve)$, since each atom (more correctly, each ion) has a charge ve, and this total charge N_Ae must just be equal to F. Hence we have the important relation

$$F = N_Ae. \tag{3.60}$$

Basically, the value of the Faraday is found by determining the mass of electrolyte liberated when a known current is passed through a solution for a measured length of time. The most reliable methods are those in which silver is dissolved in perchloric acid, and the 'iodine coulometer', in which the basic reaction is

$$3I^- \rightleftharpoons I_3^- + 2e.$$

Results obtained at the National Bureau of Standards (see Bower 1972) are

$$9{\cdot}64865 \times 10^4 \text{ C mol}^{-1} \text{ (silver)},$$

$$9{\cdot}64831 \times 10^4 \text{ C mol}^{-1} \text{ (iodine)},$$

which agree well with the value of $9{\cdot}64846 \times 10^4$ C mol^{-1} obtained indirectly from measurements of e/m_p for the proton.

Although the primary reaction in electrolysis is simply the flow of ions of positive and negative sign to the cathode and anode respectively, secondary processes may occur at the electrodes, so that different products appear there which do not correspond to the primary ions. The first product may be unstable, as in the case

$$NH_4 + H_2O \rightarrow NH_4OH + H,$$

or it may react with the solvent, the solute, or the electrode in a chemical reaction such as

$$Na + H_2O \rightarrow NaOH + H.$$

In these examples the hydrogen atoms or ions will combine to form hydrogen molecules which appear as a mass of small bubbles of gas covering the electrode. In a cell where gas appears at the electrodes, the phenomenon of 'polarization' of the electrodes is generally observed. For example, in the electrolysis of acidulated water using polished platinum electrodes, no electrolysis occurs until the applied e.m.f. V exceeds a certain critical value V', known as the decomposition potential of the electrolyte. If $V > V'$, the current I flowing obeys a modified Ohm's-law relation

$$V - V' = IR,$$

but if $V < V'$, no current flows. The cell therefore becomes irreversible owing to 'polarization' of the electrodes. This is probably due to the presence of positively charged hydrogen ions in the gas bubbles, which repel other positive ions and so give the effect of a back e.m.f. Practical cells include a 'depolarizer', a substance which reacts chemically with hydrogen ions appearing at the electrode and so prevents the formation of gas bubbles.

3.10. Voltaic cells

When two electrodes are placed in an electrolyte, it is generally found that there exists an e.m.f. between them. The combination is known as a 'voltaic cell' or battery and can be used to drive a current round an external circuit. There are a number of types in large-scale commercial use, from which many amperes of current can be drawn for short periods. These include the 'dry cell', a zinc electrode (generally forming the container), with NH_4Cl solution in the form of a paste with MnO_2 as depolarizer round a carbon rod, giving 1·5 V; the lead–dilute sulphuric acid–lead oxide (PbO_2) battery, giving 2·05 V when fully charged; and (with a more favourable stored-energy-to-weight ratio) the heavy-duty iron–nickel cell (Fe–KOH solution–Ni), giving 1·4 V.

In contrast, special cells have been developed to maintain a very constant e.m.f. for long periods, provided virtually no current is drawn; the 'Weston standard cell' is widely used as a laboratory standard for precision measurements of voltage. Its construction is rather complicated; basically, at one electrode (an amalgam of mercury and cadmium) cadmium atoms go into solution as Cd^{2+} ions in an aqueous solution of $CdSO_4$–Hg_2SO_4, while Hg^+ ions leave the electrolyte to become neutral mercury atoms at the second electrode (pure mercury). Thus the second electrode becomes positively charged, while the first becomes negatively charged because the cadmium atoms must leave two electrons behind when going into solution as Cd^{2+} ions. When current flows the changes are

$$Cd - 2e^- \rightarrow Cd^{2+},$$

$$2Hg^+ + 2e^- \rightarrow 2Hg,$$

and chemical energy is available because the latter reaction releases more energy than is required to turn a cadmium atom into a Cd^{2+} ion. Electrically two electrons are transferred in this process from the second to the first electrode; on open-circuit the e.m.f. between the two electrodes at 290 K is just over 1·018 V.

Application of thermodynamics to voltaic cells

Since in a reversible cell the chemical reactions taking place when a current is passed through it in one direction may be reversed by sending the current through it in the opposite direction, the standard equations of thermodynamics may be applied to the cell. In an ideal case we may suppose the current to be infinitely small, so that Joule heat losses, which depend on the square of the current, can be made negligibly small in comparison with the chemical-energy changes which vary with the first power of the current. Then the equation for the change in the Helmholtz function F of a cell when charge Q passes at constant temperature is

$$F = U + T(\partial F/\partial T),$$

where U is the change in internal energy and F is the energy available for external work provided the cell volume is constant (that is, no gases are liberated at the electrodes). If V is the e.m.f. of the cell, then $F = VQ$, and we have

$$\begin{aligned} V &= (U/Q)+T(\partial V/\partial T) \\ &= h+T(\partial V/\partial T), \end{aligned} \tag{3.61}$$

where $h = U/Q$ is the heat of reaction when unit charge is passed. The reason why VQ is not just equal to U is because it may be necessary for the cell to exchange heat with its surroundings in order to remain at constant temperature, and this flow of heat may be related, by the second law of thermodynamics, to the temperature coefficient of the e.m.f.

3.11. Metallic conduction: classical theory

In a theory of metallic conduction put forward by Drude in 1900 it was assumed that there are electrons (known as the 'conduction electrons') which are free to move throughout the metal in the same way as the molecules of a perfect gas move within a container. In the absence of an electric field these electrons move in random directions and with random velocities so that there is no net current flow. Under the action of an electric field they are accelerated, but the additional momentum thus acquired is dissipated by a process resembling collisions, presumably with the positive ions which are fixed in the lattice. The conduction mechanism is clearly similar to that in an ionized gas at low field strengths (§ 3.8). If the mean free time between 'collisions' is τ, Then the mobility is given by eqn (3.51) and the electrical conductivity is

$$\sigma = n(q^2/m)\tau.$$

These formulae are identical with eqns (3.10) and (3.12) if we equate the mean free time to the relaxation time.

The theory holds only if the drift velocity is small compared with the random velocity. If the mean free path L is independent of velocity then τ is inversely proportional to the average speed of the carriers. If this is appreciably altered by the additional velocity acquired between collisions through acceleration by the electric field, τ would depend on $\mathbf{E}$ and the conductivity would be dependent on field strength. No such effects are observed in metals.

Since metals are much better conductors of heat than most electrical insulators, it is plausible to assume that thermal conduction in a metal is also mainly due to the free carriers. Application of the ordinary kinetic theory formula for the thermal conductivity K of a gas to the 'electron gas' in a conductor gives

$$K = \tfrac{1}{3}ncL(\mathrm{d}\bar{W}/\mathrm{d}T), \tag{3.62}$$

where c is the random thermal velocity and $\bar{W} = \tfrac{1}{2}mc^2$ the average kinetic energy of a carrier. Taking the mean free path L as equal to $c\tau$, we obtain an expression for the ratio of the thermal to the electrical conductivity

$$\frac{K}{\sigma}=\frac{mc^2}{3q^2}\frac{d\bar{W}}{dT}=\frac{2\bar{W}}{3q^2}\frac{d\bar{W}}{dT}. \tag{3.63}$$

If we assume that the carriers are electrons ($q=-e$) and obey classical statistics so that $\bar{W}=\frac{1}{2}mc^2=\frac{3}{2}kT$, then

$$\frac{K}{\sigma T}=\frac{3}{2}\left(\frac{k}{e}\right)^2 \tag{3.64}$$

should be the same for all metals and thus a universal constant. This is in accordance with an empirical law discovered by Wiedemann and Franz in 1853, and the numerical value predicted by eqn (3.64) is in good agreement with the experimental values for copper, silver, and gold over the limited temperature range of their experiments. As in ordinary kinetic theory, the exact numerical value depends on corrections made for the velocity spread; Drude originally obtained a factor 3 rather than $\frac{3}{2}$, while Lorentz deduced a factor 2.

In spite of this remarkable success of Drude's theory, neither σ nor K can be derived separately, since on classical theory there is no way of estimating L or τ. A much more important objection arises from the heat capacity; from the average energy $\bar{W}$ assumed above, the free electrons should provide a contribution $\frac{3}{2}R$ per mole in addition to the value $3R$ from the law of Dulong and Petit. This should make the value for a metal some 50 per cent greater than that for an insulator, but experimentally there is no significant difference. This difficulty was not overcome until it was realized that electrons must be treated using quantum statistics rather than the classical Maxwell–Boltzmann statistics. In fact, since electrons have an intrinsic spin angular momentum of $\frac{1}{2}\hbar=\frac{1}{2}h/2\pi$, the correct type of statistics is that associated with the names of Fermi and Dirac. The use of these statistics is the starting point of the discussion in Chapter 11.

References

BOWER, V. E. (1972). *Atomic masses and fundamental constants 4* (Eds J. H. Sanders and A. H. Wapstra). Plenum Press, London.

CALLENDAR, H. L. (1910). *Proc. phys. Soc.* **22,** 220.

DUFFIN, W.J. (1965). *Electricity and Magnetism*. McGraw-Hill, London.

KENNARD, E.H. (1938). *The kinetic theory of gases.* McGraw-Hill, New York.

KETTERING, C. F. and SCOTT. G. C. (1944). *Phys. Rev.* **66,** 257.

LLEWELLYN JONES, F. (1953). *Ann. Rep. Progr. Phys.* **16,** 216.

MORSE, P. W. (1964). *Thermal physics.* Benjamin, New York.

SMYTHE, W. R. (1939). *Static and dynamic electricity*. McGraw-Hill, New York.

TOLMAN, R. C. and STEWART, T. D. (1917). *Phys. Rev.* **9,** 64.

Problems

3.1. For the Wheatstone's bridge circuit (Fig. 3.7) show that the current through the detector is

$$I=\frac{V(R_1R_4-R_2R_3)}{D+R_G(R_1+R_2)(R_3+R_4)},$$

where $D=R_1R_2R_3+R_2R_3R_4+R_3R_4R_1+R_4R_1R_2$.

3.2. Show from the results of the previous problem that if $R_G=\infty$ the open-circuit voltage across the terminals AB is

$$V_{AB}=V(R_1R_4-R_2R_3)/(R_1+R_2)(R_3+R_4),$$

while if the battery is short-circuited the resistance measured at the terminals AB would be

$$R_{AB}=\frac{R_1R_2}{(R_1+R_2)}+\frac{R_3R_4}{(R_3+R_4)}.$$

Verify that with detector and battery in place, the bridge behaves as a voltage generator of e.m.f. V_{AB} with internal resistance R_{AB} working into a load R_G (Thévenin's theorem).

3.3. Using the nomenclature of Callendar (Fig. 3.7) for Wheatstone's bridge, show that

$$\frac{I}{I_4}=\frac{R-R'}{R_G(1+1/n)+R'(1+m)}.$$

The ratio I/I_4 is a measure of the sensitivity which is useful when the allowed current I_4 through the unknown resistance R is limited.

3.4. Resistances P, Q, R each of 10 Ω are placed in three arms of a Wheatstone's bridge, and a resistance S is adjusted in the fourth arm so that the bridge is balanced. The resistance R is now replaced by a resistance X, and the balance is restored by shunting S with a resistance of 10·123 $k\Omega$. What is the value of X?

Discuss the advantages and disadvantages of this method of measuring resistances when high accuracy is required.

3.5. In Fig. 3.14, show that a balance condition for the bridge (that is, no current through the galvanometer) which is independent of the value of r is

$$R_1/R_2=R_3/R_4=R_5/R_6.$$

This is Kelvin's double bridge for measuring small resistances of the order of 0·01 Ω. The resistance r represents the contact resistance between the two small resistances R_1, R_2, and its value does not affect the balance. If readings are taken with the currents flowing in both directions, so that errors due to thermoelectric voltages at the junctions are avoided, an accuracy of about 0·02 per cent can be achieved.

3.6. An accurate method, due to Hamon, of obtaining a known resistance ratio is to use n equal resistances, alternately connected all in series or all in parallel. Show that the ratio of the net resistance in series to that in parallel is n^2 and that, if the fractional error in each resistance is of order δ, then the fractional error in the ratio is of order δ^2.

3.7. Six 1 Ω resistances are joined to form a regular tetrahedron, and a potential

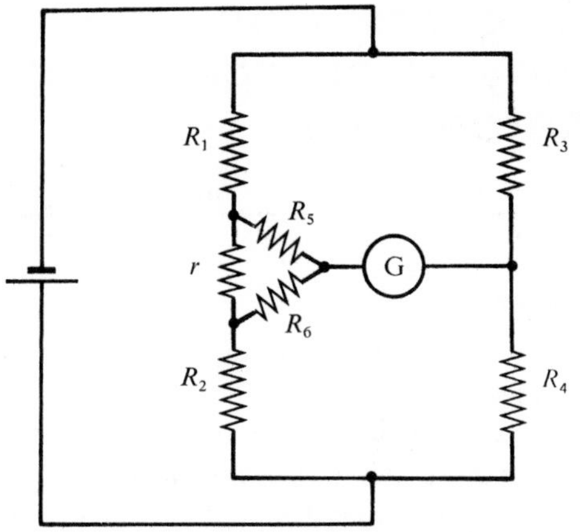

FIG. 3.14. Kelvin's double bridge.

of 1 V is maintained across one of the resistances. Find the current flowing in each conductor.

(*Answer:* 1, $\frac{1}{2}$, $\frac{1}{2}$, $\frac{1}{2}$, $\frac{1}{2}$, and 0A.)

3.8. Two long parallel circular copper rods of diameter 5·5 mm are placed with their axes 0·2 m apart in a large tank of copper-sulphate solution. If the conductivity of the solution is 4·1 Ω^{-1} m^{-1}, find the conductance between the rods of a section 1 m long.

(*Answer:* 3·0 Ω^{-1}.)

3.9. A solid sphere of mass density ρ_d and radius R is composed of hydrogen atoms. If the proton and electron differ in the magnitude of their charges by a fraction δ, show that the electric field strength at the surface of the sphere is

$$\delta(RF\rho_d/3\epsilon_0),$$

where F is the Faraday constant.

At the surface of the earth ($R = 6\times10^3$ km, $\rho_d = 5{\cdot}5\times10^3$ kg m^{-3}) the normal electric field strength is about 100 V m^{-1}. Show that this gives a value of $\delta \sim 10^{-22}$, if ascribed to such an effect.

3.10. Find the equation of motion of a free electron in an electric field strength $E = E_0 \sin \omega t$.

If the field strength E_0 is 10^4 V m^{-1}, and the frequency is 100 MHz, show that the amplitude of oscillation of the electron is 0·0045 m and that its maximum energy is 22·5 eV.

3.11. A spark passes between two electrodes 10 mm apart in air at atmospheric pressure when a uniform field of 10^6 V m^{-1} is applied across the gap. If the mean free path of an oxygen molecule in air is 6×10^{-8} m show that the time required for a singly ionized oxygen molecule to cross the gap is $4{\cdot}5\times10^{-5}$ s.

3.12. An electron of mass m collides with a molecule of mass M. Show that if the molecule is stationary the fraction of the electron energy which is transferred to the molecule in a head-on collision is $4Mm/(m+M)^2$, and evaluate this for the case $M = 200\times$mass of the proton.

(*Answer:* $\approx 1{\cdot}1\times10^{-5}$.)

4. The magnetic effects of currents and moving charges; and magnetostatics

4.1. Forces between currents

THE first experimental investigation of the interaction between coils carrying electric currents was performed by Ampère during the years 1820–5, and the work was continued by Oersted, Biot, and Savart. They found that two long parallel wires carrying currents in opposite directions repel one another, whereas when carrying currents in the same direction they attract one another, so that the direction of the force is reversed when one current is reversed. Ampère used circular coils, the leads to the coils being twisted together, and as these leads each carried equal currents in opposite directions they exerted no force on other circuits, so any forces observed were due only to the coils. He found that, if the dimensions of the coils were small compared with their distance apart, one coil exerts a force and a couple on another coil exactly similar to the force and couple which one electric dipole exerts on another. In magnitude, this force and couple are proportional to the current through the coil, the number of turns, and the area. If the plane of each coil is normal to the line joining the centre of each coil, the force is along this line. It is found also that if a coil carrying a current is placed near a magnet it experiences both a force and a couple. At distances large compared with the dimensions of either coil or magnet, this force and couple are similar in nature to those due to a second coil carrying a current. Thus both a magnet and a current-carrying coil are said to produce a magnetic field described by a flux density **B**, which exerts forces on other coils or magnets. **B** is a vector quantity and lines of **B** can be drawn whose direction at any point is that of **B**, in the same way as lines of electric force are drawn in an electric field. The number of lines of **B** passing normally through any area is known as the 'magnetic flux' through that area, and the term 'magnetic flux density' refers to the number of lines of **B** passing normally through unit area.

The force exerted on an element of wire $\mathrm{d}\mathbf{s}_1$ carrying a current I_1 at a place where the magnetic flux density is **B** can be expressed in the simple form (see Fig. 4.1)

$$\mathrm{d}\mathbf{F} = I_1(\mathrm{d}\mathbf{s}_1 \wedge \mathbf{B}). \tag{4.1}$$

This equation can be used to define a unit for **B** as that flux density which exerts unit force on unit length of a wire carrying one unit of current. In

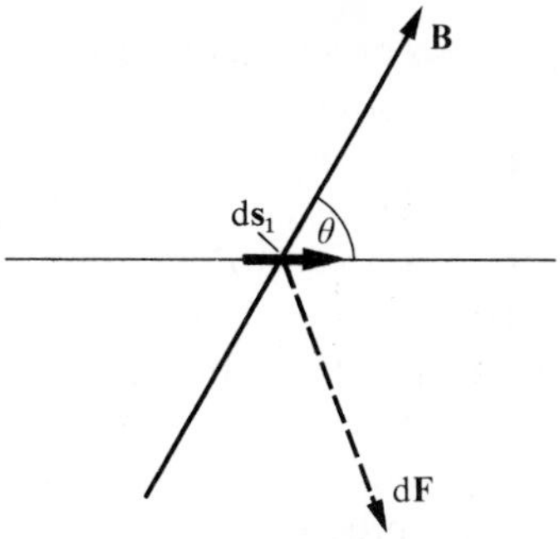

FIG. 4.1. Diagram to illustrate eqn (4.1). $\mathbf{ds}_1$ and $\mathbf{B}$ are in the plane of the paper, the angle between them being θ. $\mathbf{dF}$ is normal to the paper, towards the reader, and has magnitude $ds_1 B \sin\theta$.

S.I., the unit of force is the newton, that of length is the metre, and the unit of current is the ampere, thus the unit of **B** is the newton ampere^{-1} metre^{-1} (N A^{-1} m^{-1}); it is known as the tesla (T).

The experiments of Ampère and others showed that the force on an element $\mathbf{ds}_1$ carrying a current I_1 due to another element $\mathbf{ds}_2$ carrying a current I_2 is

$$\mathbf{dF}_1 = \frac{\mu_0 I_1 I_2}{4\pi r^3}\{\mathbf{ds}_1 \wedge (\mathbf{ds}_2 \wedge \mathbf{r})\}, \tag{4.2}$$

where μ_0 is a constant. $\mathbf{r}$ is the vector joining the two elements, being positive when drawn from $\mathbf{ds}_2$ to $\mathbf{ds}_1$, as in Fig. 4.2. The force $\mathbf{dF}_2$ on the element $\mathbf{ds}_2$ due to $\mathbf{ds}_1$ is given by a similar expression with $\mathbf{ds}_1$ and $\mathbf{ds}_2$ interchanged, and $\mathbf{r}$ must then be taken as positive when drawn from $\mathbf{ds}_1$ to $\mathbf{ds}_2$. The directions of the forces for the special case of two coplanar elements are shown in Fig. 4.2, and it will be seen that they are not equal

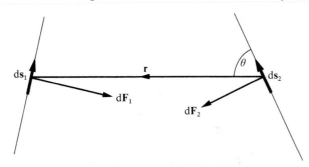

FIG. 4.2. Diagram showing the direction of the forces between two current elements. $\mathbf{ds}_1$ and $\mathbf{ds}_2$ are in the plane of the paper. The vector $(\mathbf{ds}_2 \wedge \mathbf{r})$ is normal to the plane of the paper, and the vector $\{\mathbf{ds}_1 \wedge (\mathbf{ds}_2 \wedge \mathbf{r})\}$ is in the plane of the paper, normal to $\mathbf{ds}_1$. The flux density $\mathbf{dB}$ due to $\mathbf{ds}_2$ is parallel to $(\mathbf{ds}_2 \wedge \mathbf{r}))$, and the force $\mathbf{dF}_1$ on $\mathbf{ds}_1$ due to it is parallel to $\{\mathbf{ds}_1 \wedge \mathbf{r})\}$.

and opposite unless the current elements are parallel. This apparent violation of Newton's third law of motion has caused much discussion, but Page and Adams (1945) have shown that there is no real violation, since the electromagnetic field of the current elements possesses momentum which is changing at a rate just equal to the difference of the two forces. Ampère's original formulation of the law of force between two current elements was different from eqn (4.2), but gave the correct result when integrated over a closed circuit carrying a constant current.

Comparison of eqns (4.1) and (4.2) shows that we may say that the current I_2 in the element $\mathbf{ds}_2$ produces a magnetic flux density $\mathrm{d}\mathbf{B}$ at a distance $\mathbf{r}$ given by the formula

$$\mathrm{d}\mathbf{B} = \frac{\mu_0}{4\pi r^3} I_2(\mathrm{d}\mathbf{s}_2 \wedge \mathbf{r}). \tag{4.3}$$

These equations may be used to calculate the value of $\mathbf{B}$ produced by a current in an infinite straight wire, and hence the force between currents in two parallel infinite straight wires. In Fig. 4.3 we have two such wires a distance a apart, carrying currents I_1, I_2; we choose a coordinate system where the first wire lies along the z-axis, and the second is parallel to it but passes through the point $x = a$, $y = 0$. We first calculate $\mathbf{B}$ at the point $(0, 0, 0)$ due to the current I_2 in the second wire, using eqn (4.3). Then the element $\mathrm{d}\mathbf{s}_2$ has components $(\mathrm{d}x, \mathrm{d}y, \mathrm{d}z) = (0, 0, \mathrm{d}z)$ and $\mathbf{r}$ has components $(-a, 0, -z)$ since it is defined by the coordinates of the point $(0, 0, 0)$ relative to the point $\mathrm{A}(a, 0, z)$ at which $\mathrm{d}\mathbf{s}_2$ is placed. The components of $\mathrm{d}\mathbf{s}_2 \wedge \mathbf{r}$ are $(0, -a\,\mathrm{d}z, 0)$, showing that $\mathrm{d}\mathbf{B}$ at O is antiparallel to the y-axis, wherever the point A lies along the second wire. Hence $\mathbf{B} = \int \mathrm{d}\mathbf{B}$ will also be antiparallel to the y-axis, so that $B_x = B_z = 0$, and integration yields for

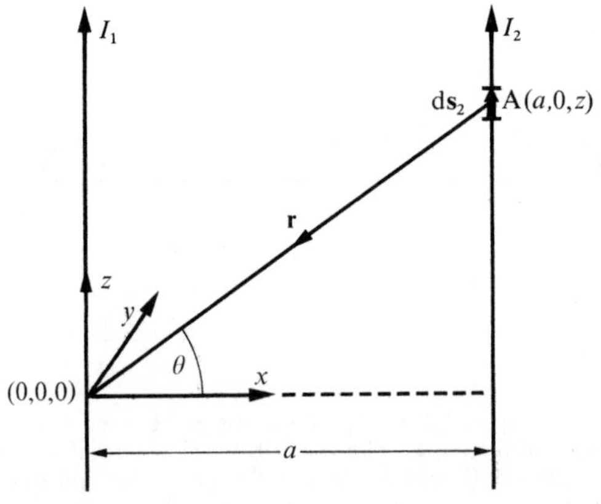

FIG. 4.3. Parallel wires carrying currents.

B_y the result

$$B_y = \frac{\mu_0 I_2}{4\pi}\int_{-\infty}^{+\infty}\frac{-a\,dz}{(a^2+z^2)^{\frac{3}{2}}} = -\frac{\mu_0 I_2}{4\pi a}\int_{-\frac{1}{2}\pi}^{+\frac{1}{2}\pi}\cos\theta\,d\theta = -\frac{\mu_0 I_2}{2\pi a}, \quad (4.4)$$

where we have used the substitution $z = a \tan\theta$. Eqn (4.4) shows that the field of a current I_2 in an infinite wire is proportional to I_2 and inversely proportional to the distance a from the wire. $\mathbf{B}$ is normal to the plane containing the wire and the radius vector $\mathbf{r}$, so that lines of constant B form closed circles centred on the wire.

We can now use eqn (4.1) to find the force $d\mathbf{F}$ on an element $d\mathbf{s}_1$ of the first wire. Since $\mathbf{B}$ is in the y direction, and $d\mathbf{s}_1$ in the z direction, the force is in the x direction, its only component being

$$dF_x = \frac{\mu_0 I_1 I_2\, ds_1}{2\pi a}. \quad (4.5)$$

If the currents are in the same direction the force is one of attraction, if the currents are opposed the force is one of repulsion, as stated above.

The value of the constant μ_0 depends on the system of units employed. We regard the ampere (or coulomb, since 1 A is a current of 1 C s^{-1}) as a standard of current (or charge) defined in an arbitrary way, a base unit similar to the kilogram and metre. The value of μ_0 is defined to be exactly $4\pi\times10^{-7}$ N A^{-2}. Eqn (4.5) then shows that for two parallel wires 1 m apart, each carrying 1 A of current, the force per metre length of wire is 2×10^{-7} N. This is a convenient way of defining the ampere. The quantity μ_0 is known as the 'permeability of a vacuum' (see § 4.4), and its unit is generally called the henry metre^{-1} (H m^{-1}) (see § 5.2) rather than the newton ampere^{-2} (N A^{-2}); the two units are equivalent.

We can use eqn (4.3) to evaluate the quantity div $d\mathbf{B}$. In this operation $d\mathbf{s}_2$ is a constant vector, and the variable is $\mathbf{r}/r^3 = -\text{grad}\,(1/r)$; the quantity div $(d\mathbf{s}_2\wedge\mathbf{r}/r^3)$ is identically zero (see eqns (A.15) and (A.19)), so that div $d\mathbf{B} = 0$. If this is true for every contribution $d\mathbf{B}$, it must also be true for the vector $\mathbf{B}$ itself. Thus the variation of the magnetic flux density in space obeys the differential equation

$$\text{div}\,\mathbf{B} = \frac{\partial B_x}{\partial x}+\frac{\partial B_y}{\partial y}+\frac{\partial B_z}{\partial z} = 0. \quad (4.6)$$

4.2. Magnetic dipole moment and magnetic shell

The investigations by Ampère of the forces between two small coils showed that they were similar to those between two dipoles. Comparison with eqns (1.14) and (1.15) shows that we should expect such a dipole, if placed in a field of uniform flux density, to experience a couple but no translational force, while in a non-uniform flux density it will experience a translational force. We now derive expressions for the couple and force, starting from eqn (4.1).

A set of elementary current circuits is shown in Fig. 4.4; they are similar but lie in three mutually orthogonal planes, each normal to one axis of a Cartesian coordinate system. We begin with the case of a uniform magnetic flux density **B**. Then, for the current circuit in Fig. 4.4(a), the forces on the sides CD, EF are clearly equal and opposite to one another, and so are those on the sides DE, FC. Thus there is no net translational force. The couple on the current element can be found as follows. The force $F_z = -I\,\mathrm{d}xB_y$ on the side EF has a moment about the x-axis equal to

$$\mathrm{d}T_x = +F_z\,\mathrm{d}y = -I\,\mathrm{d}x\mathrm{d}y\,B_y = -\mathrm{d}m_zB_y,$$

where $m_z = I\,\mathrm{d}x\,\mathrm{d}y$. Similarly, the force $F_z = -I\,\mathrm{d}yB_x$ on DE has a moment about the y-axis equal to

$$\mathrm{d}T_y = -F_z\,\mathrm{d}x = +I\,\mathrm{d}x\mathrm{d}y\,B_x = +\mathrm{d}m_zB_x,$$

while $\mathrm{d}T_z = 0$. If we write

$$\mathrm{d}\mathbf{m} = I(\mathrm{d}\mathbf{S}), \tag{4.7a}$$

the moments are the components of a couple

$$\mathrm{d}\mathbf{T} = I(\mathrm{d}\mathbf{S} \wedge \mathbf{B}) = \mathrm{d}\mathbf{m} \wedge \mathbf{B}. \tag{4.8a}$$

The quantity $\mathrm{d}x\,\mathrm{d}y$ is the area of the current loop, which can be represented by a vector normal to it, whose direction is that in which a right-handed screw would move if turned in the sense in which current flows round the loop; in our case this vector has the single component $(\mathrm{d}S)_z = \mathrm{d}x\,\mathrm{d}y$ and the vector $\mathrm{d}\mathbf{m} = I\,\mathrm{d}\mathbf{S}$ has the single component $\mathrm{d}m_z$. Similar relations hold for Fig. 4.4(b) and (c), where the components are $\mathrm{d}m_x = I'\,\mathrm{d}y\,\mathrm{d}z$ and $\mathrm{d}m_y = I''\,\mathrm{d}z\,\mathrm{d}x$. The quantity $\mathrm{d}\mathbf{m}$ is known as the

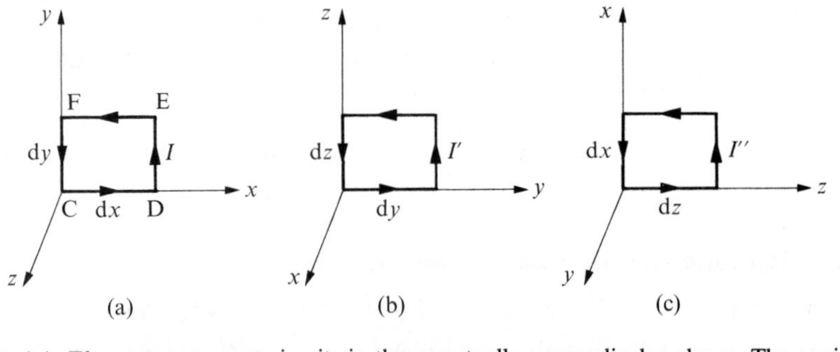

FIG. 4.4. Elementary current circuits in three mutually perpendicular planes. The equivalent magnetic dipole moments are $m_z = I\ \mathrm{d}x\ \mathrm{d}y$ in (a), $m_x = I'\ \mathrm{d}v\ \mathrm{d}z$ in (b), and $m_y = I''\,\mathrm{d}z\,\mathrm{d}x$ in (c).

'magnetic dipole moment' of the elementary current loop. For a loop of finite size we can write

$$\mathbf{m} = \int \mathrm{d}\mathbf{m} = I\int \mathrm{d}\mathbf{S} = I\mathbf{S}, \tag{4.7b}$$

where the vector **S** must be found by a vector integration, except in the case of a plane loop, and **m** is the equivalent magnetic dipole moment. In a field of uniform flux density **B** the couple is then

$$\mathbf{T} = \mathbf{m} \wedge \mathbf{B}. \tag{4.8b}$$

The translational force on a current loop in a non-uniform flux density may also be derived with the help of Fig. 4.4(a). The x-component of this force is

$$\begin{aligned} \mathrm{d}F_x &= I\left\{\mathrm{d}y\left(B_z + \frac{\partial B_z}{\partial x}\,\mathrm{d}x\right) - \mathrm{d}y B_z\right\} \\ &= I\,\mathrm{d}x\,\mathrm{d}y(\partial B_z/\partial x) = \mathrm{d}m_z(\partial B_z/\partial x), \end{aligned}$$

where B_z is the component of the flux density acting on the element FC and $(B_z + (\partial B_z/\partial x)\,\mathrm{d}x)$ that on DE. Similarly

$$\begin{aligned} \mathrm{d}F_y &= I\left\{-\mathrm{d}x B_z + \mathrm{d}x\left(B_z + \frac{\partial B_z}{\partial y}\,\mathrm{d}y\right)\right\} \\ &= \mathrm{d}m_z(\partial B_z/\partial y), \end{aligned}$$

while

$$\begin{aligned} \mathrm{d}F_z &= I\left\{\mathrm{d}x B_y - \mathrm{d}y\left(B_x + \frac{\partial B_x}{\partial x}\,\mathrm{d}x\right) - \mathrm{d}x\left(B_y + \frac{\partial B_y}{\partial y}\,\mathrm{d}y\right) + \mathrm{d}y B_x\right\} \\ &= -\mathrm{d}m_z\left\{\frac{\partial B_x}{\partial x} + \frac{\partial B_y}{\partial y}\right\}. \end{aligned}$$

Since, from eqn (4.6), div $\mathbf{B} = 0$, this last equation is equal to

$$\mathrm{d}F_z = \mathrm{d}m_z(\partial B_z/\partial z).$$

By considering the circuits in Fig. 4.4(b) and (c) it is easily shown that altogether, for a dipole **m** in a constant gradient of magnetic flux density,

$$F_x = m_x\frac{\partial B_x}{\partial x} + m_y\frac{\partial B_y}{\partial x} + m_z\frac{\partial B_z}{\partial x}, \tag{4.9}$$

with similar relations for the other force components F_y, F_z.

Comparison of eqns (4.8b) and (4.9) with (1.14) and (1.15b) shows that the magnetic dipole moment **m** plays a similar role to the electric dipole moment **p** and the magnetic flux density **B** plays the same role as the electric field strength **E**. We may therefore expect that it can be written as the gradient of a scalar potential ϕ, such that

$$\mathbf{B} = -\mu_0\,\mathrm{grad}\,\phi. \tag{4.10}$$

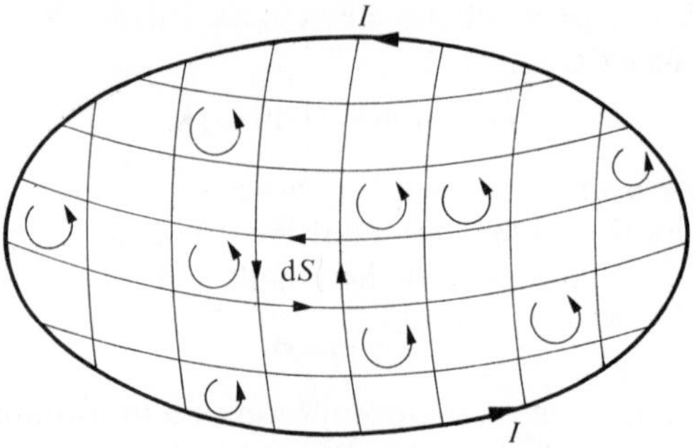

FIG. 4.5. A finite current circuit divided into a number of small current loops.

Whatever its shape, a finite current circuit can be divided up as in Fig. 4.5 into a number of current loops which lie on any surface which is bounded by the current circuit. Each element of area dS may be regarded as having a current I flowing round its edge, and summation of the currents in all the elements comprising the entire surface leaves only the current in the circuit as the resultant. If dS is infinitesimal in size, it can be regarded as a plane element and will have an associated magnetic dipole moment given by eqn (4.7a), and the couple on the whole circuit is obtained by integration of eqn (4.8a). The surface forms a magnetic 'double layer', or 'magnetic shell', with a certain dipole moment per unit area. The potential due to such a shell will now be calculated using eqn (4.3).

The field at a point P due to the current circuit is found by integration of eqn (4.3) round the circuit. In Fig. 4.6 if the point P is displaced a distance $\delta\mathbf{s}$ the change in potential will be

$$\delta\phi = -\frac{1}{\mu_0}\mathbf{B}.\delta\mathbf{s} = -I\,\delta\mathbf{s}.\int\frac{(\mathrm{d}\mathbf{a}\wedge\mathbf{r})}{4\pi r^3} = -I\int\frac{\delta\mathbf{s}.(\mathrm{d}\mathbf{a}\wedge\mathbf{r})}{4\pi r^3}, \quad (4.11)$$

where $\delta\mathbf{s}$ can be taken inside the integral sign because it is a constant during the integration. It is clear that we should obtain the same change

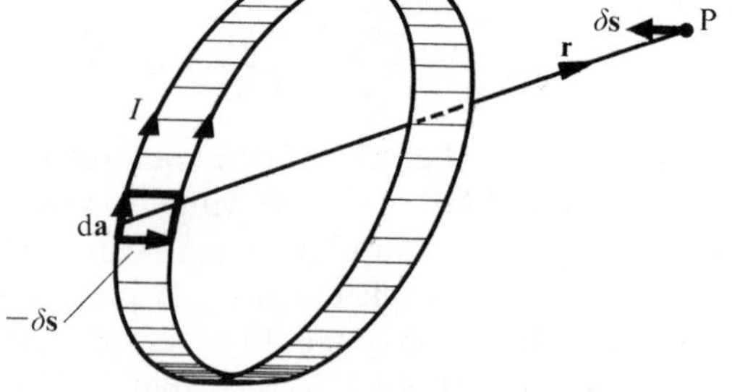

FIG. 4.6. Displacement of a current element, in order to calculate the potential at P due to a magnetic shell.

in potential if the point P were kept fixed and the circuit were displaced by an amount $-\delta\mathbf{s}$. In such a displacement the circuit element $\mathbf{da}$ sweeps out an area $-(\delta\mathbf{s}\wedge\mathbf{da})$, and this area subtends a solid angle at the point P of

$$-\frac{\mathbf{r}.(\delta\mathbf{s}\wedge\mathbf{da})}{r^3}=-\frac{\delta\mathbf{s}.(\mathbf{da}\wedge\mathbf{r})}{r^3}.$$

Hence the line integral in eqn (4.11) is the total solid angle subtended at P by the area swept out by the circuit when it is displaced by $-\delta\mathbf{s}$, and this is equal to the change $\delta\Omega$ in the solid angle due to the displacement of P by $\delta\mathbf{s}$. Hence we may write the change in potential as

$$\delta\phi=I\frac{\delta\Omega}{4\pi},$$

and the potential at P is

$$\phi=I\Omega/4\pi, \tag{4.12}$$

where Ω is the solid angle which the circuit subtends at P. But

$$\Omega=\int\frac{\mathbf{dS}.\mathbf{r}}{r^3},$$

where $\mathbf{dS}$ is an element of area of any surface bounded by the circuit, and hence

$$\phi=\int\frac{I\,\mathbf{dS}.\mathbf{r}}{4\pi r^3}=\int\frac{\mathbf{dm}.\mathbf{r}}{4\pi r^3}. \tag{4.13}$$

Here the integration is over the surface of the magnetic shell, and the potential of an individual dipole $\mathbf{m}$ must therefore be

$$\phi=\frac{\mathbf{m}.\mathbf{r}}{4\pi r^3}. \tag{4.14}$$

This is of the same form as eqn (1.10b), except that μ_0, which we might expect to replace the constant ϵ_0, does not occur here but in eqn (4.10). The reason for this choice will appear later (see eqn (4.32)).

The quantity ϕ is related to the line integral of $\mathbf{B}$ between two points, since

$$\int\mathbf{B}.\mathbf{ds}=-\mu_0\int\text{grad}\,\phi.\mathbf{ds}=-\mu_0\phi. \tag{4.15}$$

By analogy with electromotive force, which is the line integral of $\mathbf{E}$ (see eqn (3.19)), the quantity $-\phi$ is sometimes known as the 'magnetomotive force', or m.m.f.

In the electrostatic case, the work done in traversing a closed circuit is zero, and this would also be the case for a true double layer of magnetic charge. If we take the integral $\int\mathbf{B}.\mathbf{ds}$ from a point P very close to a magnetic shell round to a point P′ just on the other side of the magnetic shell, the difference $\Delta\Omega$ in the solid angle which the shell subtends at these two points is -4π, and the m.m.f. between these two points is

$$-\Delta\phi = -I\,\Delta\Omega/4\pi = I,$$

from eqn (4.12). With a real magnetic shell, if we now move from P′ to P through the shell, there would be a contribution to the m.m.f. which would just make the total zero, but with a current circuit there is no such contribution. We have therefore an important difference, that the m.m.f. increases by I every time we go round a closed path which threads the coil positively (that is, in the same direction as the lines of **B**). Thus the magnetostatic potential is not single-valued and cannot be used in a region where there are currents flowing. On the other hand, if the path does not thread a current circuit, the change in solid angle is zero, and the potential is single-valued. If the path does encircle a current I, we have

$$\oint \mathbf{B}.\mathrm{d}\mathbf{s} = \mu_0 I. \tag{4.16}$$

In a region of distributed current flow, the total current threaded by the path is $\int \mathbf{J}.\mathrm{d}\mathbf{S}$, where $\mathbf{J}$ is the current density in an element $\mathrm{d}\mathbf{S}$ of a surface bounded by the path. Hence

$$\oint \mathbf{B}.\mathrm{d}\mathbf{s} = \int \mathrm{curl}\,\mathbf{B}.\mathrm{d}\mathbf{S} = \mu_0 \int \mathbf{J}.\mathrm{d}\mathbf{S},$$

where the transformation from a line integral of **B** to a surface integral of curl **B** is an example of Stokes's theorem (see Appendix § A.8). Since the integrals must be equal over any surface, the integrands must be equal, and we have

$$\mathrm{curl}\,\mathbf{B} = \mu_0 \mathbf{J}. \tag{4.17}$$

The relation given by eqn (4.16) is known as Ampère's law, and eqn (4.17) is its representation in differential form. Since any function of the form curl(grad ϕ) is identically zero (see Appendix A, eqn (A.19)), we stress again that **B** can be derived only from a scalar potential ϕ in a region where the current density $\mathbf{J} = 0$. In such a region, we can derive a general expression for the translational force on a current loop in a non-uniform flux density. From the vector relation (A.27) in Appendix A, we can write

$$\begin{aligned} \mathbf{F} &= I\int \mathrm{d}\mathbf{s}\wedge\mathbf{B} = -I\int \mathbf{B}\wedge\mathrm{d}\mathbf{s} \\ &= -I\int (\mathrm{div}\,\mathbf{B})\,\mathrm{d}\mathbf{S} - I\int (\mathrm{curl}\,\mathbf{B})\wedge\mathrm{d}\mathbf{S} + I\int (\mathrm{d}\mathbf{S}\,.\,\mathrm{grad})\mathbf{B} \\ &= I\int (\mathrm{d}\mathbf{S}.\mathrm{grad})\mathbf{B} = \int (\mathrm{d}\mathbf{m}.\mathrm{grad})\mathbf{B}. \end{aligned} \tag{4.18a}$$

Here we have used both the relation curl $\mathbf{B}=0$ when $\mathbf{J}=0$, and also div $\mathbf{B}=0$ from eqn (4.6). If grad $\mathbf{B}$ is constant in the region occupied by the loop we have

$$\mathbf{F}=I(\mathbf{S}\cdot\text{grad})\mathbf{B}=(\mathbf{m}\cdot\text{grad})\mathbf{B}, \tag{4.18b}$$

where $\mathbf{m}=I\mathbf{S}$ is the equivalent dipole moment of the current loop of area $\mathbf{S}$. The components of the force $\mathbf{F}$ in (4.18b) do not have the same form (see Appendix A, eqn (A.10)) as those in eqn (4.9) and its analogues, but they are in fact identical when curl $\mathbf{B}=0$, since then $\partial B_y/\partial x=\partial B_x/\partial y$, etc. (cf. eqns (1.15).

Eqns (4.8b) for the couple and (4.18b) for the translational force on a magnetic dipole are exact analogues of the relations given by eqns (1.14) and (1.15c) for an electric dipole. It follows that both the torque and the force can be derived from a potential energy

$$U_P=-\mathbf{m}\cdot\mathbf{B} \tag{4.19}$$

which is the analogue of eqn (1.13). For a current circuit we have, from eqn (4.8a),

$$\mathbf{T}=I\int \mathrm{d}\mathbf{S}\wedge\mathbf{B}$$

for the couple, while the translational force is given by eqn (4.18a),

$$\mathbf{F}=I\int(\mathrm{d}\mathbf{S}\cdot\text{grad})\mathbf{B}.$$

The integrands are of the same form as the relations for electric dipoles, and the potential energy of the current circuit in the magnetic flux density $\mathbf{B}$ is

$$U_P=-I\int \mathrm{d}\mathbf{S}\cdot\mathbf{B}. \tag{4.20}$$

The quantity

$$\Phi=\int \mathrm{d}\mathbf{S}\cdot\mathbf{B}=\int\mathbf{B}\cdot\mathrm{d}\mathbf{S} \tag{4.21}$$

is the magnetic flux through the circuit, and the potential energy may be written as

$$U_P=-I\Phi. \tag{4.22}$$

The unit of flux is the weber (Wb); it is equal to 1 J A^{-1}. The unit of $\mathbf{B}$, the tesla, is equal to 1 weber metre^{-2} (1 Wb m^{-2}). The unit of magnetic dipole moment $\mathbf{m}$ can be written either as A m^2, from eqn (4.7), or J T^{-1}, from eqn (4.19). The unit of scalar magnetic potential ϕ is the ampere (A), from eqn (4.12).

4.3. Magnetostatics and magnetic media

The theory so far has been concerned with the magnetic effects of currents *in vacuo*, that is, in the absence of any magnetizable media. It is found experimentally that a material substance acquires a magnetic polarization when placed in a magnetic field, just as a dielectric medium acquires an electric polarization in an electric field. The magnetic dipole moment per unit volume is called the intensity of magnetization (or often simply the magnetization), and is represented by a vector **M**. All such magnetic effects are produced by current loops of atomic dimensions, which arise from the circulation of electric charge within the atom. The relation between the magnetic moment of such a loop and the circulating current is given by eqn (4.7b). If a magnetic medium has a magnetization **M**, which is not necessarily uniform throughout the substance, the equivalent current flow can be found by considering elementary current loops, as in Fig. 4.5. There the currents were all equal, and cancelled one another except at the perimeter, but in general this will not be the case. In Fig. 4.7 both **M** and I are functions of the space coordinates. We consider an element of volume $\mathrm{d}\tau = \mathrm{d}x\,\mathrm{d}y\,\mathrm{d}z$ at the point (x, y, z), for which the magnetic moment has a component $M_z\,\mathrm{d}x\,\mathrm{d}y\,\mathrm{d}z$ in the z direction. This is equivalent to a current flowing round the loop, the strength of the current being

$$I = (M_z\,\mathrm{d}x\,\mathrm{d}y\,\mathrm{d}z)/(\mathrm{d}x\,\mathrm{d}y) = M_z\,\mathrm{d}z,$$

since the area of the loop is $\mathrm{d}x\,\mathrm{d}y$. The adjacent loop at the point $(x+\mathrm{d}x, y, z)$ has a current

$$I' = I+(\partial I/\partial x)\,\mathrm{d}x = M_z\,\mathrm{d}z+(\partial M_z/\partial x)\,\mathrm{d}x\,\mathrm{d}z.$$

Hence the net current flow on the interface between the two elements has a component in the y direction of magnitude

$$I-I' = -(\partial M_z/\partial x)\,\mathrm{d}x\,\mathrm{d}z$$

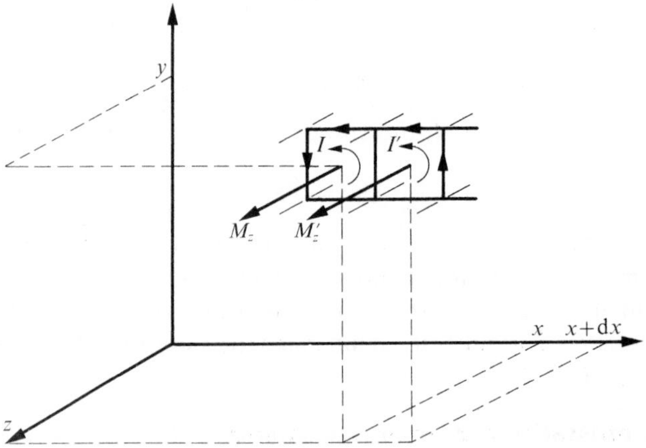

FIG. 4.7. Representation of a non-uniform magnetization by circulating currents,

$$M_z' = M_z + \frac{\partial M_z}{\partial x}\,\mathrm{d}x.$$

and if **J** is the current density, this component must equal $J_y\,dx\,dz$. By considering similar current loops in the yz plane, we find there is another current component in the y direction of magnitude $(\partial M_x/\partial z)\,dx\,dz$ associated with the component of magnetization M_x. Hence the total y component of the current density is $J_y=(\partial M_x/\partial z)-(\partial M_z/\partial x)$, with similar expressions for the other components of **J**. These are the components of the vector curl **M**, so that we can write for the current density $\mathbf{J}_m$ associated with a magnetization **M**

$$\mathbf{J}_m=\text{curl}\,\mathbf{M}. \tag{4.23}$$

It follows that if **M** is uniform in space, $\mathbf{J}_m=0$, so that an equivalent current flow exists only in regions where **M** is varying.

In a medium which is both electrically conducting and magnetizable, the total current density will be the sum of the real current density **J** and the equivalent magnetization current density $\mathbf{J}_m$, both of which must be counted in Ampère's law. Hence eqn (4.17), which was derived for a vacuum, must be replaced by

$$\text{curl}\,\mathbf{B}=\mu_0(\mathbf{J}+\mathbf{J}_m)=\mu_0(\mathbf{J}+\text{curl}\,\mathbf{M}),$$

or

$$\text{curl}(\mathbf{B}-\mu_0\mathbf{M})=\mu_0\mathbf{J}. \tag{4.24}$$

This is the form which Ampere's law takes in the presence of a magnetizable medium, and the quantity $(\mathbf{B}-\mu_0\mathbf{M})$, whose curl is related to the flow of real current, is used to define a new vector such that

$$\mathbf{B}-\mu_0\mathbf{M}=\mu_0\mathbf{H},$$

or

$$\mathbf{B}=\mu_0(\mathbf{H}+\mathbf{M}). \tag{4.25}$$

The vector **H** is called the 'magnetic field strength'. It is clear from eqn (4.25) that the dimensions of **H** are different from those of magnetic flux density **B**; it is important to remember this distinction, because magnetic flux density is often loosely referred to as 'the field **B**'.

Ampère's law now takes the simple form

$$\text{curl}\,\mathbf{H}=\mathbf{J}, \tag{4.26}$$

which is more general than eqn (4.17), since it holds both *in vacuo* and in a medium. Obviously, *in vacuo* $\mathbf{B}=\mu_0\mathbf{H}$, so that eqn (4.17) is a special case of (4.26), which is the general differential form of Ampère's law. Similarly, eqn (4.16) must be replaced by the more general equation

$$\int\mathbf{H}.d\mathbf{s}=\int\text{curl}\,\mathbf{H}.d\mathbf{S}=\int\mathbf{J}.d\mathbf{S}=I. \tag{4.27}$$

It is clear from this equation that the dimensions of **H** must be amperes metre^{-1} (A m^{-1}), since the line integral of **H** round a circuit is equal to the total current threading the circuit. From eqn (4.25), **M** must have the same dimensions as **H**, and this can be readily verified, since **M** = magnetic moment per unit volume, and magnetic moment = current×area, from eqn (4.7).

The process by which we have introduced a new vector **H** in modifying our equations to allow for the presence of a polarizable medium is analogous to that in electrostatics, where a new vector **D** was introduced. There, this vector followed from the modification of Gauss's theorem needed in the presence of a polarizable medium; the force vector **E** is related by Gauss's theorem to the sum of the real charge density and the polarization charge density, and the advantage of **D** is that div **D** is related only to the real charge density. In the magnetic case, the force vector **B** is related by Ampère's law to the sum of the real current density and the magnetization current density, and the advantage of **H** is that curl **H** is determined solely by the real current density.

For a volume distribution of current, eqn (4.3) may be written

$$d\mathbf{B} = \frac{\mu_0}{4\pi}\left(\frac{\mathbf{J}\wedge\mathbf{r}}{r^3}\right)d\tau = -\frac{\mu_0}{4\pi}\left(\mathbf{J}\wedge\operatorname{grad}\frac{1}{r}\right)d\tau. \tag{4.28}$$

Since the only change is the replacement of $I\,d\mathbf{s}_2$ by $\mathbf{J}\,d\tau$, it is clear that the proof of eqn (4.6) given in §4.2 is again valid, and div $\mathbf{B} = 0$. This relation was derived only for a current *in vacuo*, but we have shown that any magnetization **M** present can be replaced by an equivalent current density $\mathbf{J}_m$ for which it will also be true that div $\mathbf{B} = 0$. Hence eqn (4.6) holds also in a magnetizable medium.

This equation is similar to that derived for div **D** in electrostatics, except that div $\mathbf{D} = \rho_e$, where ρ_e is the density of true electric charge, while div $\mathbf{B} = 0$ because we have no true magnetic charges. Again, as in electrostatics, we can use Gauss's theorem applied to an elementary flat box surrounding the boundary between two magnetic media as in Fig. 4.8 to show that $\int \mathbf{B}.d\mathbf{S} = 0$ over the surface of the box. If the height of the

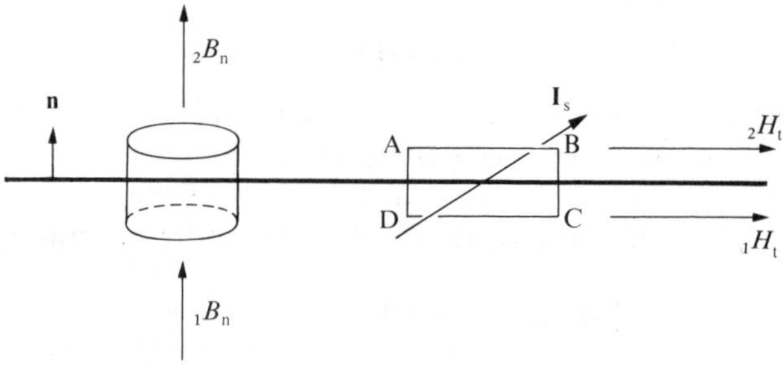

FIG. 4.8. Boundary conditions at the surface between two media. I_s is the surface current per unit width normal to the plane of the circuit ABCDA.

box is very small compared with its cross-section, the only contributions to $\int \mathbf{B}.\mathrm{d}\mathbf{S}$ come from the components of **B** normal to the boundary. Hence we have

$$ {}_1B_n = {}_2B_n. \tag{4.29}$$

The boundary conditions for **H** are found by using Ampère's law applied to a small rectangular circuit ABCDA whose sides BC, AD are very small compared with AB, CD. If there is a surface current I_s per unit length of the surface normal to the circuit, then eqn (4.27) leads immediately to the result

$$ {}_2H_t - {}_1H_t = I_s$$

for the components of H_t normal to I_s, while the components of H_t parallel to I_s are continuous. If the surface current is represented by a vector $\mathbf{I}_s$, the boundary conditions can be expressed as

$$\mathbf{n} \wedge ({}_2\mathbf{H} - {}_1\mathbf{H}) = \mathbf{I}_s, \tag{4.30}$$

where **n** is a unit vector normal to the surface as in Fig. 4.8. Similarly, eqn (4.29) can be written as

$$\mathbf{n}.({}_2\mathbf{B} - {}_1\mathbf{B}) = 0. \tag{4.31}$$

Thus the normal components of **B** are always continuous, while the tangential components of **H** are continuous if there is no surface current. These boundary conditions are similar to those in electrostatics, but note that the formal equivalence here is between **B** and **D**, and between **H** and **E**. This equivalence can be carried a stage further, since if no currents are present, we have curl $\mathbf{H} = 0$, and we can therefore write

$$\mathbf{H} = -\text{grad}\ \phi, \tag{4.32}$$

which is analogous to $\mathbf{E} = -\text{grad}\ V$. Eqn (4.32) is true both *in vacuo* and in a magnetizable medium, our earlier eqn (4.10) being limited to the special case of a vacuum.

4.4. Solution of magnetostatic problems

In many materials it is found that the magnetization **M** is linearly proportional to the magnetic field strength **H**, so that we can write

$$\mathbf{M} = \chi \mathbf{H}. \tag{4.33}$$

Here χ is known as the magnetic susceptibility; if we wish to distinguish it from the electric susceptibility (§ 1.5) we may write them as χ_m and χ_e respectively, but where there is no danger of confusion the subscripts may be omitted. Representative values of χ for different substances vary widely, and will be discussed in Chapter 6. At ordinary temperatures χ is small and independent of **H** for most substances, the exceptions being ferromagnetics, where χ is large and very dependent on field strength; **M** may even be non-zero when $\mathbf{H}=0$.

From eqn (4.25) we have

$$\mathbf{B}=\mu_0(\mathbf{H}+\mathbf{M})=\mu_0\mathbf{H}(1+\chi)$$
$$=\mu_r\mu_0\mathbf{H}=\mu\mathbf{H}, \tag{4.34}$$

where the quantity μ is known as the magnetic permeability at the medium, and μ_0 is the magnetic permeability of vacuum. The quantity μ_r given by

$$\mu/\mu_0=\mu_r=1+\chi \tag{4.35}$$

is known as the relative magnetic permeability of the medium. It is clear that μ_r plays a similar role in magnetostatics to that played by the relative permittivity ϵ_r in electrostatics.

Since $\operatorname{div}\mathbf{B}=0$, we have, when μ is independent of position,

$$\operatorname{div}\mathbf{B}=\operatorname{div}(\mu\mathbf{H})=-\mu\operatorname{div}\operatorname{grad}\phi$$
$$=-\mu\nabla^2\phi=0,$$

or

$$\nabla^2\phi=0, \tag{4.36}$$

showing that the magnetostatic potential obeys Laplace's equation. The theory of Chapter 2 may therefore be adapted to magnetostatic problems, and we shall illustrate this by treating a special case.

The problem of a polarizable sphere in a uniform electric field was solved by means of spherical harmonics in § 2.4. The corresponding magnetic problem may be approached in the same way, but we shall extend it slightly by assuming that **M** is not necessarily proportional to **H**, though still parallel to it. Then, from eqn (4.25),

$$\operatorname{div}\mathbf{B}=-\mu_0\operatorname{div}\operatorname{grad}\phi+\mu_0\operatorname{div}\mathbf{M}=0,$$

whence

$$\nabla^2\phi=\operatorname{div}\mathbf{M}. \tag{4.37}$$

In the corresponding electrostatic case we found that the sphere was uniformly polarized, and we shall assume that this is true also in the magnetic case. Then $\operatorname{div}\mathbf{M}=0$, and the potentials required are solutions of Laplace's equation.

In Fig. 4.9, the potentials inside and outside the sphere are assumed to be

$$\phi_1=-H_1r\cos\theta \qquad (r<a),$$
$$\phi_2=-H_0r\cos\theta+Ar^{-2}\cos\theta \qquad (r>a).$$

As in the electrostatic case, there can be no term in $r^{-2}\cos\theta$ inside the sphere, since it would become infinite at $r=0$; thus the field inside is

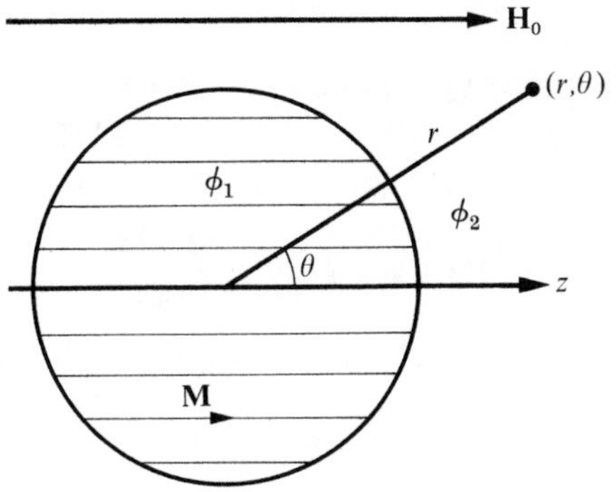

FIG. 4.9. A magnetizable sphere of radius a in a uniform field H_0. μ_1, μ_2 are the relative permittivities inside and outside the sphere.

uniform and equal to H_1. Outside the sphere the field at large distances is uniform and equal to H_0; we take H_0 to be given, so that H_1 and A are the unknowns to be determined from the boundary conditions.

The first boundary condition, that the tangential components of **H** be continuous at the boundary, is equivalent to making $\phi_1 = \phi_2$ at $r = a$, giving

$$H_1 = H_0 - Aa^{-3}.$$

We now assume that the magnetization of the sphere is parallel to $\mathbf{H}_0$, but that it has a permanent component $\mathbf{M}_0$ as well as an induced component $(\mu_1 - 1)\mathbf{H}_1$. Then the magnetization inside the sphere is

$$\mathbf{M}_1 = (\mu_1 - 1)\mathbf{H}_1 + \mathbf{M}_0,$$

so that $\mathbf{B}_1 = \mu_1\mu_0\mathbf{H}_1 + \mu_0\mathbf{M}_0$, while outside $\mathbf{B}_2 = \mu_2\mu_0\mathbf{H}_2$. Then the radial components of **B** are

$$-\mu_1\mu_0(\partial\phi_1/\partial r) + \mu_0 M_0 \cos\theta = \mu_0(\mu_1 H_1 + M_0)\cos\theta \quad \text{(inside)}$$

and

$$-\mu_2\mu_0(\partial\phi_2/\partial r) = \mu_2\mu_0(H_0 + 2Ar^{-3})\cos\theta \quad \text{(outside)}.$$

On equating the two at $r = a$ and eliminating A, we have

$$\mathbf{H}_1 = \mathbf{H}_0 \frac{3\mu_2}{\mu_1 + 2\mu_2} - \mathbf{M}_0 \frac{1}{\mu_1 + 2\mu_2}. \tag{4.38a}$$

Here the first term is the analogue of eqn (2.43) and the second arises from the permanent moment $\mathbf{M}_0$. The difference between $\mathbf{H}_1$ and $\mathbf{H}_0$ can be ascribed (cf. Appendix B) to a 'demagnetizing field' $\mathbf{H}_d = \mathbf{H}_1 - \mathbf{H}_0$ which is proportional to the difference between the actual magnetization $\mathbf{M}_1$ and the value $\mathbf{M}_2 = (\mu_2 - 1)\mathbf{H}_0$ which would have existed there if the sphere were absent and replaced by the medium μ_2. The difference between the two is

$$\mathbf{M}_1-\mathbf{M}_2=\mathbf{H}_0(\mu_1-\mu_2)\frac{1+2\mu_2}{\mu_1+2\mu_2}+\mathbf{M}_0\frac{1+2\mu_2}{\mu_1+2\mu_2}, \tag{4.38b}$$

and on combining this with eqn (4.38a) we obtain

$$\mathbf{H}_d=\mathbf{H}_1-\mathbf{H}_0=-\frac{\mathbf{M}_1-\mathbf{M}_2}{1+2\mu_2}. \tag{4.38c}$$

This result is the analogue of eqn (2.46), where no permanent moment was assumed. If the sphere is in a vacuum, then $\mathbf{M}_2=0$ and $\mu_2=1$, so that $\mathbf{H}_d=-\mathbf{M}_1/3$. These results obviously hold both in the special cases of no induced moment ($\mu_1=1$) or no permanent moment ($\mathbf{M}_0=0$).

We conclude the discussion of magnetostatics by finding a general expression for the magnetic potential due to a magnetized substance. In an element $d\tau$ the dipole moment is $\mathbf{M}\,d\tau$, and the potential eqn (4.13) may be written in the form

$$\phi=\frac{1}{4\pi}\int \mathbf{M}.\operatorname{grad}(1/r)\,d\tau, \tag{4.39}$$

where the differentiation is here with respect to the coordinates of the volume element $d\tau$ (cf. eqn (1.11b)). Then, by a vector transformation similar to that used in deriving eqn (1.16), we find

$$\phi=\frac{1}{4\pi}\int\frac{1}{r}(\mathbf{M}.d\mathbf{S})-\frac{1}{4\pi}\int\frac{1}{r}(\operatorname{div}\mathbf{M}\,d\tau), \tag{4.40}$$

showing that the potential can be attributed to an apparent surface distribution of magnetic charge of surface density $M\cos\theta$, where θ is the angle $\mathbf{M}$ makes with the normal $d\mathbf{S}$ to the surface, and an apparent volume distribution of volume density $-\operatorname{div}\mathbf{M}$, which, since $\operatorname{div}\mathbf{B}=0$, is equal to $+\operatorname{div}\mathbf{H}$. Thus the field lines of $\mathbf{H}$ terminate on the polarization charges, while the field lines of $\mathbf{B}$ are all closed loops since there are no real magnetic charges. If the substance is uniformly magnetized, $\operatorname{div}\mathbf{M}=0$ and there are no volume charges, but there is a surface distribution corresponding to the 'magnetic poles' of classical magnetic theory.

4.5. Steady currents in uniform magnetic media

In § 4.3 the effects of the presence of a magnetizable medium were considered, and it was shown that Ampère's law takes the simple form

$$\operatorname{curl}\mathbf{H}=\mathbf{J} \tag{4.26}$$

or in integral form

$$\int \mathbf{H}.d\mathbf{s}=I. \tag{4.27}$$

It follows from these equations that in an infinite uniform magnetizable medium of relative permeability μ_r the value of the magnetic field strength **H** is unaltered by the presence of the medium, provided the current flow is unaltered, and is independent of μ_r. Returning to eqn (4.3), which holds *in vacuo* where $\mathbf{B}=\mu_0\mathbf{H}$, we see that it may be rewritten in terms of **H** as

$$\mathrm{d}\mathbf{H}=\frac{1}{4\pi r^3}I(\mathrm{d}\mathbf{s}\wedge\mathbf{r}) \tag{4.41}$$

$$=\frac{1}{4\pi r^3}(\mathbf{J}\wedge\mathbf{r})\,\mathrm{d}\tau, \tag{4.42}$$

where the first form refers to a current I in an element $\mathrm{d}\mathbf{s}$ and the second to a current density **J** in an element $\mathrm{d}\tau$. From the preceding remarks it is obvious that these equations are unaltered in a magnetizable medium; they are known as the law of Biot and Savart.

Since $(\mathbf{r}/r^3)=-\mathrm{grad}(1/r)$, we may also express these relations in the form

$$\mathrm{d}\mathbf{H}=-\frac{1}{4\pi}I\{\mathrm{d}\mathbf{s}\wedge\mathrm{grad}(1/r)\}=-\frac{1}{4\pi}\{\mathbf{J}\wedge\mathrm{grad}(1/r)\}\,\mathrm{d}\tau. \tag{4.43}$$

From the vector identity (A.13) of Appendix A,

$$\mathrm{curl}(\mathbf{J}/r)=(1/r)\mathrm{curl}\,\mathbf{J}-\mathbf{J}\wedge\mathrm{grad}(1/r)=-\mathbf{J}\wedge\mathrm{grad}(1/r), \tag{4.44}$$

since the curl operator acts only on the field point and **J** is invariant in this operation. Hence we can write in integral form

$$\mathbf{H}=\frac{1}{4\pi}\int\mathrm{curl}(\mathbf{J}/r)\,\mathrm{d}\tau \tag{4.45}$$

$$=\frac{1}{4\pi}\mathrm{curl}\int(\mathbf{J}/r)\,\mathrm{d}\tau, \tag{4.46}$$

where the order of the curl operation and the integration can be interchanged because the integration is over the current distribution, while the curl operation refers to the field point.

We have seen that in a region where there is no current flow, **H** can be derived from a scalar potential (eqn (4.32)). It is useful to find another potential function which is not subject to this limitation and which is valid in regions where **J** is finite. Such a potential must satisfy the fundamental relation $\mathrm{div}\,\mathbf{B}=0$ given by eqn (4.6), which, as we have seen, is valid both in the presence of currents and of magnetizable media. From the vector identity $\mathrm{div\,curl}\equiv 0$, (see eqn (A.20) of Appendix A), it is clear that (4.6) is automatically satisfied if we write

$$\mathbf{B}=\mathrm{curl}\,\mathbf{A}, \tag{4.47}$$

where **A** is known as the 'magnetic vector potential'. Since **A** is essentially derived from **B** by an integration, this definition is not complete, for we could add another term (equivalent to a constant of integration) such as grad ψ, and still have, using eqn (A.19),

$$\operatorname{curl}(\mathbf{A}+\operatorname{grad}\psi)=\operatorname{curl}\mathbf{A}=\mathbf{B}. \tag{4.48}$$

We therefore add a supplementary condition, and define $\mathbf{A}$ by the relations

$$\operatorname{curl}\mathbf{A}=\mathbf{B}, \qquad \operatorname{div}\mathbf{A}=0. \tag{4.49}$$

The magnetic vector potential enables us, using Stokes's theorem, to derive expressions in terms of line integrals for some quantities which hitherto have appeared only as surface integrals. The magnetic flux Φ through a circuit (eqn (4.21)) becomes

$$\Phi=\int(\mathbf{B}.\mathrm{d}\mathbf{S})=\int(\operatorname{curl}\mathbf{A}).\mathrm{d}\mathbf{S}=\int(\mathbf{A}.\mathrm{d}\mathbf{s}), \tag{4.50}$$

and, if the circuit carries a current I, its potential energy is

$$U_{\mathrm{P}}=-I\Phi=-I\oint(\mathbf{A}.\mathrm{d}\mathbf{s}). \tag{4.51}$$

For a volume current density $\mathbf{J}$, using the relation $\mathbf{I}\,\mathrm{d}\mathbf{s}=\mathbf{J}\,\mathrm{d}\tau$, we have

$$U_{\mathrm{P}}=-\int(\mathbf{A}.\mathbf{J})\,\mathrm{d}\tau. \tag{4.52}$$

We shall use these relations in the next chapter.

A differential equation for $\mathbf{A}$ can be found by combining eqns (4.26) and (4.48). In a medium where $\mathbf{B}=\mu\mathbf{H}$ we obtain

$$\mathbf{J}=\operatorname{curl}\mathbf{H}=\operatorname{curl}(\mathbf{B}/\mu)=\operatorname{curl}\left(\frac{1}{\mu}\operatorname{curl}\mathbf{A}\right). \tag{4.53}$$

If the medium is uniform, so that μ does not vary with position, then using the vector relation (A.22) (Appendix A), we have

$$\mu\mathbf{J}=\operatorname{curl}(\operatorname{curl}\mathbf{A})=\operatorname{grad}\operatorname{div}\mathbf{A}-\nabla^2\mathbf{A},$$

and since $\operatorname{div}\mathbf{A}=0$,

$$\nabla^2\mathbf{A}=-\mu\mathbf{J}. \tag{4.54}$$

This equation is similar to Poisson's equation (2.1), except that the operand is a vector instead of a scalar quantity. This should not cause any difficulty if one remembers that eqn (4.54) implies that each of the components of the vector separately must satisfy the differential equation. A formal solution similar to eqn (2.6) can then be found for each of the components

$$A_x=\frac{\mu}{4\pi}\int\frac{J_x}{r}\,\mathrm{d}\tau,\text{ etc.},$$

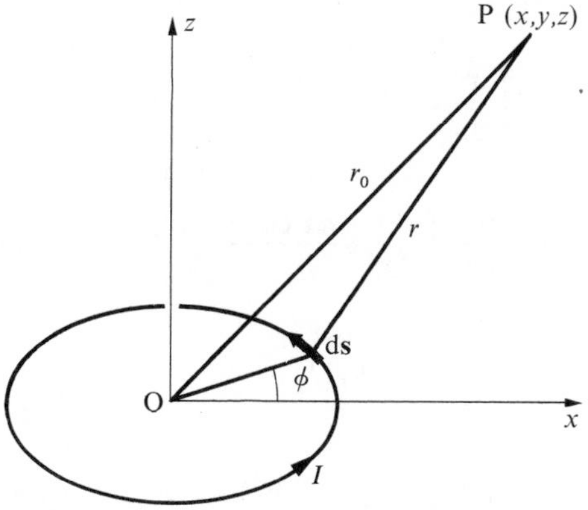

FIG. 4.10. The magnetic vector potential due to a plane circular coil carrying a current I.

which may be expressed in vector form as

$$\mathbf{A} = \frac{\mu}{4\pi}\int \frac{\mathbf{J}}{r}\,\mathrm{d}\tau. \tag{4.55}$$

Since $\mathbf{H} = \mathbf{B}/\mu = (1/\mu)\mathrm{curl}\,\mathbf{A}$ if μ is constant in space, we have a solution for $\mathbf{A}$ which is obviously consistent with eqn (4.46).

For a current circuit carrying a current I, the solution for $\mathbf{A}$ is

$$\mathbf{A} = \frac{\mu I}{4\pi}\int \frac{\mathrm{d}\mathbf{s}}{r}, \tag{4.56}$$

and we can use this to find the value of the magnetic vector potential for an elementary current circuit and hence for a point dipole. We take a small circular current of radius a and calculate the value of $\mathbf{A}$ at a point P. For convenience we use Cartesian coordinates, whose z-axis is normal to the plane of the coil and whose origin is at the centre of the coil. Then P is at $r_0 \equiv (x, y, z)$, and r is the distance of P from an element $\mathrm{d}\mathbf{s}$ of the circuit. In terms of the azimuthal angle ϕ (see Fig. 4.10), the components of $\mathrm{d}\mathbf{s}$ are $(-a\,\mathrm{d}\phi \sin\phi,\ a\,\mathrm{d}\phi\cos\phi,\ 0)$ so that $A_z = 0$. Since

$$r^2 = (x - a\cos\phi)^2 + (y - a\sin\phi)^2 + z^2$$

$$= r_0^2 - 2ax\cos\phi - 2ay\sin\phi + a^2,$$

and $a \ll r$, we have

$$\frac{1}{r} = \frac{1}{r_0} + \frac{ax\cos\phi + ay\sin\phi}{r_0^3} + \ldots$$

and hence

$$A_x = \frac{\mu I}{4\pi}\int_0^{2\pi} \frac{-a\sin\phi\,d\phi}{r}$$

$$= -\frac{a\mu I}{4\pi}\int_0^{2\pi}\left(\frac{1}{r_0}+\frac{ax\cos\phi+ay\sin\phi}{r_0^3}\right)\sin\phi\,d\phi$$

$$= -\frac{\mu}{4\pi}\left\{\pi a^2 I\left(\frac{y}{r_0^3}\right)\right\}.$$

Similarly,

$$A_y = \frac{\mu I}{4\pi}\int_0^{2\pi}\frac{a\cos\phi\,d\phi}{r} = +\frac{\mu}{4\pi}\left\{\pi a^2 I\left(\frac{x}{r_0^3}\right)\right\}.$$

The dipole moment equivalent to the current circuit can be represented by a vector **m** of size $\pi a^2 I$ directed along the z-axis, and the components of **A** are then proportional to those of the vector $\mathbf{m}\wedge\mathbf{r}_0$. Hence, dropping the subscript on $\mathbf{r}_0$, we can write

$$\mathbf{A} = \frac{\mu}{4\pi r^3}(\mathbf{m}\wedge\mathbf{r}) = -\frac{\mu}{4\pi}\left(\mathbf{m}\wedge\mathrm{grad}_P\frac{1}{r}\right), \tag{4.57}$$

where the subscript P denotes that O is fixed and P is the moveable point. From eqns (A.15) and (A.20) it can be seen that eqn (4.57) satisfies the relation $\mathrm{div}\,\mathbf{A}=0$. It can readily be verified that the lines of constant A are circles about the z-axis, and it is generally true that for simple current circuits the lines of **A** are similar in their geometry to those of the current flow, as in the case just discussed.

Eqn (4.57) may be compared with eqn (4.14). The vector potential is proportional to the vector product $(\mathbf{m}\wedge\mathbf{r})$, the scalar potential to the scalar product $(\mathbf{m}\cdot\mathbf{r})$; in addition the quantity μ appears in the vector potential but not in the scalar potential because the former is connected with **B** and the latter with **H**.

In § 2.3 a general expression was found for the equivalent electric dipole moment of a distributed charge, and some applications on the atomic scale were given. A similar expression may be found for the equivalent magnetic dipole moment of a current distribution, and it is convenient to do this from the formula for the energy in a uniform flux density **B**, using the magnetic vector potential. For a uniform **B**,

$$\mathbf{A} = \tfrac{1}{2}(\mathbf{B}\wedge\mathbf{r}) \tag{4.58}$$

as can readily be verified by calculating the components of curl **A** in Cartesian coordinates. Then from eqn (4.52) the potential energy of a distributed current is

$$U_P = -\int(\mathbf{A}.\mathbf{J})\,\mathrm{d}\tau = -\tfrac{1}{2}\int\{(\mathbf{B}\wedge\mathbf{r}).\mathbf{J}\}\,\mathrm{d}\tau$$

$$= -\tfrac{1}{2}\int\mathbf{B}.(\mathbf{r}\wedge\mathbf{J})\,\mathrm{d}\tau = -\mathbf{B}.\int\tfrac{1}{2}(\mathbf{r}\wedge\mathbf{J})\,\mathrm{d}\tau,$$

where $\mathbf{B}$ can be taken out of the integral because it is constant and independent of the space coordinates. The equivalent magnetic dipole moment may be found by equating this expression for the energy to that in eqn (4.19), $U_P = -\mathbf{m}.\mathbf{B}$, giving

$$\mathbf{m} = \int\tfrac{1}{2}(\mathbf{r}\wedge\mathbf{J})\,\mathrm{d}\tau. \tag{4.59}$$

We may check that this agrees with our earlier definition of the dipole moment equivalent to a current circuit since, for the latter, eqn (4.59) becomes

$$\mathbf{m} = I\int\tfrac{1}{2}(\mathbf{r}\wedge\mathrm{d}\mathbf{s}) = I\int\mathrm{d}\mathbf{S} = I\mathbf{S},$$

in agreement with eqn (4.7b).

The formulae derived in this chapter (with the exception of the relation $U = \int\tfrac{1}{2}(\mathbf{B}.\mathbf{H})\,\mathrm{d}\tau$ which is derived in § 5.5) are summarized in Table 4.1 in a form which gives a ready comparison with electrostatics. It is assumed that the permeability μ is uniform, isotropic and independent of $\mathbf{H}$ and that there is no spontaneous magnetization.

Non-linear magnetic media

In a ferromagnetic medium where μ is not independent of $\mathbf{H}$, formulae must be derived using the relation $\mathbf{B} = \mu_0(\mathbf{H}+\mathbf{M})$ instead of $\mathbf{B} = \mu\mathbf{H}$. Instead of eqns (4.53) and (4.54) we have

$$\text{grad div}\,\mathbf{A} - \nabla^2\mathbf{A} = \mu_0\,\text{curl}(\mathbf{H}+\mathbf{M}) = \mu_0\mathbf{J} + \mu_0\,\text{curl}\,\mathbf{M},$$

so that

$$\nabla^2\mathbf{A} = -\mu_0\mathbf{J} - \mu_0\,\text{curl}\,\mathbf{M}. \tag{4.60}$$

This has the formal solution

$$\mathbf{A} = \frac{\mu_0}{4\pi}\int\frac{\mathbf{J}}{r}\,\mathrm{d}\tau + \frac{\mu_0}{4\pi}\int\frac{\text{curl}\,\mathbf{M}}{r}\,\mathrm{d}\tau, \tag{4.61}$$

where, from eqn (4.23), we see that the second term is equivalent to

$$\frac{\mu_0}{4\pi}\int\frac{\mathbf{J}_\mathrm{m}}{r}\,\mathrm{d}\tau$$

and is thus similar in form to the first term, $\mathbf{J}_\mathrm{m}$ being the equivalent current density of the magnetization. By using the vector relations (A.13)

TABLE 4.1
Comparison of various formulae

Electrostatics	Magnetostatics	Currents
$\mathbf{D}=\epsilon_0\mathbf{E}+\mathbf{P}$	$\mathbf{B}=\mu_0(\mathbf{H}+\mathbf{M})$	$\mathbf{H}=(\mathbf{B}/\mu_0)-\mathbf{M}$
$=\epsilon_r\epsilon_0\mathbf{E}=\epsilon\mathbf{E}$	$=\mu_r\mu_0\mathbf{H}=\mu\mathbf{H}$	$=\mathbf{B}/\mu_r\mu_0=\mathbf{B}/\mu$
$\operatorname{div}\mathbf{D}=\rho_e$	$\operatorname{div}\mathbf{B}=0$	$\operatorname{div}\mathbf{B}=0$
$\operatorname{curl}\mathbf{E}=0$	$\operatorname{curl}\mathbf{H}=0$	$\operatorname{curl}\mathbf{H}=\mathbf{J}$
$\mathbf{E}=-\operatorname{grad}V$	$\mathbf{H}=-\operatorname{grad}\phi$	$\mathbf{B}=\operatorname{curl}\mathbf{A}$
$\nabla^2V=-\rho_e/\epsilon$	$\nabla^2\phi=0$	$\nabla^2\mathbf{A}=-\mu\mathbf{J}$
$V=\dfrac{1}{4\pi\epsilon}\displaystyle\int\frac{\rho_e\,d\tau}{r}$		$\mathbf{A}=\dfrac{\mu}{4\pi}\displaystyle\int\frac{\mathbf{J}\,d\tau}{r}$
$\mathbf{D}=\dfrac{q\mathbf{r}}{4\pi r^3}$		$\mathbf{H}=\dfrac{I\,d\mathbf{s}\wedge\mathbf{r}}{4\pi r^3}$
$\mathbf{p}=\displaystyle\int\rho_e\mathbf{r}\,d\tau$		$\mathbf{m}=\displaystyle\int\tfrac{1}{2}(\mathbf{r}\wedge\mathbf{J})\,d\tau$
$V=\dfrac{\mathbf{p}.\mathbf{r}}{4\pi\epsilon r^3}$	$\phi=\dfrac{\mathbf{m}.\mathbf{r}}{4\pi r^3}$	$\mathbf{A}=\dfrac{\mu}{4\pi r^3}(\mathbf{m}\wedge\mathbf{r})$
$U_P=-\mathbf{p}.\mathbf{E}$	$U_P=-\mathbf{m}.\mathbf{B}$	$U_P=-I\Phi=-I\displaystyle\int\mathbf{A}.d\mathbf{s}$
$U_P=\displaystyle\int\rho_eV\,d\tau$		$U_P=-\displaystyle\int(\mathbf{A}.\mathbf{J})\,d\tau$
$U=\displaystyle\int\tfrac{1}{2}\mathbf{D}.\mathbf{E}\,d\tau$		$U=\displaystyle\int\tfrac{1}{2}\mathbf{B}.\mathbf{H}\,d\tau$
$\mathbf{F}=q\mathbf{E}=\displaystyle\int\rho_e\mathbf{E}\,d\tau$		$\mathbf{F}=I\displaystyle\int(d\mathbf{S}\wedge\mathbf{B})=\int(\mathbf{J}\wedge\mathbf{B})\,d\tau$
$\mathbf{F}=(\mathbf{p}.\operatorname{grad})\mathbf{E}$	$\mathbf{F}=(\mathbf{m}.\operatorname{grad})\mathbf{B}$	$\mathbf{F}=I\displaystyle\int(d\mathbf{S}.\operatorname{grad})\mathbf{B}$
$\mathbf{T}=\mathbf{p}\wedge\mathbf{E}$	$\mathbf{T}=\mathbf{m}\wedge\mathbf{B}$	$\mathbf{T}=I\displaystyle\int(d\mathbf{S}\wedge\mathbf{B})$

and (A.24) (from Appendix A) we have

$$\frac{\mu_0}{4\pi}\int\frac{\operatorname{curl}\mathbf{M}}{r}\,d\tau=\frac{\mu_0}{4\pi}\int\operatorname{curl}\left(\frac{\mathbf{M}}{r}\right)d\tau+\frac{\mu_0}{4\pi}\int\mathbf{M}\wedge\operatorname{grad}_O\left(\frac{1}{r}\right)d\tau$$

$$=\frac{\mu_0}{4\pi}\int\frac{\mathbf{M}\wedge d\mathbf{S}}{r}+\frac{\mu_0}{4\pi}\int\mathbf{M}\wedge\operatorname{grad}_O\left(\frac{1}{r}\right)d\tau. \qquad (4.62)$$

In the second term, the vector $\mathbf{r}$ is drawn from a fixed field point P to a variable point O where the element of dipole moment $\mathbf{M}\,d\tau$ is located, and the integration is over the magnetization. This is indicated by the subscript O in $\operatorname{grad}_O(1/r)$. In fact the second term is equivalent to eqn (4.57), with $\mathbf{M}\,d\tau$ instead of $\mathbf{m}$, since there P is the variable point and O a fixed point, and

$$\operatorname{grad}_O(1/r)=-\operatorname{grad}_P(1/r).$$

This change of sign is exactly the same as that in § 1.4, eqns (1.11a) and (1.11b); we note also that in the latter equations for the scalar potential V we have the scalar product ($\mathbf{p}.\text{grad}$), while in eqn (4.62) for the vector potential $\mathbf{A}$ we have the vector product ($\mathbf{M}\wedge\text{grad}$).

The first term in eqn (4.62) is the contribution to $\mathbf{A}$ from a surface layer of magnetization, which is equivalent to a surface current density $(\mathbf{I}_s)_m$ tangential to the surface and normal to $\mathbf{M}$, which can be written as

$$(\mathbf{I}_s)_m = \mathbf{M}\wedge\mathbf{n}, \tag{4.63}$$

where $\mathbf{n}$ is a unit vector normal to the surface. In such a surface layer the magnetization must be parallel to the surface (unlike the magnetic shell considered in § 4.2), a phenomenon which does not often arise in practice.

Boundary conditions

The relations which express the boundary conditions at a surface between two dielectric or two magnetic media are summarized in Table 4.2. They include the cases where there is a surface density σ_e of true electric charge, and a true surface current of $\mathbf{I}_s$ per unit width. The boundary conditions for $\mathbf{D}$ and $\mathbf{E}$ come from eqns (1.26), (1.27), and (1.28) expressed in vectorial form (see Problem 1.13) with $\mathbf{n}$ as a unit vector normal to the surface; those for $\mathbf{B}$ and $\mathbf{H}$ come from eqns (4.30) and (4.31). Since $\mathbf{B} = \text{curl}\,\mathbf{A}$, the boundary conditions for $\mathbf{B}$ and $\mathbf{H}$ give relations only for the spatial derivatives of $\mathbf{A}$. However, it can be shown

TABLE 4.2
Summary of boundary conditions

Surface charge, σ_e per unit area (single layer)	$\mathbf{n}.({}_2\mathbf{D}-{}_1\mathbf{D}) = \sigma_e$ $\mathbf{n}\wedge({}_2\mathbf{E}-{}_1\mathbf{E}) = 0$ ${}_2V-{}_1V = 0$
Surface current, $\mathbf{I}_s$ per unit width (single layer)	$\mathbf{n}.({}_2\mathbf{B}-{}_1\mathbf{B}) = 0$ $\mathbf{n}\wedge({}_2\mathbf{H}-{}_1\mathbf{H}) = \mathbf{I}_s$ ${}_2\mathbf{A}-{}_1\mathbf{A} = 0$
Electric double layer, dipole moment $\mathbf{P}_s$ per unit area	${}_2V-{}_1V = (\mathbf{P}_s.\mathbf{n})/\varepsilon_0$
Magnetic double layer, magnetic moment $\mathbf{M}_s$ per unit area	${}_2\phi-{}_1\phi = \mathbf{M}_s.\mathbf{n}$ ${}_2\mathbf{A}-{}_1\mathbf{A} = \mu_0(\mathbf{M}_s\wedge\mathbf{n})$

The unit vector $\mathbf{n}$, the electric dipole moment $\mathbf{P}_s$, and the magnetic dipole moment $\mathbf{M}_s$ per unit area of the double layer are directed from side 1 to side 2.

(see Stratton 1941) that both the normal and tangential components of **A** are continuous through a single layer of surface current; for a magnetic double layer (a 'magnetic shell') this is true for the normal components of **A**, and also for the tangential components of **A** if the magnetic moment is normal to the surface. On the other hand the scalar potential ϕ has different values on the two sides of a magnetic shell (see § 4.2 and cf. Problem 2.19), the difference being equal to the magnetic dipole moment per unit area.

4.6. Calculation of the magnetic field strengths of simple circuits

The magnetic field strength of a circuit of simple shape may be found in a number of ways, the chief of which are:

(1) use of eqn (4.27). This is possible only when the field strength has a high degree of symmetry as in the rather similar use of Gauss's theorem in electrostatics;

(2) use of the potential of the equivalent magnetic shell, eqn (4.12);

(3) use of the Biot–Savart law (eqn (4.41));

(4) use of the magnetic vector potential (eqns (4.49) and (4.56)).

Simple illustrations will be given of the use of the various methods.

The field strength due to an infinite straight wire carrying a current I was calculated by method (3) in §4.1, but is quickly found by method (1). By symmetry, **H** is a function only of the radial distance from the wire, and by applying eqn (4.27) to a circle of radius r about the wire we find $\int \mathbf{H}.d\mathbf{s} = 2\pi rH = I$. Hence the azimuthal component of **H** is $I/2\pi r$; since this depends only on r, the lines of force are concentric circles about the wire, and no other components of **H** exist.

If the wire has radius a, and the current density is uniform, the field inside the wire can be found by a similar application of eqn (4.27). In this case the current threading a circle of radius r is $I(r^2/a^2)$, so that $\int \mathbf{H}.d\mathbf{s} = 2\pi rH = I(r^2/a^2)$ and $H = Ir/2\pi a^2$, showing that H increases linearly from the centre to the surface of the wire.

The case of a straight wire serves also as a simple example where the vector potential can be found by solving eqn (4.54). Taking the axis of the wire to be along the z-axis, it is obvious that the only component of current density is J_z, and hence the only component of **A** is A_z, so that the lines of **A** are parallel to the wire. Inside the wire (permeability $\mu_r\mu_0 = \mu_1$)

$$\nabla^2 A_z = -\mu_1 J_z = -\mu_1 I/\pi a^2.$$

Since J_z is independent of z and θ (making use of cylindrical coordinates r, θ, z), so also is A_z and the differential equation becomes

$$\frac{1}{r}\frac{\partial}{\partial r}\left(r\frac{\partial A_z}{\partial r}\right) = -\frac{\mu_1 I}{\pi a^2}.$$

Integration gives

$$r\left(\frac{\partial A_z}{\partial r}\right) = -\frac{\mu_1 I r^2}{2\pi a^2},$$

where the constant of integration vanishes, because $r(\partial A_z/\partial r) = 0$ at $r = 0$ since we cannot have a discontinuity in A_z on crossing the axis. A second integration gives

$$A_z = \frac{\mu_1 I}{4\pi}\left(1 - \frac{r^2}{a^2}\right) \quad \text{(inside)}, \tag{4.64}$$

where for convenience we make $A_z = 0$ at $r = a$.

Outside the wire (assuming a medium of permeability μ_2)

$$\frac{1}{r}\frac{\partial}{\partial r}\left(r\frac{\partial A_z}{\partial r}\right) = 0,$$

whence $\partial A_z/\partial r = c/r$, and

$$A_z = c \ln(r/a) \quad \text{(outside)},$$

where the second constant of integration is chosen to make A_z continuous at the boundary, that is, $A_z = 0$ at $r = a$. The constant c is determined by the boundary condition for $\partial A_z/\partial r$ at $r = a$. By writing $r^2 = x^2 + y^2$ and finding the components of curl $\mathbf{A}$, it can be verified that

$$B_x = (\partial A_z/\partial r)(y/r), \qquad B_y = -(\partial A_z/\partial r)(x/r),$$

so that $B_\theta = -\partial A_z/\partial r$. Since B_θ is purely tangential (the other components are zero), the boundary condition is that H_θ must be continuous at $r = a$, and it is easily shown then that

$$A_z = -\frac{\mu_2 I}{2\pi}\ln\left(\frac{r}{a}\right) \quad \text{(outside)}. \tag{4.65}$$

Methods (2) and (3) may be compared in finding the magnetic field strength on the axis of a plane circular coil. We assume that the coil has n turns each carrying a current I, and the radius of the coil is a. Then, at a point on the axis a distance z away, the solid angle subtended by the coil is

$$2\pi\left\{1 - \frac{z}{(z^2+a^2)^{\frac{1}{2}}}\right\}$$

(this formula may be verified by the integration $\Omega = \int (\mathrm{d}\mathbf{S}\cdot\mathbf{r}/r^3)$, taken over the plane surface bounded by the coil). Hence

$$H_z = -\mathrm{d}\phi/\mathrm{d}z = -(nI/4\pi)(\mathrm{d}\Omega/\mathrm{d}z) = \tfrac{1}{2}nIa^2/(z^2+a^2)^{\frac{3}{2}}. \tag{4.66}$$

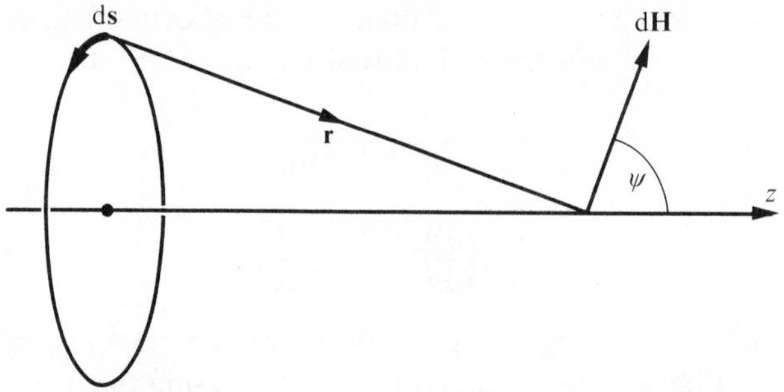

FIG. 4.11. The magnetic field due to a circular coil at a point on its axis.

When $a \ll z$, this formula is the same as that for the field of a dipole of moment $\mathbf{m} = nI(\pi a^2)$ at a point on its axis (cf. eqn (4.7b)).

In applying the formula of Biot and Savart we consider first the field strength $\mathbf{dH}$ due to an element of wire $\mathbf{ds}$, as in Fig. 4.11. Since $\mathbf{dH}$ is normal both to $\mathbf{ds}$ and to $\mathbf{r}$, it will have the direction shown in the figure. On integrating round the coil it is clear that the sum of all the components normal to the axis will be zero. The components parallel to the axis sum to

$$H_z = nI\int(\cos\psi/4\pi r^2)\,\mathrm{d}s = 2\pi anI\cos\psi/4\pi r^2 = \tfrac{1}{2}nIa^2/(z^2+a^2)^{\frac{3}{2}}.$$

This formula may be extended to the case of a solenoid with m turns per unit length, uniformly wound round a cylinder of radius a (Fig. 4.12). If the turns are closely wound, we may regard them as being equivalent to a uniform current flowing round the cylinder, so that an element $\mathrm{d}z$ of it forms a plane coil with a current $mI\,\mathrm{d}z$. At the point O this gives a field along the axis equal to

$$\mathrm{d}H_z = \tfrac{1}{2}mIa^2\,\mathrm{d}z/(z^2+a^2)^{\frac{3}{2}} = -\tfrac{1}{2}mI\sin\phi\,\mathrm{d}\phi,$$

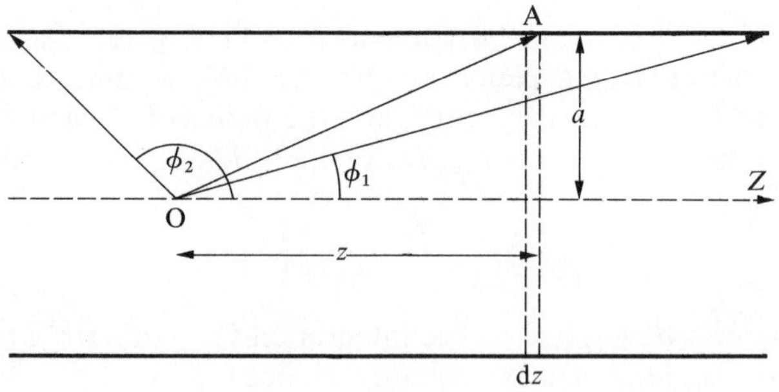

FIG. 4.12. Magnetic field on the axis of a solenoid.

where ϕ is the angle AOZ. Hence

$$H_z = -\tfrac{1}{2}mI\int_{\phi_2}^{\phi_1} \sin\phi \, d\phi = \tfrac{1}{2}mI(\cos\phi_1 - \cos\phi_2). \tag{4.67}$$

For an infinite solenoid, $\phi_1 = 0$ and $\phi_2 = \pi$, so that

$$H_z = mI \tag{4.68}$$

and is uniform inside the solenoid.

We end this section by calculating the force between two small plane circular coils, each of one turn of radius a carrying a current I, with a common axis, and separated by a distance z ($a \ll z$). From eqn (4.66) the field strength on the axis at the centre of the second coil is

$$H_z = \tfrac{1}{2}Ia^2/z^3.$$

Since $a \ll z$, H_z will not vary appreciably for a small distance off the axis, and the flux through the second coil is therefore

$$\Phi = \int \mathbf{B}.d\mathbf{S} = \mu(\tfrac{1}{2}Ia^2/z^3)(\pi a^2) = \tfrac{1}{2}\mu\pi Ia^4/z^3.$$

The potential energy of the second coil is $U_P = -\Phi I$, and hence the force on it is

$$F = -dU_P/dz = -\tfrac{3}{2}\mu\pi a^4 I^2/z^4. \tag{4.69}$$

It is instructive to see just how this force arises. Since div $\mathbf{B} = 0$, in the region around the second coil we have

$$\frac{\partial B_x}{\partial x} + \frac{\partial B_y}{\partial y} + \frac{\partial B_z}{\partial z} = 0,$$

where x and y are normal to the common axis of the coils. By symmetry, $\partial B_x/\partial x = \partial B_y/\partial y$, and hence each equals $-\tfrac{1}{2}(\partial B_z/\partial z)$. At a small distance a from the axis there will be a radial component of field equal to $a(\partial B_x/\partial x) = -\tfrac{1}{2}a(\partial B_z/\partial z) = 3\mu a^3 I/4z^4$. There will therefore be a force $I(d\mathbf{s} \wedge \mathbf{B})$ on each element of the coil, of which the components due to B_z are radial and sum to zero over the whole coil, while the force components due to the radial component of B all act in the negative z direction (assuming the currents in each coil flow in the same sense). These sum to $-I(2\pi a)\mu(3a^3I/4z^4)$, which gives the same result as in eqn (4.69) (see also Problem 4.4).

4.7. Moving charges in electric and magnetic fields

The field strength of the magnetic field produced by a current density $\mathbf{J}$ is given by eqn (4.42). If the current density arises from the movement of an electrical charge density ρ_e with velocity $\mathbf{v}$, then $\mathbf{J}=\rho_e\mathbf{v}$ and

$$\mathbf{H}=\int\frac{(\mathbf{J}\wedge\mathbf{r})}{4\pi r^3}\,\mathrm{d}\tau=\int\frac{(\mathbf{v}\wedge\mathbf{r})}{4\pi r^3}\,\rho_e\,\mathrm{d}\tau.$$

Similarly the force $\mathbf{F}$ on a moving charge density is

$$\mathbf{F}=\int(\mathbf{J}\wedge\mathbf{B})\,\mathrm{d}\tau=\int(\mathbf{v}\wedge\mathbf{B})\rho_e\,\mathrm{d}\tau.$$

For a point charge q these equations reduce to

$$\mathbf{H}=\frac{q(\mathbf{v}\wedge\mathbf{r})}{4\pi r^3},\tag{4.70}$$

and

$$\mathbf{F}=q(\mathbf{v}\wedge\mathbf{B}).\tag{4.71}$$

If an electric field of strength $\mathbf{E}$ is also present, the total force is

$$\mathbf{F}=q(\mathbf{E}+\mathbf{v}\wedge\mathbf{B}).\tag{4.72}$$

It may be remarked that though eqn (4.71) has here been introduced as an additional postulate, it follows as a consequence of eqn (1.3) when we apply the special theory of relativity. An observer in whose system a charge is at rest will ascribe the forces on it to a purely electrostatic field of strength $\mathbf{E}$. On applying the laws for the transformation of mechanical force we find that a moving observer would measure a force of the type given by eqn (4.72), that is, he would ascribe the effects to the action of both electric and magnetic fields. In a similar manner, eqn (4.70) can be deduced from the electrostatic formula for $\mathbf{D}$, (eqn (1.23)).

The motion of charged particles, usually electrons or positive ions, under the action of electric and magnetic fields is the basis of many fundamental experiments in physics, a few of which will be used as illustrations here. The motion in purely electrostatic fields has already been discussed (§ 3.6), and we shall begin by considering the motion of a charge in a uniform magnetic flux density $\mathbf{B}$. If the charge is initially moving in a plane normal to $\mathbf{B}$, then the force on it (assuming $\mathbf{E}=0$) is also in this plane and normal to its direction of motion. Thus no work is done on the particle, since $\mathbf{F}\cdot\mathbf{v}=q\mathbf{v}\cdot(\mathbf{v}\wedge\mathbf{B})=0$, and its velocity remains constant in magnitude. The charge will therefore move in a circle in this plane, the force towards the centre being

$$Mv^2/r=qvB,$$

where M is the mass of the charged particle, and r the radius of the circle. Hence we have

$$r=Mv/qB\quad\text{and}\quad\omega_c=v/r=B(q/M),\tag{4.73}$$

where ω_c is the angular velocity. This equation shows that ω_c, and hence the time taken to make one revolution, is independent of the velocity of the particle (so long as the relativistic change of mass with velocity can be neglected). This fact is made use of in many applications.

Magnetic focusing

If the initial velocity of the charge is not normal to **B**, but makes an angle θ with the direction of **B**, then we can resolve the velocity into a component $v \cos \theta$ parallel to **B**, and a component $v \sin \theta$ normal to **B**. The vector product $\mathbf{v} \wedge \mathbf{B}$ has no component parallel to **B**, and the component of velocity $v \cos \theta$ will therefore continue unaltered. The projection of the motion on a plane normal to **B** will be a circle of radius $r = Mv \sin \theta / qB$, and the actual path of the particle will be a helix. One revolution of the helix is completed in a time $2\pi/\omega = 2\pi M/qB$, and the particle has then moved a distance $z = 2\pi Mv \cos \theta / qB$ in the direction of **B**. For small values of θ, this distance is independent of θ in the first approximation (since $\cos \theta \approx 1 - \frac{1}{2}\theta^2$), and this is the principle used in magnetic focusing.

In Fig. 4.13 electrons leave a point F, and it is desired to focus them so that they all reach a point P a distance z away. If the electrons emerge from a gun with electrostatic focusing, they all have closely the same velocity v, but are not moving quite parallel to the line FP. By means of a solenoid, a magnetic field is applied in the direction FP, and the current in the solenoid adjusted so that the time taken to reach P is equal to one or more periods of revolution in the helical motion caused by the magnetic field. It is often impracticable to use a long solenoid, and one or more short solenoids, encased in iron with a small annular gap round the inner circumference, as shown in Fig. 4.13, are used instead. Such coils give a localized, non-uniform field, which acts like a thin lens; their design is largely empirical.

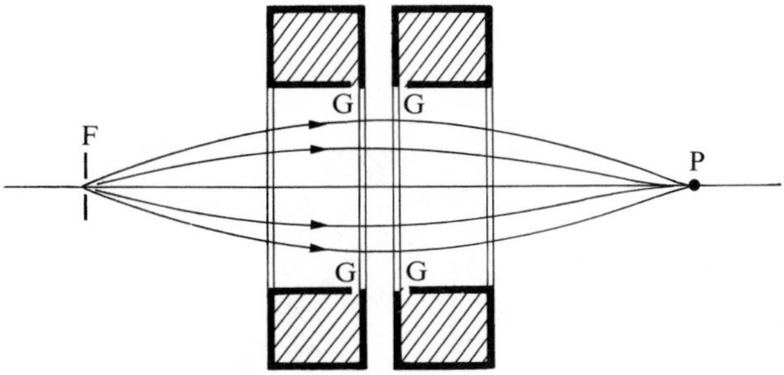

FIG. 4.13. Double solenoid encased in iron, for magnetic focusing. G, G are small annular gaps in the iron casing.

Measurement of specific charge

Determination of the ratio of charge to mass (or 'specific charge') of atomic particles is of prime importance in atomic physics. All such particles carry a charge (positive or negative) equal to the electronic charge e, or a small integral multiple of it, and the ratios of the charge to the mass for the electron and the proton are fundamental constants. From eqn (4.73) it will be seen that an accurate measurement of ω_c and B for particles moving in a circle suffices to determine q/M, and recent methods based on this principle are described in Chapter 24. For positive ions of heavier nuclei, the main interest lies in the measurement of the mass, and instruments for measuring the specific charge q/M for this purpose are known as 'mass spectrometers'. In general they make use of both electric and magnetic fields, and two measurements are required since both the velocity of the particle and the value of q/M are unknown. The positive ions are usually formed in a gaseous discharge and more than one type of ion with varying velocity may be present; the instruments are therefore designed to sort these out, and bring all particles with the same specific charge to a common focus.

Many such instruments have been designed, but since our purpose here is just to illustrate the principles, we shall describe only one, due to Bainbridge. It makes use of a 'velocity selector', formed by the flat plates P_1, P_2 in Fig. 4.14, which have a very small separation. Ions enter these plates from a source through slits S_1, S_2, so that they are travelling with velocity v parallel to the z-axis of the coordinate system shown in the figure. A voltage is maintained between the plates, so that there is an

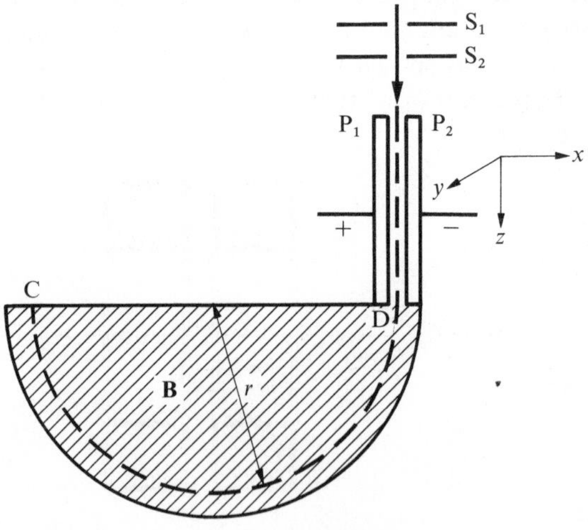

FIG. 4.14. Diagram illustrating the principle of Bainbridge's mass spectrometer.

electric field of strength $\mathbf{E}$ in the x direction. By means of a pair of Helmholtz coils (see Problem 4.2) a uniform magnetic flux density $\mathbf{B}$ is maintained in the y direction, and the total force on an ion between the plates is therefore $q(E-vB)$ in the x direction. If the plates are long and close together, only ions for which this force is zero will emerge, and their velocity must therefore be $v=E/B$. The device thus selects ions of a particular velocity determined by this ratio. On emerging from the plates the ions travel in a semicircular path under the influence of the field $\mathbf{B}$ alone until they strike a detector at C. The distance DC is twice the radius of the orbit and is thus

$$2r=\frac{2Mv}{qB}=2(M/q)\frac{E}{B^2}. \tag{4.74}$$

Hence the distance DC is linearly proportional to the mass of the ion. In Problem 4.9 it is shown that the distance DC is independent (to the first order) of the angle to the z-axis at which an ion emerges from the plates, provided this is small, so that we have 'first-order' focusing of the ions with a given value of q/M.

References

PAGE, L. and ADAMS, N. I. (1945). *Am. J. Phys.* **13,** 141.

ROBINSON, F. N. H. (1973). *Macroscopic electromagnetism*, pp. 95–6. Pergamon Press, Oxford.

STRATTON, J. A. (1941). *Electromagnetic theory*. McGraw–Hill, New York.

Problems

4.1. Two short magnets are attached to a cork so that they float on water with their axes horizontal. One magnet lies with its centre on the axis of the other, but with its own axis perpendicular to the line joining them. Assuming that the distance between the magnets is large compared with their lengths, so that they can be treated as point dipoles, calculate the force and the couple on each magnet, and satisfy yourself that there is no resultant force or couple on the system as a whole.

4.2. Two identical circular coils, each of n turns of radius a, are placed with their planes parallel and normal to the line joining them, a distance r apart. Calculate the magnetic field on the axis midway between the coils due to a current I through each coil. Show that, if $r=a$, the field midway between the coils is uniform over a considerable region; that is, $\partial H/\partial r$, $\partial^2 H/\partial r^2$ and $\partial^3 H/\partial r^3$ are all zero at the midpoint. Such an arrangement of coils was used by Helmholtz for a galvanometer, a small magnetic needle suspended by a torsion fibre at the centre being deflected by the current through the coils.

4.3. Two infinite cylindrical conductors are placed parallel to one another at a distance $2a$ apart. They carry equal and opposite currents. Show that in the equatorial plane the gradient of the magnetic field is greatest at a distance $a/\sqrt{3}$ from the plane through the axes of the cylinders.

4.4. Deduce eqn (4.69) by treating each coil as a point dipole of moment $I(\pi a^2)$, and using the formula for the force on a dipole in a non-uniform field.

4.5. A uniform flux density $\mathbf{B}_0$ is present in a liquid of relative permeability μ_2. Show that inside a circular cylinder of relative permeability μ_1, inserted in the liquid with its axis normal to $\mathbf{B}_0$, the flux density is

$$\mathbf{B}_1 = \frac{2\mu_1}{\mu_1+\mu_2}\mathbf{B}_0.$$

Important. Note that if the cylinder carries a current I, the force on it per unit length is $(\mathbf{I}\wedge\mathbf{B}_0)$ and not $(\mathbf{I}\wedge\mathbf{B}_1)$ (see, for example, Robinson 1973).

4.6. A spherical shell has radii a and b respectively $(b>a)$, and is made of a material of permeability μ_r. It is placed in a uniform field of strength H. Show that the field inside the shell is

$$H_i = \frac{9\mu_r H}{(2\mu_r+1)(\mu_r+2)-2(\mu_r-1)^2(a/b)^3}$$

and that for large values of μ_r, this approximates to

$$H_i = 9H/\{2\mu_r(1-a^3/b^3)\}.$$

Thus if μ_r is large, H_i is much smaller than H, and an instrument can be shielded from stray magnetic fields by placing it in an iron case. Magnetic shielding is much less efficient than electrostatic shielding (especially if a/b is close to unity), for the effective value of ϵ_r in the equivalent expression for a conductor is infinite.

4.7. A hollow sphere of internal radius a, external radius b, has a uniform spontaneous magnetization $\mathbf{M}$ per unit volume. Show that the field strength in the internal cavity $(r<a)$ is zero, and that the external field strength $(r>b)$ is the same as that of a dipole moment $\mathbf{m}=4\pi\mathbf{M}(b^3-a^3)/3$, the total moment of the hollow sphere.

Show also that the square of the field strength outside the sphere at a point (r, θ), measured from the centre of the sphere and with respect to the direction of magnetization, is

$$H^2 = (3\cos^2\theta+1)\left\{\frac{M(b^3-a^3)}{3r^3}\right\}^2.$$

If the angle of dip δ is defined as the angle which the lines of force at a point on the external surface of the sphere make with the tangent at that point, show that

$$\tan\delta = 2\cot\theta.$$

4.8. A particle of mass M and charge q is rotating in a circular orbit of radius r with angular velocity ω. Show that a magnetic dipole moment $\mathbf{m}$ is associated with the motion of the charge, such that

$$\mathbf{m} = (q/2M)\mathbf{G},$$

where $\mathbf{G}=Mr^2\omega$ is the angular momentum of the particle. (The orbit may be regarded as a small circuit carrying a current $I=q\times$the frequency at which the charge passes any point in the orbit per unit time.)

4.9. In Bainbridge's mass spectrometer the ions emerge in a wedge-shaped beam of small semi-vertical angle θ (see Fig. 4.14). If the resolving power of the instrument as a mass spectrometer is defined as the reciprocal of the smallest fractional change of mass which will produce non-overlapping traces on the plane CD, show that the resolving power is $2/\theta^2$.

4.10. A charged particle starts from rest at the origin of coordinates in a region where there is a uniform electric field of strength **E** parallel to the x-axis, and a uniform magnetic flux density **B** parallel to the z-axis. Show that the coordinates of the particle at a time t later will be

$$x = (E/\omega B)(1-\cos \omega t),$$

$$y = -(E/\omega B)(\omega t-\sin \omega t),$$

$$z = 0,$$

where $\omega = qB/m$. (The path of the particle is a cycloid.)

A charge $-q$ is liberated with zero velocity from the negative plate of a parallel-plate capacitor, to which is applied a flux density **B** parallel to the plates. Show that it will not reach the positive plate if the plate separation d is greater than $2mE/qB^2$, where **E** is the field strength between the plates.

5. Electromagnetic induction and varying currents

5.1. Faraday's laws of electromagnetic induction

THE experiments of Oersted and others showed that 'electricity can produce magnetism', and established the laws governing the magnetic field set up by a current. Many experiments were devised to detect the inverse effect, the flow of electric current due to a magnetic field, without success, mainly because a steady current flow was looked for. In 1831 Faraday found that a transient flow of current occurred in a closed circuit when the magnetic flux through the circuit was changed. The change of flux could be brought about in a number of ways: in his first experiment two coils of wire were wound on a ring of soft iron as in Fig. 5.1. The presence of a current in the second coil was detected by connecting it to another coil near a small suspended magnet. When the first coil was connected to a battery, a momentary oscillation of the magnet occurred, after which it settled in its original position. A similar oscillation, though with an initial kick in the opposite direction, was observed on disconnecting the battery. In other experiments Faraday showed that similar effects were observed if a permanent magnet was moved near the second coil, or if the coil was moved in the neighbourhood of a magnet. His results were summed up in the two laws:

(1) when the magnetic flux through a circuit is changing, an electromotive force is induced in the circuit;

(2) the magnitude of this e.m.f. is proportional to the rate of change of the flux.

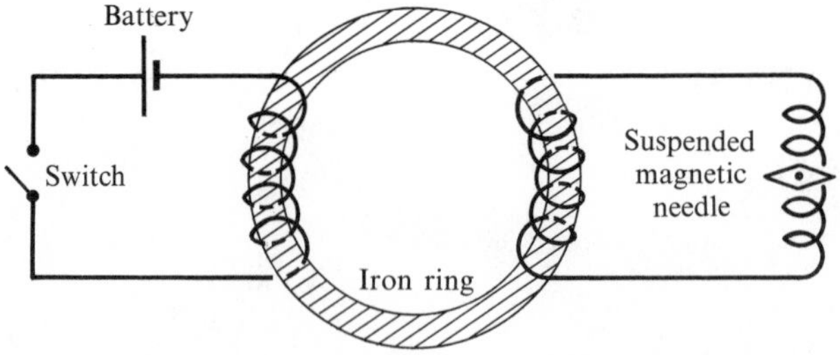

FIG. 5.1. Faraday's experiment on electromagnetic induction.

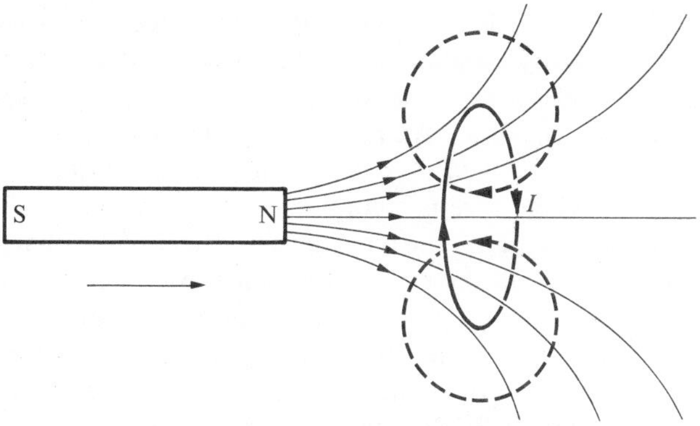

FIG. 5.2. Current induced in a loop by a moving magnet (broken lines represent lines of magnetic field produced by the induced current when the magnet moves towards the loop).

The sign of the e.m.f. is given by Lenz's law, which states that it is such that any current flow is in the direction which would oppose the flux change causing the e.m.f. Thus, in Fig. 5.2, if the magnet is moved towards the closed loop of wire so that the magnetic flux through the coil is increased, the induced current will flow in such a direction that its own flux opposes the increased flux of the magnet threading the loop.

These laws are expressed in the equation

$$V = -\mathrm{d}\Phi/\mathrm{d}t, \tag{5.1}$$

where V is the e.m.f. round the circuit, and Φ is the instantaneous value of the magnetic flux through the circuit. Now $\Phi = \int \mathbf{B}.\mathrm{d}\mathbf{S}$ and $V = \int \mathbf{E}.\mathrm{d}\mathbf{s}$, where the former integral is taken over any area bounded by the circuit and the latter integral is taken round the circuit. Hence we have

$$\int \mathbf{E}.\mathrm{d}\mathbf{s} = -\frac{\mathrm{d}}{\mathrm{d}t}\int \mathbf{B}.\mathrm{d}\mathbf{S}. \tag{5.2}$$

Using the transformation $\int \mathbf{E}.\mathrm{d}\mathbf{s} = \int \operatorname{curl} \mathbf{E}.\mathrm{d}\mathbf{S}$, and the fact that the time and space coordinates are independent variables, this can be rewritten in the form

$$\int \operatorname{curl} \mathbf{E}.\mathrm{d}\mathbf{S} = -\int \frac{\mathrm{d}\mathbf{B}}{\mathrm{d}t}.\mathrm{d}\mathbf{S},$$

and since this must hold over any surface area, the integrands must be equal, giving the differential form

$$\operatorname{curl} \mathbf{E} = -\mathrm{d}\mathbf{B}/\mathrm{d}t. \tag{5.3}$$

At first sight it would have been expected that eqn (5.1) would have contained a multiplying constant to be determined either experimentally or from theory. This constant is in fact unity, as can be seen in the following way. Let us assume we have a very thin conductor carrying no current, which is moved with a velocity **v** in a uniform field of magnetic flux density **B**. Since the wire is a conductor, it carries charges (electrons) which are free to move along the wire; let the velocity of a charge q in the conductor be **u** relative to the conductor. Since the conductor is very thin **u** must be parallel to the direction of the wire at any point. The velocity of the charge q relative to the observer is **v**+**u**, and the force on it will therefore be $\mathbf{F}=q(\mathbf{v}+\mathbf{u})\wedge\mathbf{B}$. If the charge moves a distance $\mathrm{d}\mathbf{r}$ along the wire, the work done is $\mathbf{F}.\mathrm{d}\mathbf{r}$; since $(\mathbf{u}\wedge\mathbf{B}).\mathrm{d}\mathbf{r}=0$ because **u**, d**r** are parallel, this is the same as if there existed in d**r** an e.m.f.

$$\mathrm{d}V=(\mathbf{v}\wedge\mathbf{B}).\mathrm{d}\mathbf{r}. \tag{5.4}$$

For a closed circuit in a uniform flux density **B**, since $(\mathbf{v}\wedge\mathbf{B})$ is constant, the total e.m.f. $V=\int(\mathbf{v}\wedge\mathbf{B}).\mathrm{d}\mathbf{r}=0$. If the flux density is not uniform, the flux through the circuit will change as the circuit moves, and we can relate V to the rate of change of flux. Consider a small rectangular circuit ABCD (Fig. 5.3) with sides a, b parallel to the x-, y-axes of a Cartesian system. The e.m.f. in an anticlockwise direction (the sense in which a right-handed screw would turn to advance along the z-axis) is

$$V=a(v_yB_z-v_zB_y)+b\left\{v_z\left(B_x+a\frac{\partial B_x}{\partial x}\right)-v_x\left(B_z+a\frac{\partial B_z}{\partial x}\right)\right\}-$$

$$-a\left\{v_y\left(B_z+b\frac{\partial B_z}{\partial y}\right)-v_z\left(B_y+b\frac{\partial B_y}{\partial y}\right)\right\}-b(v_zB_x-v_xB_z)$$

$$=-ab\left\{v_x\frac{\partial B_z}{\partial x}+v_y\frac{\partial B_z}{\partial y}-v_z\left(\frac{\partial B_x}{\partial x}+\frac{\partial B_y}{\partial y}\right)\right\}$$

$$=-ab\left\{v_x\frac{\partial B_z}{\partial x}+v_y\frac{\partial B_z}{\partial y}+v_z\frac{\partial B_z}{\partial z}\right\} \tag{5.5}$$

since $\operatorname{div}\mathbf{B}=0$. But $abB_z=\Phi$, the flux through the circuit, and since $v_x=\mathrm{d}x/\mathrm{d}t$, etc.,

$$V=-\left(\frac{\partial\Phi}{\partial x}v_x+\frac{\partial\Phi}{\partial y}v_y+\frac{\partial\Phi}{\partial z}v_z\right)=-\frac{\mathrm{d}\Phi}{\mathrm{d}t},$$

in agreement with eqn (5.1). The unit of magnetic flux Φ is the weber, and an e.m.f. of 1 V is generated in a circuit where the flux is changing at the rate of $1\ \mathrm{Wb\,s^{-1}}$.

Eqn (5.3) may be combined with the relation $\mathbf{B}=\operatorname{curl}\mathbf{A}$ (eqn (4.47)) to give

$$\operatorname{curl}\mathbf{E}=-\frac{\partial}{\partial t}\operatorname{curl}\mathbf{A}=-\operatorname{curl}(\partial\mathbf{A}/\partial t),$$

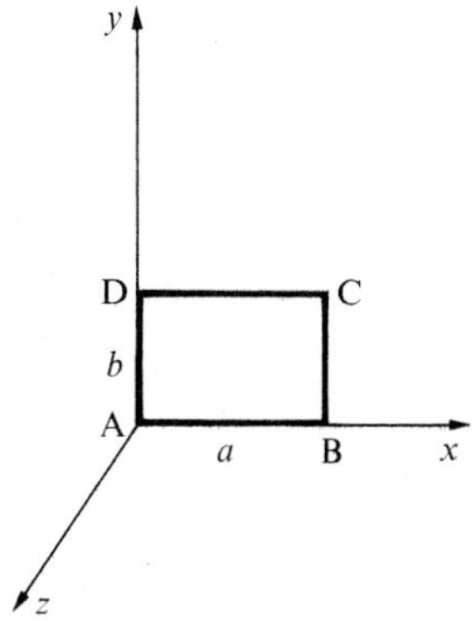

FIG. 5.3. Relation between Faraday's law and the force on a moving conductor.

whence

$$\mathbf{E} = -\frac{\partial \mathbf{A}}{\partial t} + \text{constant} = -\frac{\partial \mathbf{A}}{\partial t} - \text{grad } V. \tag{5.6}$$

This is a more general equation than eqn (1.6) which applies only to steady fields. As an example of its use we consider a particle accelerator such as the betatron, in which a particle of charge q is constrained to move in a circle of radius R through which there is a changing magnetic flux. If the flux density, apart from varying with time, depends only on the radius, then the lines of **A** are concentric circles, and from eqn (4.50), $A = A_\theta = \Phi/2\pi R$. If grad $V = 0$, the rate of change of momentum of the particle is

$$\mathrm{d}\mathbf{p}/\mathrm{d}t = -q(\mathrm{d}\mathbf{A}/\mathrm{d}t)$$

and

$$\mathbf{p} - \mathbf{p}_0 = -q\mathbf{A} = -(q/2\pi R)\Phi, \tag{5.7}$$

if the momentum was $\mathbf{p}_0$ when Φ was zero. The negative sign reminds us that the magnetic flux must link the circuit in the correct sense in order to produce an increase and not a decrease in momentum.

5.2. Self-inductance and mutual inductance

When a current I flows in a circuit, a magnetic field is set up and there will be a magnetic flux Φ through the circuit arising from its own magnetic field. The magnetic field strength at any point is proportional to the current I; so also is the flux Φ, and we may therefore write

$$\Phi = LI, \tag{5.8}$$

where L is a constant which depends on the geometry of the circuit and the permeability of the medium in which it is immersed. L is called the self-inductance of the circuit, and is equal to the total flux through the circuit when unit current is flowing. A circuit has unit self-inductance (one henry) if it is threaded by one weber of flux when one ampere of current is flowing.

If a second coil is brought near to a coil carrying a current I, there will in general be a magnetic flux Φ_2 through the second coil due to the current in the first coil. Since Φ_2 is again linearly proportional to I_1, we may write

$$\Phi_2 = L_{21} I_1, \tag{5.9a}$$

where L_{21} is called the mutual inductance between the two circuits. The unit of mutual inductance is again the henry. There will also be a flux Φ_1 through the first circuit due to a current I_2 in the second circuit, given by

$$\Phi_1 = L_{12} I_2. \tag{5.9b}$$

The coefficients L_{12} and L_{21} are equal, as can be seen from energy considerations. The potential energy of the system can be found from the flux of either coil due to the field of the other; from eqn (4.51)

$$U_P = -\Phi_2 I_2 = -L_{21} I_1 I_2 = -\Phi_1 I_1 = -L_{12} I_2 I_1,$$

showing that

$$L_{12} = L_{21}. \tag{5.10}$$

By using eqns (4.51) and (4.55) we can derive a formula for the mutual inductance, since

$$U_P = -I_1 \int \mathbf{A}_{21} \cdot d\mathbf{s}_1 = -I_1 \int \left(\frac{\mu}{4\pi} \int \frac{I_2\, d\mathbf{s}_2}{r} \right) \cdot d\mathbf{s}_1$$

$$= -\frac{\mu}{4\pi} I_1 I_2 \int\int \frac{d\mathbf{s}_1 \cdot d\mathbf{s}_2}{r} = -L_{12} I_1 I_2, \tag{5.11}$$

where, by symmetry,

$$M = L_{12} = L_{21} = \frac{\mu}{4\pi} \int\int \frac{d\mathbf{s}_1 \cdot d\mathbf{s}_2}{r}. \tag{5.12}$$

This result is known as Neumann's formula. Since the unit of mutual inductance is the henry, this formula shows that the dimensions of μ_0 are henry metre^{-1} (H m^{-1}).

If the two coils are closely wound, so that all the flux generated by the first coil passes through the second, and vice versa, then the ratio of the two fluxes Φ_1 and Φ_2 will just be equal to the ratio of the number of turns n_1, n_2 on the two coils. For a current I_1 in the first coil we have

$$\Phi_1/\Phi_2 = (L_1 I_1)/(M I_1) = L_1/M = n_1/n_2,$$

while for the flux generated by a current I_2 in the second coil

$$\Phi_2/\Phi_1 = (L_2I_2)/(MI_2) = L_2/M = n_2/n_1.$$

Hence

$$L_1/M = M/L_2 = n_1/n_2 \quad \text{and} \quad M^2 = L_1L_2. \tag{5.13}$$

If the flux through the two coils is changing, the voltages induced in the two coils will be in the ratio

$$V_1/V_2 = (\mathrm{d}\Phi_1/\mathrm{d}t)/(\mathrm{d}\Phi_2/\mathrm{d}t) = n_1/n_2 = 1/n.$$

Hence such a device may be used as a transformer, since if a changing voltage V_1 is applied to the 'primary' coil, a changing voltage of different magnitude will be induced in the 'secondary' coil. The voltage transformation ratio is $n = V_2/V_1$, the 'turns ratio' of secondary to primary. In general not all the flux of one circuit passes through the other, and M is less than $(L_1L_2)^{\frac{1}{2}}$; it may be written as

$$M = k(L_1L_2)^{\frac{1}{2}} \quad (0 \leq k \leq 1), \tag{5.14}$$

where k is called the 'coefficient of coupling'. The theory of transformers is considered further in § 7.5.

The magnitude of an inductance may be calculated from first principles by computing the field produced by a given current in the coil, and then finding the total flux through the same or another coil, according to whether the self-inductance of the first coil or the mutual inductance between the two coils is required. The calculations are illustrated below for a number of simple shapes of coil.

Long solenoid

For an infinitely long solenoid, wound with m turns per unit length and carrying a current I, the magnetic field strength inside is uniform and given by eqn (4.68):

$$H = mI.$$

If the core of the solenoid has permeability μ, the flux through each turn is $\Phi' = \mu AmI$, where A is the cross-sectional area of the solenoid. The self-inductance per unit length is therefore $L' = m\Phi'/I = \mu m^2A$; for a solenoid of length l, large compared with its diameter, this formula is still very nearly correct, and we may write for the total self-inductance

$$L = \mu m^2 Al. \tag{5.15}$$

If a second short coil of n turns, insulated from the first, is wound on the solenoid, as in Fig. 5.4, the mutual inductance is

$$M = \mu mnA. \tag{5.16}$$

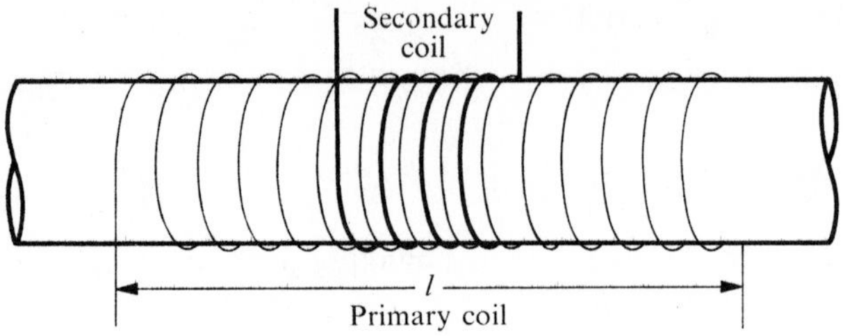

FIG. 5.4. Solenoid with primary and secondary coils.

Two coaxial coils

Another simple case is that of two plane coaxial coils A and B as in Fig. 5.5, of radii a and b, and total numbers of turns n_1 and n_2 respectively, whose centres are a distance z apart, where $z \gg a, b$. The field strength at the centre of B due to a current I in A is, from eqn (4.66),

$$H = \tfrac{1}{2}Ia^2n_1/z^3,$$

and the total flux through B is $\pi b^2 n_2(\mu H)$, since the field through the coil will be uniform in the first approximation when the inequality $z \gg a, b$ holds. Hence the mutual inductance is

$$M = \frac{\mu\pi a^2 b^2 n_1 n_2}{2z^3}. \tag{5.17}$$

Pair of coaxial circular cylinders

An important method of carrying radio-frequency alternating currents is by means of a pair of coaxial cylinders of radii a, b $(b > a)$, as in Fig. 5.6. At any point the current in the inner cylinder is I, while that in the outer cylinder is $-I$, that is, it is exactly equal in magnitude but flowing in the opposite direction. The magnetic field at a distance r from the axis when $r < b$ is the same as that due to a straight wire, so that

$$H = I/2\pi r.$$

Application of the same equation shows that there will be no field outside the larger cylinder, since any circuit drawn round it is threaded by two

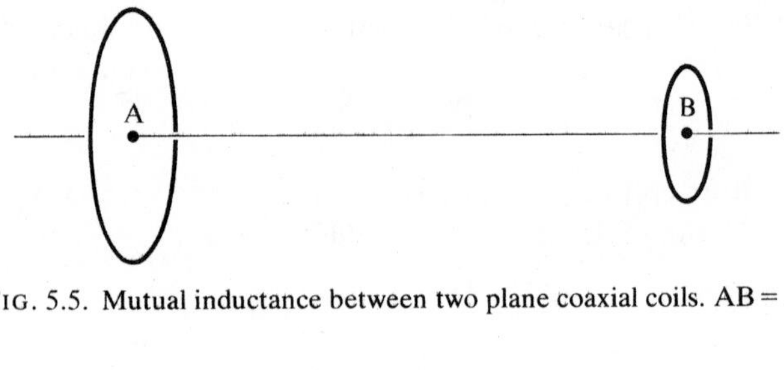

FIG. 5.5. Mutual inductance between two plane coaxial coils. AB = z.

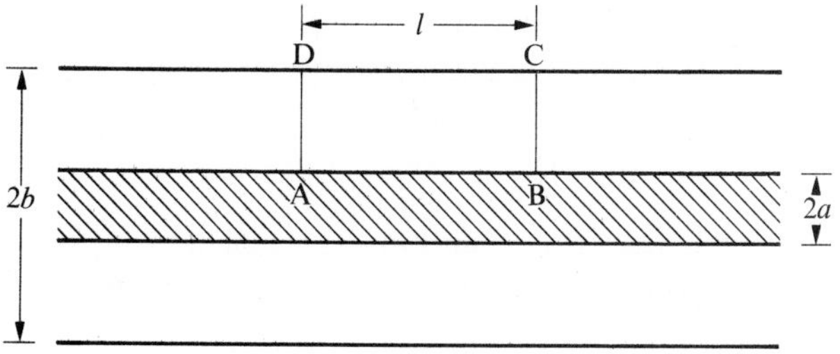

FIG. 5.6. Self-inductance of coaxial circular cylinders.

equal and opposite currents. To compute the self-inductance of a length l, we find the flux through a circuit such as ABCD in Fig. 5.6. Instead of calculating this directly from the equation above for H, we illustrate an alternative method using the magnetic vector potential $\mathbf{A}$. Since the current flow is entirely parallel to the z-axis, the vector potential has just the single component A_z, whose value in the space between the cylinders is, from eqn (4.65),

$$A_z = -(\mu I/2\pi)\ln(r/a).$$

The flux through the circuit ABCD of length l is therefore

$$\int \mathbf{A}\cdot\mathrm{d}\mathbf{s} = l(A_z)_{r=a} - l(A_z)_{r=b} = (l\mu I/2\pi)\ln(b/a)$$

and the inductance L per unit length is

$$L = \frac{\{(A_z)_{r=a} - (A_z)_{r=b}\}}{I} = \frac{\mu}{2\pi}\ln(b/a). \tag{5.18}$$

These equations may be compared with the corresponding equation (1.30) for the capacity per unit length of such a pair of coaxial cylinders,

$$C = \frac{Q}{(V_a - V_b)} = \frac{2\pi\epsilon}{\ln(b/a)}.$$

On combining the two sets of equations we obtain

$$LC = \mu\epsilon.$$

This rather surprising result (to which we shall return in § 9.3 when discussing the propagation of electromagnetic waves along parallel cylinders) is not accidental. It arises when we are dealing with pairs of parallel conductors whose dimensions do not change in one direction (in this case, the z-axis), and the electric and magnetic fields are entirely transverse to this direction. If the current flow is confined to this direction, then the sole component of **A** is, from eqn (4.56),

$$A_z = \frac{\mu I}{4\pi}\int_{-\infty}^{+\infty}\frac{dz}{r}. \tag{5.19a}$$

Clearly this is independent of z, and in the xy plane it varies in the same way as the electric potential V, which from eqn (2.6) is

$$V = \frac{Q}{4\pi\epsilon}\int_{-\infty}^{+\infty}\frac{dz}{r}, \tag{5.19b}$$

where Q is the charge per unit length on the outer cylinder (also independent of z). As a result, similar geometrical factors appear in the calculation of L and $1/C$, so that they disappear in the product LC (see also Problem 5.2).

5.3. Transient currents in circuits containing inductance, resistance, and capacitance

If a circuit containing a battery of e.m.f. V and a resistance R is connected to a coil in which the magnetic flux Φ is changing, the net e.m.f. V' applied to R will be the sum of the battery voltage V and the e.m.f. developed in the coil. Hence $V' = V - d\Phi/dt = RI$, or

$$V = RI + d\Phi/dt. \tag{5.20a}$$

We can apply this equation to a number of problems, the first being a circuit containing a self-inductance L and a resistance R, as shown in Fig. 5.7, which is connected at time zero to a battery of constant e.m.f. V. Since $d\Phi/dt = L(dI/dt)$, we have

$$V = IR + L(dI/dt), \tag{5.20b}$$

which is the fundamental differential equation relating the current I to the voltage V. Integration of this equation, with the condition $I = 0$ at $t = 0$, gives

$$I = \frac{V}{R}(1 - e^{-(R/L)t}), \tag{5.21}$$

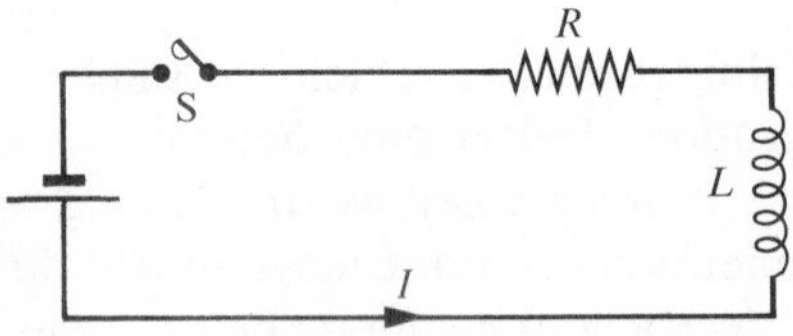

FIG. 5.7. Battery driving current through R and L.

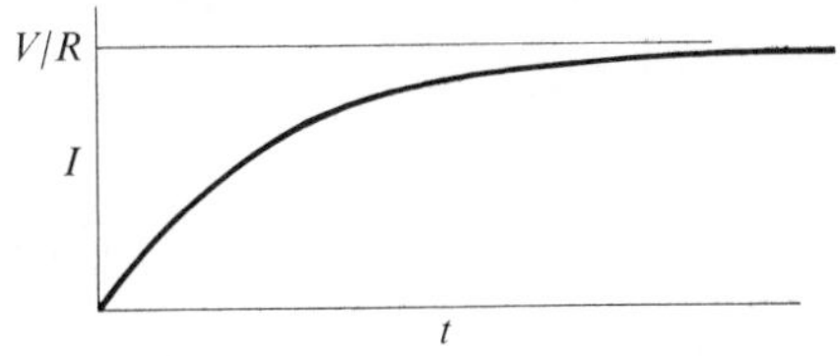

FIG. 5.8. Rise of current in circuit of Fig. 5.7.

showing that the current approaches exponentially the value V/R which it would have if there were no inductance present (see Fig. 5.8). The rate of approach to this steady value depends on the ratio of resistance to inductance. If $R = 0$, the steady state, corresponding to infinite current, is never reached, but the current rises linearly according to the equation $I = (V/L)t$, obtained by direct integration of eqn (5.20b) with $R = 0$. When R is finite, the initial rate of rise of current, given by the tangent at the origin in Fig. 5.8, is $dI/dt = V/L$, but as the current through the resistance increases, the voltage across the inductance falls, with a corresponding decrease in dI/dt.

The converse problem, in which a battery has been connected to the circuit for a long time so that a steady current I_0 is flowing, and then at time zero the battery is replaced by a short-circuit, leads to the same differential eqn (5.20b), but with $V = 0$. Its solution is

$$I = I_0 e^{-(R/L)t}, \tag{5.22}$$

showing that the effect of the inductance is to prevent the current from falling instantaneously to zero. If the battery is suddenly open-circuited, the sudden cessation of the current produces a large impulse voltage $-L(dI/dt)$ in the inductance, which may be sufficient to cause a spark across the point at which the circuit is broken. With large inductances such as are found in electromagnets (see § 6.5) very high voltages may arise in this way which can damage the insulation if the circuit is broken suddenly.

In both the cases considered above the exponential is of the form $e^{-t/\tau}$, and the exponential rate of change of the current is the same; the quantity $\tau = L/R$ is called the time constant of the circuit.

Circuit with capacitance and resistance

An analogous problem is that of a capacitance C in series with a resistance R, to which a battery of e.m.f. V is connected at time zero (see Fig. 5.9). At any instant the charge on the capacitor is Q, and the voltage across it is Q/C. Since the current $I = dQ/dt$, we have

$$V = Q/C + RI = Q/C + R(dQ/dt). \tag{5.23}$$

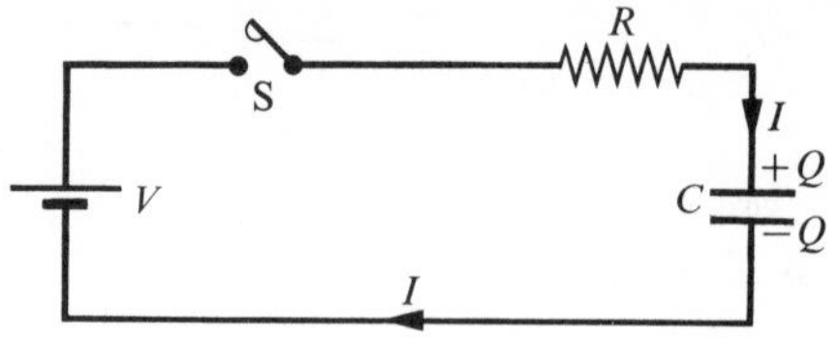

FIG. 5.9. Battery charging capacitance C through resistance R.

The solution of this equation gives

$$Q = CV(1-e^{-t/RC}) \tag{5.24}$$

showing that the charge on the capacitor builds up in a manner similar to the current in the previous problem. The time constant of the circuit is now $\tau = RC$, and the current at any instant is

$$I = dQ/dt = \frac{V}{R}e^{-t/RC} = \frac{V}{R}e^{-t/\tau}. \tag{5.25}$$

It therefore falls exponentially from its initial value (V/R) to zero, as shown in Fig. 5.10. The voltage across the capacitor increases from zero to its steady value V, when no more current can flow in the circuit.

If we have an isolated capacitor initially at a voltage V_0, and a resistance R is then connected across it at time $t = 0$, the differential equation for the charge at a subsequent time is given by eqn (5.23) with $V = 0$. The solution is

$$Q = CV_0e^{-t/RC} = CV_0e^{-t/\tau}, \tag{5.26}$$

showing that the charge decays exponentially to zero.

Circuit with inductance, capacitance, and resistance

An inductance, a capacitance, and a resistance are connected in series, as in Fig. 5.11, and the circuit is closed at an instant $t = 0$ when the charge on the capacitor is Q_0. Since the total voltage in the circuit is always zero,

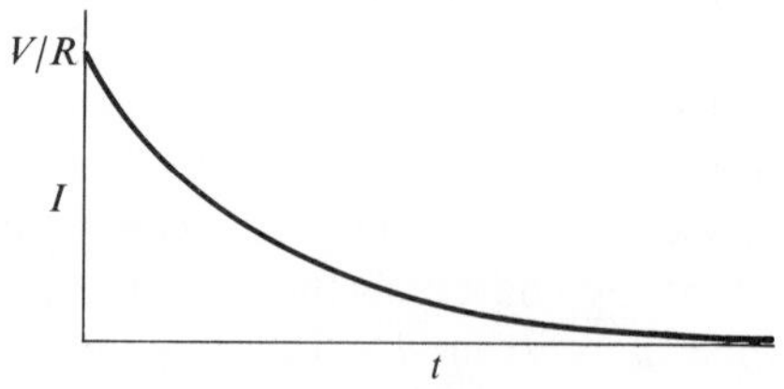

FIG. 5.10. Current in circuit of Fig. 5.9 after switch is closed.

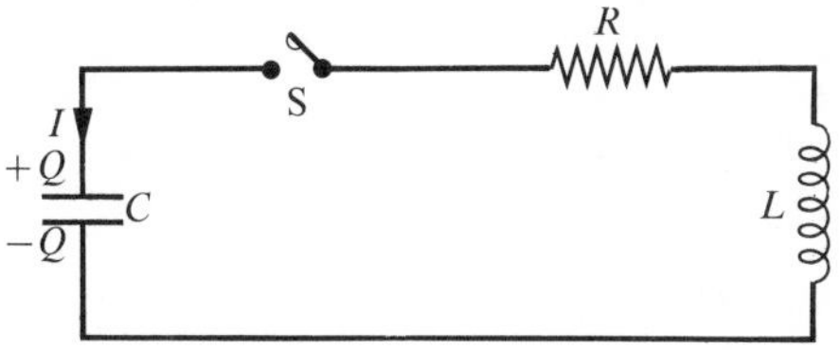

FIG. 5.11. Discharge of a capacitance C through resistance R and inductance L.

we have

$$L\frac{dI}{dt}+IR+\frac{Q}{C}=0.$$

Since $I=dQ/dt$, Q may be eliminated by differentiation, giving

$$L\frac{d^2I}{dt^2}+R\frac{dI}{dt}+\frac{I}{C}=0. \tag{5.27}$$

This equation has a general solution of the form

$$I=e^{-(R/2L)t}(Ae^{nt}+Be^{-nt}), \tag{5.28}$$

where

$$n^2=(R/2L)^2-(1/LC)$$

and A, B are constants determined by the initial conditions. Since $I=0$ at time $t=0$, we must have $B=-A$, and the value of A can be found by integration of (5.28) and setting $Q=Q_0$ at $t=0$.

The nature of the solution depends on whether n is real or imaginary. Three cases can be distinguished.

(1) n real, that is, $(R/2L)>1/(LC)^{\frac{1}{2}}$. The discharge of the capacitor is aperiodic, as shown in Fig. 5.12.

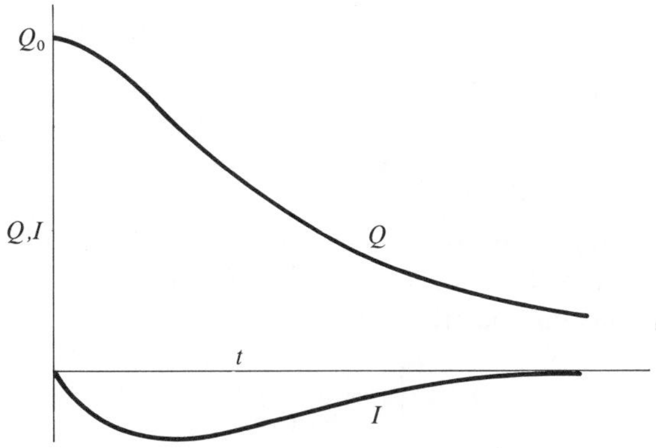

FIG. 5.12. Charge on capacitor and current in circuit of Fig. 5.11 (non-oscillatory discharge).

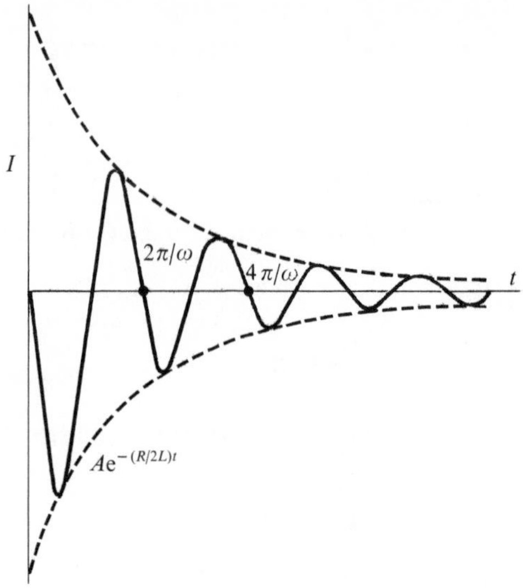

FIG. 5.13. Current in circuit of Fig. 5.11 (oscillatory discharge).

(2) $n = 0$, that is, $(R/2L) = 1/(LC)^{\frac{1}{2}}$. In this case the solution is of the form

$$Q = Q_0(1 + tR/2L)e^{-(R/2L)t} \tag{5.29}$$

and the aperiodic discharge is most rapid.

(3) n imaginary, that is, $(R/2L) < 1/(LC)^{\frac{1}{2}}$. The discharge is now oscillatory, as in Fig. 5.13, and the current may be written as

$$I = Ae^{-(R/2L)t} \sin \omega t, \tag{5.30}$$

where $\omega^2 = 1/LC - R^2/4L^2$. If the resistance is small, so that

$$(R/2L) \ll 1/(LC)^{\frac{1}{2}} \quad (\text{that is, } \tfrac{1}{2}R(C/L)^{\frac{1}{2}} \ll 1),$$

the angular frequency ω will be close to $1/(LC)^{\frac{1}{2}}$. We may call

$$f_0 = \frac{1}{2\pi(LC)^{\frac{1}{2}}}$$

the natural frequency of oscillation of the circuit in the absence of damping. When the damping is small the amplitude of the oscillations decays slowly, and the maxima, which lie on the exponential curve

$$I = A\left\{1 + \left(\frac{R}{2\omega L}\right)^2\right\}^{-\frac{1}{2}} e^{-(R/2L)t},$$

occur very nearly at the points where $\sin \omega t = 1$. These points occur at intervals of time T, where $T = 2\pi/\omega = 2\pi(LC)^{\frac{1}{2}}$ (where we have again

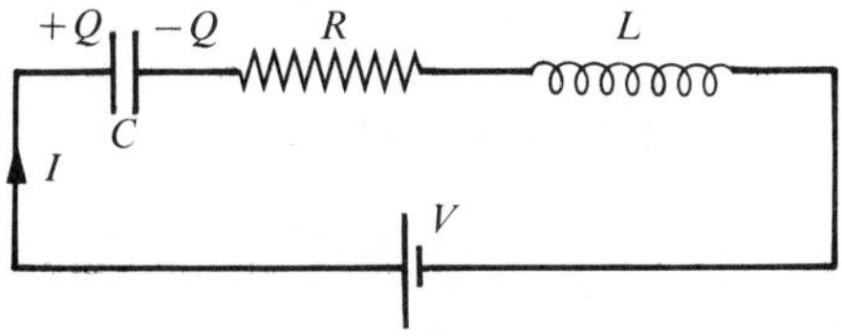

FIG. 5.14. Battery of e.m.f. V charging capacitance C through resistance R and inductance L.

assumed that the damping is small), and the amplitude of successive maxima (of the same sign) therefore decreases by the constant ratio $e^{-TR/2L}$. If we denote such successive maxima by I_m, I_{m+1}, the quantity

$$\ln(I_m/I_{m+1}) = TR/2L = \pi R(C/L)^{\frac{1}{2}} = \pi/Q_F \tag{5.31}$$

is known as the 'logarithmic decrement' of the circuit. The quantity

$$Q_F = \frac{1}{R}\left(\frac{L}{C}\right)^{\frac{1}{2}}$$

is known as the 'quality factor' of the circuit. Either of these quantities serves as an important criterion for the performance of an oscillatory circuit, but in electricity it is customary to use the quality factor Q_F rather than the logarithmic decrement. Further relations involving Q_F will be obtained in Chapter 7.

The converse problem, where a battery of e.m.f. V is connected at time zero to a circuit containing L, C, R in series (Fig. 5.14), is left as an exercise for the reader. The differential equation is the same as (5.27), and the solutions are similar to those above, with an aperiodic or oscillatory approach to the equilibrium state where no current flows and the voltage across the capacitance is equal to V. An important point is that the approach to the equilibrium state in the L, C, R circuit is most rapid at the change-over point from an aperiodic to an oscillatory condition, that is, when $(R/2L) = (LC)^{-\frac{1}{2}}$ or $Q_F = (L/C)^{\frac{1}{2}}/R = \frac{1}{2}$.

5.4. Magnetic energy and mechanical forces in inductive circuits

In the circuit of Fig. 5.14 the relation between the applied voltage V and the current is given by the equation

$$V = L\frac{dI}{dt} + RI + \frac{Q}{C},$$

where Q is the charge on the capacitance. The rate at which work is done is found by multiplying by I, giving

$$VI = LI\frac{dI}{dt} + RI^2 + \left(\frac{dQ}{dt}\right)\frac{Q}{C}$$

since $I = dQ/dt$. The work done in a time interval t, in which the current changes from I_1 to I_2, and the charge on the capacitance from Q_1 to Q_2, will be

$$W = \int_0^t VI\,dt = \tfrac{1}{2}L(I_2^2 - I_1^2) + \int_0^t RI^2\,dt + \frac{1}{2C}(Q_2^2 - Q_1^2).$$

Now W is the total work done by the battery, and the integral of RI^2 is the energy dissipated as heat in the resistance, which is always positive. The last term, $\tfrac{1}{2}(Q_2^2 - Q_1^2)/C$, represents the change in the stored energy of the capacitance, and we interpret the first term $\tfrac{1}{2}L(I_2^2 - I_1^2)$ as the change in the energy stored in the inductance. We note that this change is reversible, since if the current is first increased to I_2 and then returned to its initial value I_1, the change in the stored energy is zero. If $I_1 = 0$, and a current I is established in the inductance, an energy $\tfrac{1}{2}LI^2$ will be associated with it.

An expression for the energy stored in a series of inductances will now be derived in a more general way. If Φ_k is the flux through the kth circuit, the voltage induced in it is $V_k = d\Phi_k/dt$, and the rate of doing work is $I_k V_k = I_k(d\Phi_k/dt)$. We assume that currents in all the circuits were initially zero and increased proportionately with time, so that at an intermediate instant t' the values are $I'_k = \alpha I_k$, $\Phi'_k = \alpha\Phi_k$. Then the total work done is

$$\int_0^t I'_k V'_k\,dt' = I_k\Phi_k \int_0^t \alpha(d\alpha/dt')\,dt'$$

$$= I_k\Phi_k \int_0^1 \alpha\,d\alpha = \tfrac{1}{2}I_k\Phi_k. \qquad (5.32)$$

On summing over all the circuits, we have

$$U = \tfrac{1}{2}\sum_k I_k\Phi_k. \qquad (5.33)$$

If each coil has both self- and mutual inductance,

$$\Phi_k = L_k I_k + \sum_{j\neq k} L_{kj} I_j. \qquad (5.34)$$

Hence eqn (5.33) can be written in the form

$$U = \tfrac{1}{2}\sum_k L_k I_k^2 + \tfrac{1}{2}\sum_j \sum_k^{j\neq k} L_{kj} I_k I_j. \qquad (5.35)$$

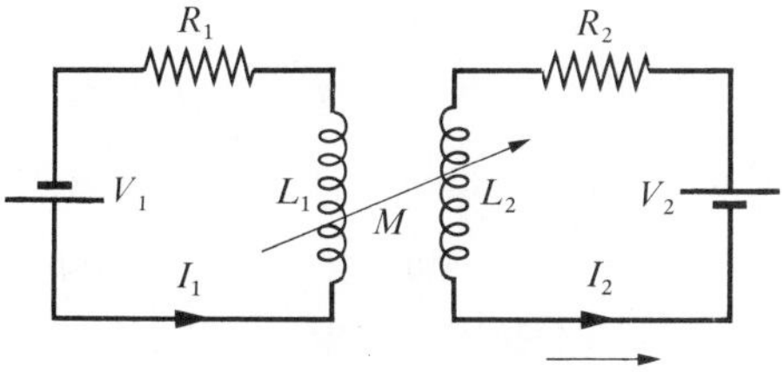

FIG. 5.15. Two circuits with a mutual inductance M. Circuit (2) is moving relative to circuit (1) so that M is varying, but L_1, L_2 and I_1, I_2 are constant.

Since $L_{kj} = L_{jk}$, each term in the second summation occurs twice. This is clearly seen if we consider just two coils, for which (writing M for L_{12})

$$\Phi_1 = L_1 I_1 + M I_2, \qquad \Phi_2 = L_2 I_2 + M I_1,$$

so that the energy becomes

$$\begin{aligned} U &= \tfrac{1}{2} I_1 (L_1 I_1 + M I_2) + \tfrac{1}{2} I_2 (L_2 I_2 + M I_1) \\ &= \tfrac{1}{2}(L_1 I_1^2 + L_2 I_2^2) + M I_1 I_2. \end{aligned} \tag{5.36}$$

It is important to realize the distinction between this stored energy U and the potential energy of one circuit carrying an invariant current I_1 in the field of a second circuit carrying an invariant current I_2, given by eqn (5.11). The difference lies in the fact that the formula for U_P assumes that the currents are invariant, and no account is taken of any work done by the batteries in maintaining the currents, whereas the stored energy U includes the work done by the batteries in setting up the current flow, starting from zero current. Thus U_P is the potential-energy function from which any mechanical force can be calculated by the usual formulae such as

$$F_x = -\mathrm{d}U_P/\mathrm{d}x \tag{5.37}$$

for the x-component of the force, under the condition that the currents are kept constant.

The distinction between U and U_P can be seen from a simple example, that of two rigid circuits as in Fig. 5.15, one of which is moving with velocity $\mathrm{d}x/\mathrm{d}t$ with respect to the other. Then the rate at which work is done by the batteries is

$$\mathrm{d}W/\mathrm{d}t = (\mathrm{d}U/\mathrm{d}t) + F_x(\mathrm{d}x/\mathrm{d}t) + R_1 I_1^2 + R_2 I_2^2, \tag{5.38}$$

where $\mathrm{d}U/\mathrm{d}t$ is the rate at which the stored energy changes. Now, from eqn (5.37), $F_x(\mathrm{d}x/\mathrm{d}t) = -(\mathrm{d}U_P/\mathrm{d}x)(dx/dt) = -\mathrm{d}U_P/\mathrm{d}t$ and hence

$$\mathrm{d}W/\mathrm{d}t = (\mathrm{d}U/\mathrm{d}t) - (\mathrm{d}U_P/\mathrm{d}t) + R_1 I_1^2 + R_2 I_2^2.$$

The flux through each circuit is

$$\Phi_1 = L_1I_1 + MI_2, \qquad \Phi_2 = L_2I_2 + MI_1,$$

and the rate at which work is done by the batteries

$$\begin{aligned} \mathrm{d}W/\mathrm{d}t &= I_1(R_1I_1 + \mathrm{d}\Phi_1/\mathrm{d}t) + I_2(R_2I_2 + \mathrm{d}\Phi_2/\mathrm{d}t) \\ &= 2(\mathrm{d}M/\mathrm{d}t)I_1I_2 + R_1I_1^2 + R_2I_2^2, \end{aligned}$$

since I_1, I_2 are constant. Comparing these two equations, we see that

$$\mathrm{d}U/\mathrm{d}t = \mathrm{d}U_\mathrm{P}/\mathrm{d}t + 2(\mathrm{d}M/\mathrm{d}t)I_1I_2. \tag{5.39}$$

From eqn (5.36), since only M is changing,

$$\mathrm{d}U/\mathrm{d}t = (\mathrm{d}M/\mathrm{d}t)I_1I_2$$

and hence

$$\mathrm{d}U_\mathrm{P}/\mathrm{d}t = -(\mathrm{d}M/\mathrm{d}t)I_1I_2 = -\mathrm{d}U/\mathrm{d}t.$$

Thus the component of the mechanical force is

$$F_x = -\mathrm{d}U_\mathrm{P}/\mathrm{d}x = +\mathrm{d}U/\mathrm{d}x. \tag{5.40}$$

From eqn (5.39) we see that the difference in sign between the two expressions for the force in (5.40) arises from the fact that the batteries do work at the rate $2(\mathrm{d}M/\mathrm{d}t)I_1I_2$ (apart from the irreversible Joule heating represented by the terms $R_1I_1^2$, etc.) which is just twice the rate of increase ($\mathrm{d}U/\mathrm{d}t$) of the stored energy. This situation with regard to the magnetic energy, and the calculation of the mechanical forces in a system where the currents are kept constant, is similar to that in electrostatics when changes take place in a system where the conductors are maintained at constant potential. It was shown in § 1.7 that in a change where the stored energy increases by $\mathrm{d}U$, the batteries do work $2\,\mathrm{d}U$, so that the net amount of external work $\mathrm{d}W$ done on the system is $-\mathrm{d}U$. Hence the component of the force in the x direction is given by

$$F_x = -\mathrm{d}U_\mathrm{P}/\mathrm{d}x = +(\mathrm{d}U/\mathrm{d}x)_V,$$

where the subscript V denotes that this formula is to be used under the condition $V = \text{constant}$. Similarly, we may emphasize that in equations such as (5.40) the current is kept constant by writing them in the form

$$F_x = +(\mathrm{d}U/\mathrm{d}x)_I, \text{ etc.}$$

As we should expect, the torque on a coil can be shown to be

$$T = -(\mathrm{d}U_\mathrm{P}/\mathrm{d}\theta) = +(\mathrm{d}U/\mathrm{d}\theta)_I = (\mathrm{d}M/\mathrm{d}\theta)I_1I_2, \tag{5.41}$$

where θ is the angle which the coil makes with some fixed axis.

5.5. Magnetic energy in magnetic media

Since for any circuit

$$\Phi=\int \mathbf{B}.\mathrm{d}\mathbf{S}=\int \mathrm{curl}\,\mathbf{A}.\mathrm{d}\mathbf{S}=\oint \mathbf{A}.\mathrm{d}\mathbf{s} \tag{5.42}$$

we can transform eqn (5.33) as follows:

$$U=\tfrac{1}{2}\sum_k I_k\Phi_k=\tfrac{1}{2}\sum_k I_k\oint \mathbf{A}.\mathrm{d}\mathbf{s}$$

For a system of distributed currents, we write $I\,\mathrm{d}\mathbf{s}=\mathbf{J}\,\mathrm{d}\tau$, and take the integral over all space, since contributions arise only from regions where **J** is finite, giving

$$U=\tfrac{1}{2}\int(\mathbf{A}.\mathbf{J})\,\mathrm{d}\tau. \tag{5.43}$$

However, $\mathbf{J}=\mathrm{curl}\,\mathbf{H}$, and using the vector identity (eqn (A.15) of Appendix A)

$$\mathrm{div}(\mathbf{A}\wedge\mathbf{H})=\mathbf{H}.\mathrm{curl}\,\mathbf{A}-\mathbf{A}.\mathrm{curl}\,\mathbf{H},$$

we have

$$U=\tfrac{1}{2}\int(\mathbf{H}.\mathrm{curl}\,\mathbf{A})\,\mathrm{d}\tau-\tfrac{1}{2}\int\mathrm{div}(\mathbf{A}\wedge\mathbf{H})\,\mathrm{d}\tau$$

$$=\tfrac{1}{2}\int(\mathbf{H}.\mathbf{B})\,\mathrm{d}\tau-\tfrac{1}{2}\int(\mathbf{A}\wedge\mathbf{H})\,\mathrm{d}\mathbf{S}.$$

If the volume integral is extended over all space, the surface integral is taken over the sphere at infinity. For a finite system of closed circuits, the field strength **H** and the flux density **B** will fall off at large distances at least as rapidly as those of dipoles; hence $\mathbf{H}\sim r^{-3}$ and $\mathbf{A}\sim r^{-2}$, so that the integrand diminishes as r^{-5} and the integral vanishes as $r\to\infty$. Thus we have

$$U=\tfrac{1}{2}\int(\mathbf{H}.\mathbf{B})\,\mathrm{d}\tau. \tag{5.44}$$

This formula shows that the energy may be regarded as distributed throughout the region occupied by the field with density $\tfrac{1}{2}\mathbf{H}.\mathbf{B}$, which is clearly analogous to the result $\tfrac{1}{2}\mathbf{D}.\mathbf{E}$ in electrostatics, eqn (1.36). In each case it has been derived under the assumption that **B**(**D**) is linearly proportional to **H**(**E**), that is, that the media have values of $\mu(\epsilon)$, which are independent of field strength. We may relax this assumption by considering infinitesimal changes of flux, for which $\delta U=\sum_k I_k\,\delta\Phi_k$. Then by applying transformations exactly similar to those above, we find

$$\delta U=\int\mathbf{J}.\delta\mathbf{A}\,\mathrm{d}\tau=\int\mathbf{H}.\delta\mathbf{B}\,\mathrm{d}\tau, \tag{5.45}$$

a result which can then be integrated over any finite change in **B** if we know how **H** and **B** are related at every point during the change. If **B** is linearly proportional to **H**, this result clearly gives again eqn (5.44).

We close this discussion by finding the work done when the magnetization changes by $\delta\mathbf{M}$ in a magnetic flux density $\mathbf{B}_0$ which is due to other sources; that is, $\mathbf{B}_0$ does not include the effect of the field due to the magnetic material itself. A simple example would be a piece of magnetizable material inside a solenoid; $\mathbf{B}_0$ is then just equal to the field $\mu_0\mathbf{H}_0$ which the solenoid would produce in the absence of the magnetizable substance, and can be calculated from the standard formulae. We start by considering two circuits, for which, from eqn (5.39), in an infinitesimal change,

$$\delta U = \delta U_{\mathrm{P}} + 2\delta(MI_1I_2).$$

But $U_{\mathrm{P}} = -\mathbf{m}_2.\mathbf{B}_1 = -I_2\int\mathbf{B}_1.\mathrm{d}\mathbf{S}_2$, and the flux through the second circuit is $MI_1 = \int\mathbf{B}_1.\mathrm{d}\mathbf{S}_2$. Hence

$$\delta U = +\delta I_2\int\mathbf{B}_1.\mathrm{d}\mathbf{S}_2 = \delta(\mathbf{B}_1.\mathbf{m}_2),$$

where $\mathbf{m}_2$ is the equivalent magnetic moment of the second coil, and we assume that $\mathbf{B}_1$ is constant over the area of this coil. If $\mathbf{B}_1$ is fixed, and $\mathbf{m}_2$ increases by $\delta\mathbf{m}$, then $\delta U = \mathbf{B}_1.\delta\mathbf{m}$. Clearly it does not matter what is the source of $\mathbf{B}_1$, so that we can write in general

$$\delta U = \mathbf{B}_0.\delta\mathbf{m} = \int(\mathbf{B}_0.\delta\mathbf{M})\,\mathrm{d}\tau, \tag{5.46}$$

where $\mathbf{B}_0$ excludes the field contribution from the magnetized substance itself. This formula has an important limitation; it assumes that $\mathbf{B}$ is the same inside the magnetized region as it was in the absence of the magnetized body. This neglects the demagnetizing field, and is a valid approximation only for weakly magnetic substances. In general the energy, and hence also any force or torque on the body (see Appendix B), depends on its shape.

5.6. Galvanometers and galvanometer damping

A galvanometer is an instrument for detecting and measuring electric current; if provided with a scale already calibrated, it is called an ammeter. The only type now in common use has a 'moving coil', usually rectangular in shape, through which the current passes while it is suspended in a uniform and constant magnetic flux density $\mathbf{B}$. The suspension is adjusted so that the plane of the coil is parallel to the lines of magnetic flux when no current is passing. Flow of current through the coil gives it a magnetic moment, on which there is a couple due to $\mathbf{B}$; this is balanced by the restoring torque of the suspension, equal to $c\theta$, where θ is the angle through which the coil turns. If the coil has N turns, each of area A, and the current is I, the magnetic moment $|\mathbf{m}| = NAI$, and at the

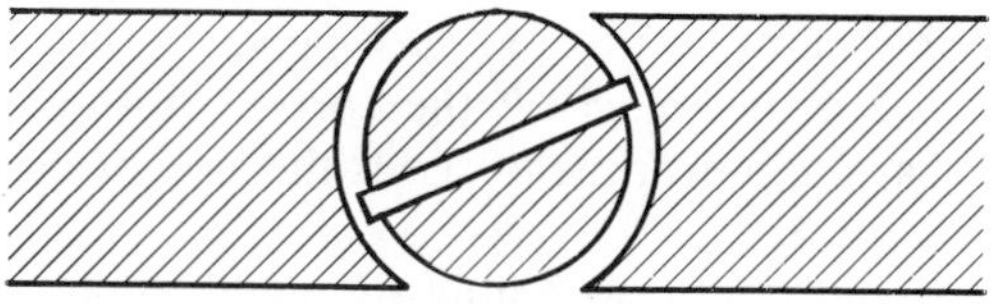

FIG. 5.16. Pole pieces of a permanent magnet in a moving-coil galvanometer.

equilibrium deflection

$$c\theta = |\mathbf{m} \wedge \mathbf{B}| = (NAB)I\cos\theta. \tag{5.47}$$

In practice the lines of **B** are produced by a small permanent magnet, with pole faces shaped as in Fig. 5.16. The centre of the gap is filled with iron, so that the flux lines in the gap are as nearly radial as possible. The term $\cos\theta$ will then be unity, so that the equilibrium deflection θ will be linearly proportional to the current.

In another class of moving-coil instruments, known as dynamometers, the flux density B is provided by a second set of fixed coils carrying a current I_1 surrounding the moving coil which carries a current I_2. If M is the mutual inductance between the fixed and moving coils, by eqn (5.41) the torque is $I_1I_2(\mathrm{d}M/\mathrm{d}\theta)$, and hence the equilibrium deflection will be given by

$$c\theta = I_1I_2(\mathrm{d}M/\mathrm{d}\theta),$$

where $c\theta$ is the restoring torque due to the suspension (usually a spiral spring). The factor $(\mathrm{d}M/\mathrm{d}\theta)/c$ is found by calibration with a known current.

If a current I to be measured is allowed to flow through the fixed and moving coils in series, so that $I = I_1 = I_2$, the instrument will act as a milliammeter with a deflection proportional to I^2; with a pointer-type dynamometer the full-scale reading is about 10 mA. A more important application is its use as a wattmeter; the coils are then connected as follows. In Fig. 5.17, the moving coil is connected in series with a resistance R, large compared with the load, so that the current I_1 is nearly equal to I, the current through the load. The current $I_2 = V/R$, where V is the voltage across the load, and the deflection is proportional to $I_2I_2 = IV/R = P/R$, where P is the power being dissipated in the load. Thus the

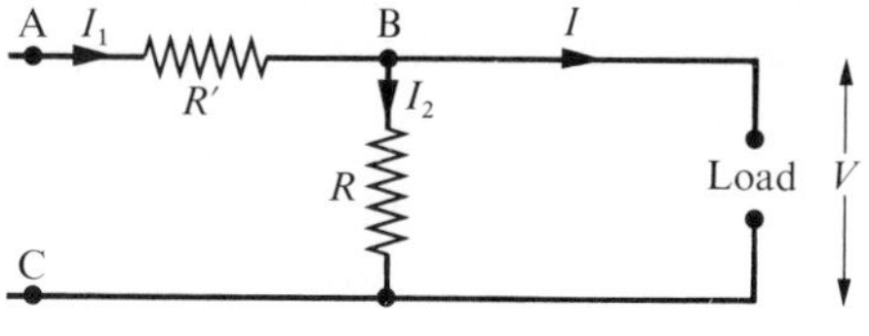

FIG. 5.17. Circuit of a wattmeter.

instrument may be calibrated to measure power directly. Note that with the coils connected as in Fig. 5.17, the current through the fixed coil is really $I+I_2$, so that the power measured is the sum of that in the load and in R (see Problem 5.17).

So far we have considered the equilibrium deflection of a galvanometer only when a steady current flows; the behaviour during the time interval between switching on the current and the attainment of the final deflection must also be discussed. When the current starts to flow through the galvanometer coil, a couple is exerted on it which gives it an angular acceleration. This couple diminishes as the coil nears its equilibrium position, owing to the reverse couple exerted by the suspension, but if the damping is small, the coil may overshoot and oscillate about its final position. If the damping is large there is no overshoot, and the coil approaches the final position very slowly. In either case some time must elapse before the equilibrium deflection can be observed, and the effect of the damping is thus of considerable importance. Though air resistance and other sources of energy loss contribute to the damping, the most important source is electromagnetic damping, due to the motion of the coil in the magnetic field of the permanent magnet. With the notation of eqn (5.47), the flux through the coil is $\Phi = NAB \sin \theta$, and there will therefore be an e.m.f. induced in the coil equal to

$$-\mathrm{d}\Phi/\mathrm{d}t = -NAB \cos \theta (\mathrm{d}\theta/\mathrm{d}t).$$

If θ is small, or if the lines of magnetic flux are always radial (as in Fig. 5.16), we can put $\cos \theta = 1$. If V is the external source of e.m.f. and R is the total resistance in the circuit (including the galvanometer coil resistance), then the current I will be given by the equation

$$V - \frac{\mathrm{d}\Phi}{\mathrm{d}t} = RI$$

at any instant. The equation of motion of the coil will therefore be

$$\Im \frac{\mathrm{d}^2\theta}{\mathrm{d}t^2} + b\frac{\mathrm{d}\theta}{\mathrm{d}t} + c\theta = NABI = \frac{NAB}{R}\left(V - \frac{\mathrm{d}\Phi}{\mathrm{d}t}\right) = \frac{NAB}{R}V - \frac{(NAB)^2}{R}\frac{\mathrm{d}\theta}{\mathrm{d}t}, \tag{5.48}$$

where $\Im$ is the moment of inertia of the coil, $b(\mathrm{d}\theta/\mathrm{d}t)$ the torque due to air damping, etc., and $c\theta$ the restoring torque due to the suspension. Rearranging the equation gives

$$\Im \frac{\mathrm{d}^2\theta}{\mathrm{d}t^2} + \left\{b + \frac{(NAB)^2}{R}\right\}\frac{\mathrm{d}\theta}{\mathrm{d}t} + c\theta = \frac{NAB}{R}V, \tag{5.49}$$

from which it is seen that the effect of the e.m.f. induced by the motion of the coil is to increase the damping torque by an amount

$$(NAB)^2(\mathrm{d}\theta/\mathrm{d}t)/R.$$

The differential equation is similar to that for the oscillatory circuit of § 5.3, and the same analysis may be applied. There are three types of solution, according to whether the damping is large or small.

(1) *Small damping.* $b' < 2(\Im c)^{\frac{1}{2}}$, where $b' = b + (NAB)^2/R$. The motion is a damped harmonic motion, of period nearly equal to $2\pi(\Im/c)^{\frac{1}{2}}$. The coil overshoots its equilibrium deflection $\theta = (NAB)V/cR$, and then oscillates with diminishing amplitude about this position.

(2) *Critical damping.* $b' = 2(\Im c)^{\frac{1}{2}}$; the coil approaches its equilibrium without overshoot.

(3) *High damping.* $b' > 2(\Im c)^{\frac{1}{2}}$; the coil approaches its final position without overshoot, but more slowly than in case (2).

In general the electromagnetic damping is much greater than other sources of damping, unless the total resistance R of the galvanometer circuit is very large. As R decreases the electromagnetic damping increases, and critical damping will be obtained with a certain value of resistance R_c given by the relation

$$R_c = \tfrac{1}{2}(NAB)^2/(\Im c)^{\frac{1}{2}}, \tag{5.50}$$

assuming that b is negligible under these conditions. It is important to operate a galvanometer at or near critical damping since it then takes up its equilibrium deflection (or, more precisely, comes to a deflection within a certain fraction of its equilibrium value) in the shortest possible time. If the galvanometer is either highly overdamped or very much underdamped it will take much longer to settle down, and this reduces the rate at which readings can be taken. If the source of voltage varies within the time required for the galvanometer to settle, the reading will always lag behind the true value; in detecting the balance point of a bridge, quite misleading indications can be obtained if the bridge arms are altered too rapidly.

Since the critical-damping resistance is of such importance, it is always specified for a galvanometer, together with the coil resistance, the sensitivity, and the period. As a rough rule, the critical-damping resistance is 10–20 times the coil resistance in a reflecting galvanometer. In a pointer instrument the coil may carry a short-circuited turn, or the coil former may be made of metal, in order to give adequate damping.

5.7. The ballistic galvanometer and fluxmeter

The ballistic galvanometer is a suspended moving-coil instrument with very light damping, which can be used for the measurement of charge by observing the maximum deflection in its oscillatory motion. For this purpose the charge must pass through the galvanometer in a time short compared with its period of swing, so that the coil does not deflect appreciably during the passage of the charge. If the current at any instant during this passage is I, the couple exerted on the coil is $NABI$, and the impulse of angular momentum given to the coil will be

$$\int NABI\,\mathrm{d}t = NABQ,$$

where Q is the total charge flowing through the coil. The effect of this impulse is to give the coil an initial angular velocity $(\mathrm{d}\theta/\mathrm{d}t)_{t=0} = NABQ/\mathfrak{I}$, where $\mathfrak{I}$ is the moment of inertia of the coil. We analyse the subsequent motion of the coil assuming it to remain connected to a circuit of total resistance R (including the coil resistance) but with no external e.m.f. The differential equation of the motion will be the same as (5.49) but with $V = 0$, so that we have

$$\mathfrak{I}\frac{\mathrm{d}^2\theta}{\mathrm{d}t^2} + b'\frac{\mathrm{d}\theta}{\mathrm{d}t} + c\theta = 0, \tag{5.51}$$

where $b' = b + (NAB)^2/R$ is the total effective damping, which we shall assume to be small. Then the solution is

$$\theta = D\exp(-b't/2\mathfrak{I})\sin\omega t,$$

where $\omega = (c/\mathfrak{I})^{\frac{1}{2}}$ to a good approximation. The constant D is found by differentiation, and setting the initial angular velocity equal to $NABQ/\mathfrak{I}$, giving

$$NABQ/\mathfrak{I} = (\mathrm{d}\theta/\mathrm{d}t)_{t=0} = D\omega,$$

whence

$$\theta = (NABQ/\mathfrak{I}\omega)\exp(-b't/2\mathfrak{I})\sin\omega t. \tag{5.52}$$

If there were no damping the deflection would oscillate between maximum and minimum values $\pm\theta_0$ occurring at the instants when $\sin\omega t = 1$, and the charge Q would be given by

$$Q = \frac{\mathfrak{I}\omega\theta_0}{NAB} = \frac{c\theta_0}{\omega NAB} = \theta_0\frac{\tau}{2\pi}\cdot\frac{c}{NAB}, \tag{5.53}$$

where we have used in succession the relations $(\mathfrak{I}/c)\omega^2 = 1$ and $\omega = 2\pi/\tau$, where τ is the period of swing of the galvanometer. This can be measured experimentally, and the constant c/NAB can be found from the deflection produced by passing a steady current through the galvanometer.

The effect of the damping is to make successive deflections smaller (cf. Fig. 5.13), and also to make the first throw θ_1 (at $t = \frac{1}{4}\tau$) smaller than θ_0,

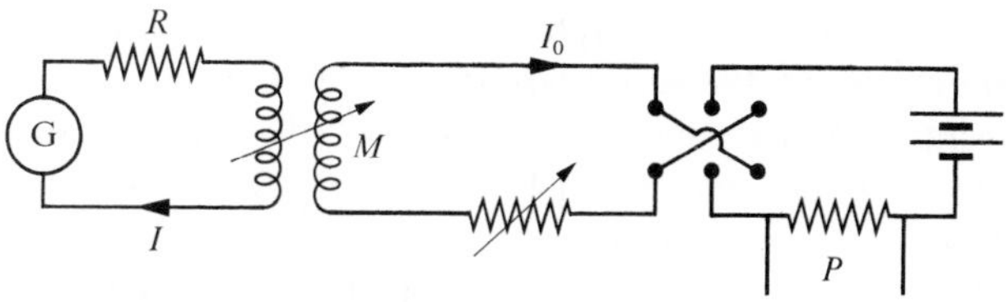

FIG. 5.18. Apparatus for calibrating a ballistic galvanometer.
G galvanometer,
M standard mutual inductance,
P potentiometer connected across standard 1 Ω resistance.

since from eqn (5.52)

$$\theta_1 = \theta_0 \exp(-b'\tau/8\mathfrak{I}) \approx \theta_0(1-\tfrac{1}{4}\Lambda). \tag{5.54}$$

The size of the correction can be found by measuring the logarithmic decrement Λ of successive swings on the same side, since

$$\Lambda = \ln(\theta_1/\theta_2) = \ln(\theta_2/\theta_3) = \ldots = b'\tau/2\mathfrak{I}.$$

For accurate use the damping of the ballistic galvanometer should be small and the resistance of the external circuit must therefore be high. If used to measure the charge on a capacitor by connecting the capacitor to the galvanometer and allowing it to discharge through the galvanometer, this condition is fulfilled because the capacitor is effectively an open-circuit for the slowly varying induced e.m.f. produced by the galvanometer swing. A convenient method of calibrating the ballistic galvanometer directly is by discharging through it a known capacitance charged to a known voltage. An alternative method is by means of a standard mutual inductance, using the circuit of Fig. 5.18. The secondary of the mutual inductance is connected to the galvanometer, and a known current is reversed in the primary coil. If L is the inductance of the secondary windings and R the total resistance in the secondary circuit, the secondary current I flowing at any instant is given by the equation

$$L\frac{\mathrm{d}I}{\mathrm{d}t} + RI = -\frac{\mathrm{d}\Phi}{\mathrm{d}t},$$

where $\mathrm{d}\Phi/\mathrm{d}t$ is the rate at which the flux is changing through the secondary coil. Integrating over the duration of time occupied by the flux change (assuming that the time constant R/L is very short compared with the period of the galvanometer) gives

$$\int -\frac{\mathrm{d}\Phi}{\mathrm{d}t}\,\mathrm{d}t = \Phi_1 - \Phi_2 = L\int \frac{\mathrm{d}I}{\mathrm{d}t}\,\mathrm{d}t + R\int I\,\mathrm{d}t =$$

$$= L[I]_0^0 + RQ = RQ, \tag{5.55}$$

showing that the charge measured by the galvanometer is just equal to the total flux change in the secondary divided by the total resistance. If I_0 is the current reversed in the primary, $\Phi_1 - \Phi_2 = 2MI_0$, and so $Q = 2MI_0/R$. The current I_0 may either be measured by an accurate ammeter, or by means of the potential drop it produces across a known resistance. To reduce the damping of the ballistic galvanometer, either R must be large, which in general will mean adding considerable resistance to the circuit and thus reducing the sensitivity of the galvanometer, or the secondary circuit must be broken immediately after the primary current is reversed and before the galvanometer has deflected appreciably.

Once the ballistic galvanometer has been calibrated, it may be used to measure a flux change or an unknown mutual inductance if the resistance R is known. An alternative method is to use a fluxmeter, which gives a direct reading of the flux change. This instrument consists of a moving coil, suspended in such a way as to give almost zero restoring torque, in a strong magnetic flux density of a permanent magnet. The electromagnetic damping is very strong, and if other forms of damping can be neglected, the equation of motion is

$$\mathfrak{I}\frac{\mathrm{d}^2\theta}{\mathrm{d}t^2} + \frac{(NAB)^2}{R}\frac{\mathrm{d}\theta}{\mathrm{d}t} = NABI = \frac{NAB}{R}\left(-\frac{\mathrm{d}\Phi}{\mathrm{d}t} - L\frac{\mathrm{d}I}{\mathrm{d}t}\right), \tag{5.56}$$

where I is the instantaneous current produced by the e.m.f. $-\mathrm{d}\Phi/\mathrm{d}t$ induced by the flux change to be measured, and L is the self-inductance. The effect of this current is to impart an angular momentum to the fluxmeter coil, which is then brought to rest under the action of the electromagnetic damping. Integration of eqn (5.56) over the entire time occupied by the motion of the coil gives

$$\mathfrak{I}\left[\frac{\mathrm{d}\theta}{\mathrm{d}t}\right]_0^0 + \frac{(NAB)^2}{R}[\theta]_0^{\theta_0} = \frac{NAB}{R}(\Phi_1 - \Phi_2) - \frac{NABL}{R}[I]_0^0,$$

and, since the angular velocity and current are zero both initially and finally,

$$NAB\theta_0 = \Phi_1 - \Phi_2, \tag{5.57}$$

where θ_0 is the ultimate deflection of the fluxmeter. The flux change through the fluxmeter coil is just $NAB\theta_0$, and so this equation shows that the total flux threading the system is the same at the beginning and the end. The fluxmeter is generally calibrated directly in terms of flux; since there is no restoring torque on the coil it has no stable position of equilibrium but can rest anywhere in neutral equilibrium. It is therefore necessary to level the instrument carefully to prevent the coil drifting during the measurement. As a rough rule, a fluxmeter is accurate to about 1 per cent, while a ballistic galvanometer, properly calibrated, is accurate to about 0·1 per cent.

5.8. Absolute measurements

Relative measurements of electrical quantities can be made using bridges or potentiometers to an accuracy of some parts in 10^5. In national Standards Laboratories the legal standard of current has usually been based on the mass of an electrolyte deposited in electrolysis in a given time (see § 3.9), while the legal standard of e.m.f. is based on a standard cell (see § 3.10). A standard resistance can then be established by passing a known current through it and comparing the potential drop across it with that of a standard cell.

For absolute measurements it is necessary to use methods in which an electrical quantity can be determined in terms of mechanical quantities using the fundamental laws of electromagnetism. One such method, for the absolute determination of current, is based on the equations of § 4.1—a current is passed through a pair of conductors, and the force between them is measured. The method of doing this, by means of a 'current balance', is illustrated in Fig. 5.19. A, B, C, D, E, F are six single-layer coils wound on marble formers and connected in series. Coils A, C, D, and F are fixed, while B and E are carried on a balance arm; the current flows through the various coils in such a direction that the force on E is upwards while that on B is downwards. The balance arm is brought back to its equilibrium position by moving a standard mass along a calibrated scale on the arm. From the distance of the mass from the fulcrum, the torque and hence the force between the coils can be evaluated. If M is the mutual inductance between B or E and either of the coils A and C or D and F, then the total torque due to the coils when

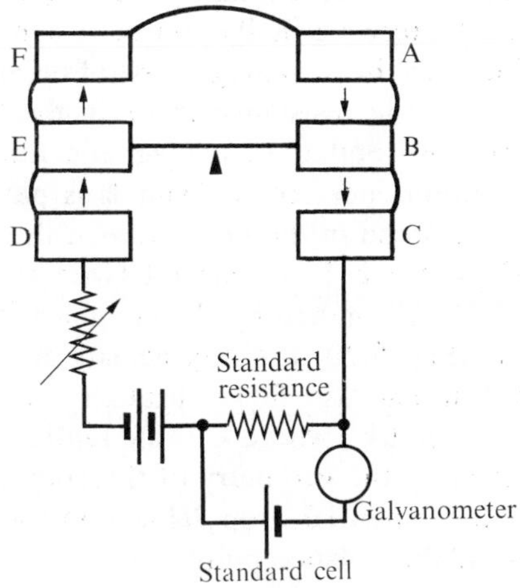

FIG. 5.19. Current balance.

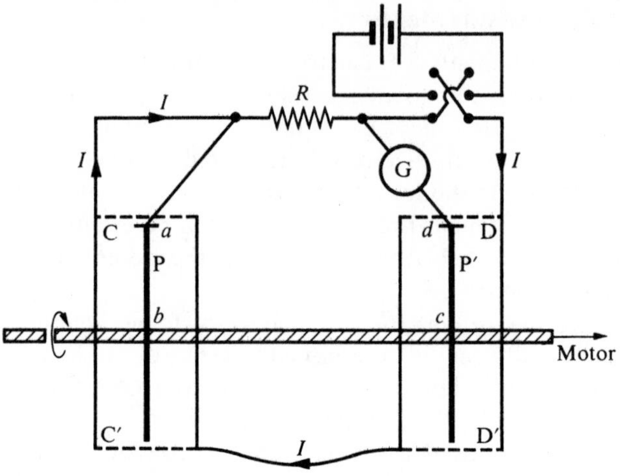

FIG. 5.20. Lorenz's apparatus for determining a resistance R in terms of μ_0 and the metre. CC′, DD′ are coils through which I flows (in opposite senses) producing the magnetic fields in which the plates PP′ are rotated.

carrying a current I is $4I^2(\mathrm{d}M/\mathrm{d}\theta)$, where θ is the angle defining the rotation of the balance arm. Since $\mathrm{d}M/\mathrm{d}\theta$ is calculated from the geometry, there must be no iron near the apparatus. Effects due to the more remote coils (that is, for example, the force between coil A and coil E) are eliminated by repeating the readings with the current through all the coils on one side reversed. An accuracy of one part in 10^5 can be achieved, but the measurements are very laborious.

The resistance of a coil was determined by Rayleigh in terms of the constant μ_0 and the standard of length, by rotating the coil in a magnetic field (see Problem 5.7). Here we shall describe in some detail the method used by Lorenz, which has the advantages first of being a null method and secondly of not needing the resistance to be in the form of a coil. If a disc-shaped conductor is rotated in a magnetic flux density $\mathbf{B}$ with a frequency of f revolutions per second, and $\mathbf{B}$ is parallel to the axis of rotation, a voltage is induced in the disc between the axis and the rim of magnitude $V = kfB$, where k is a constant (with the dimensions of an area: see Problem 5.6) determined by the geometry of the apparatus. In Lorenz's experiment (Fig. 5.20) the flux density $\mathbf{B}$ is produced by coils carrying a current I which also flows through the resistance R to be standardized. Then $B = \mu_0 I/k'$, where k' is a quantity with the dimensions of a length, again fixed by the geometry of the apparatus. The voltage V is balanced against the potential drop RI across the resistance, so that $RI = (k/k')\mu_0 fI$, and at the balance point

$$R = (k/k')\mu_0 f. \tag{5.58}$$

The frequency f can be measured accurately, $\mu_0 = 4\pi \times 10^{-7}\,\mathrm{H\,m^{-1}}$ by definition and k/k' has the dimensions of a length. In fact $\mu_0(k/k')$ is the mutual inductance between the field coil and the rotating conductor, which can be calculated from the dimensions and the geometry. Fundamentally the comparison is between the resistance and the product inductance × frequency, the rotating conductor being a device by which a steady voltage is developed across the inductance for balance against the voltage drop in the resistance.

The arrangement of the apparatus is shown in Fig. 5.20. The rotating conductor is formed by the discs P and P′ mounted on a shaft driven by a motor. The galvanometer G reads zero when the voltage induced in the circuit *abcd* is exactly equal to the potential across the resistance R, which is in series with the two coils CC′ and DD′. The current flows through these coils in the opposite sense so that the voltages induced in the two discs add round the circuit *abcd.* Contact with the discs at a and d is made by brushes, and the coils are designed so that the magnetic flux density at these contacts is zero; this minimizes any error in determining the actual radius of the disc, that is, the distance from the centre at which contact is made. Some trouble arises from thermal e.m.f.s at the brush contacts, which would give a finite galvanometer reading at the true balance point. This difficulty is overcome by reversing the current throughout the whole system, which reverses any true unbalance current but not that due to the thermal e.m.f. Hence the correct balance point is when the galvanometer reading remains unchanged on reversing the current, and this procedure also eliminates the effect of any voltages induced by stray magnetic fields. With adequate precautions it is possible to obtain an accuracy of a few parts in 10^5; other resistances can then be determined relative to the standard by means of a Wheatstone's bridge.

In these methods the value of a resistance is found in absolute units by comparing it with an inductance whose value is μ_0 times a factor with the dimensions of a length. The capacitance C of a capacitor is $\epsilon_0 l$, where l is a factor with the dimensions of a length which is determined by the geometry of the capacitor. Thus if l is calculated, and the capacitance C is determined by comparison with a standard resistance R, the value of the constant ϵ_0 can be found. An accurate experiment of this type was carried out by Rosa and Dorsey in 1907; the apparatus consists essentially of the bridge circuit shown in Fig. 5.21. The contact X vibrates between P and Q at a known frequency, and the capacitor of capacitance C is alternately charged and discharged f times a second. The voltage V across it is Q/C, and the charging of the condenser f times a second sends a current fQ through the arm AB, so that the effective resistance of this arm is $V/(fQ) = 1/(fC)$. Hence at the balance point, $R_2/R_1 = fCR$. Rosa and Dorsey used both spherical and cylindrical capacitors, and took many

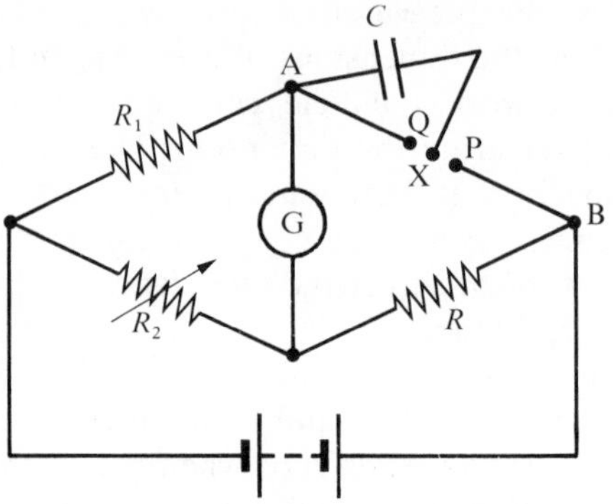

FIG. 5.21. Determination of a capacitance C in terms of a standard resistance R.

readings to correct for a large variety of possible errors. Their final result was $\epsilon_0 = (8{\cdot}8547)10^{-12}$ F m^{-1}, with an accuracy of about 4 parts in 10^5.

Reference

HEINE, V. (1956). *Proc. Camb. Phil. Soc.* **52,** 546.

Problems

5.1. Two infinite coaxial cylinders carry equal and opposite currents I. Calculate the self-inductance L per unit length (eqn (5.18)) by equating the stored energy per unit length to $\frac{1}{2}LI^2$.

5.2. Prove that the inductance per unit length of two infinite parallel wires of radius a separated by a distance $2d$ ($2d \gg a$) is

$$L = (\mu/\pi)\ln(2d/a).$$

Verify that $LC = \mu\epsilon$, where C is the capacitance per unit length (§ 2.6).

5.3. Prove that the inductance of a long solenoid of length l, radius a, and with m turns per unit length is approximately

$$L = \pi m^2 a^2 \mu\{(l^2+a^2)^{\frac{1}{2}} - a\}$$

(assume that the field is uniform over any cross-section: if $l > 10a$, this formula is accurate within 2 per cent)

5.4. Show that the mutual inductance between two coplanar coaxial coils of radii a and b ($b \gg a$), with turns n_1 and n_2 respectively, is approximately (*in vacuo*)

$$M = \mu_0 \pi a^2 n_1 n_2 / 2b.$$

Use this result to show that the flux through the larger coil due to a small magnet of magnetic dipole moment $\mathfrak{m}$ placed at its centre and pointing along its axis is

$$\Phi = \mu_0 \mathfrak{m} n_2 / 2b.$$

5.5. If the magnet in the last problem is withdrawn along the axis at a uniform velocity v, show that the e.m.f. induced in the coil when the magnet is at a distance z from the centre of the coil is

$$V = 3\mu_0 m n_2 z v b^2 / 2(b^2+z^2)^{\frac{5}{2}}.$$

If the resistance of the coil is R, show that the total charge which flows when the magnet is removed from the centre to infinity is

$$Q = \mu_0 m n_2 / 2bR.$$

5.6. A plane circular disc of radius a rotates at a speed of f revolutions per second about an axis through its centre normal to its plane. A uniform flux density **B** exists parallel to this axis. Show from first principles that there is an e.m.f. between the centre of the disc and its rim of magnitude $V = fB\pi a^2$. (Lorenz's method of determining the unit of resistance depends on this result; see § 5.8)

5.7. A circular coil of n turns of radius a, total resistance R, and no self-inductance is rotated with uniform angular velocity ω about a vertical diameter in a horizontal flux density **B**. Prove that the mean power required to maintain the coil in motion is $\frac{1}{2}n^2\pi^2a^4B^2\omega^2/R$, and that this is equal to the power dissipated in the resistance of the coil.

A small magnetic needle, which is free to turn slowly in a horizontal plane, is placed at the centre of the coil. Show that it will set at an angle ϕ to **B**, where

$$\cot \phi = 4R/\pi n^2 \mu_0 \omega a.$$

(Rayleigh's method of determining the unit of resistance is based on an experiment of this type.)

5.8. A torsional pendulum consists of two spheres of 1 cm diameter at either end of a thin rod 10 cm long suspended at its midpoint. It swings in a horizontal plane so that its instantaneous angular deflection is $\pi \cos \frac{1}{2}\pi t$. Find the magnitude and direction of the current flowing in the rod at any instant, assuming that the capacity of each sphere is the same as if each were isolated in space, and the rod and suspension have negligible resistance. Show that there will be damping of the swing if the suspension has a finite resistance, but not otherwise. (Vertical component of earth's magnetic flux density $= 6 \times 10^{-5}$ T.)

(*Answer:* $3{\cdot}2 \times 10^{-19} \cos \frac{1}{2}\pi t$ A.)

5.9. An aeroplane is in level flight at a ground speed of 300 km h^{-1}. Its metal propeller measures 3 m from tip to tip and rotates at 3000 revolutions per minute. Find an expression for the potential difference between the ends of the propeller when the aeroplane is flying along the magnetic meridian. (Vertical component of earth's magnetic flux density $= 6 \times 10^{-5}$ T.)

(*Answer:* $0{\cdot}015 \cos 100\pi t$ V)

5.10. Show that for a uniformly magnetized spherical permanent magnet, $\int \frac{1}{2}\mathbf{B}.\mathbf{H}\, d\tau$, integrated over the volume inside the sphere, is just equal and opposite to the value of the integral over the volume outside the sphere. Thus $\int \frac{1}{2}\mathbf{B}.\mathbf{H}\, d\tau$ over the whole of space is zero. (This follows because there are no currents—see the derivation of $U = \int \frac{1}{2}\mathbf{B}.\mathbf{H}\, d\tau$ in § 5.5.)

5.11. The magnetostatic energy of a permanent magnet is $\int \frac{1}{2}\mu_0 H^2\, d\tau$ taken over all space. A solid spherical permanent magnet of radius b is uniformly magnetized with a magnetization M per unit volume. Show that the total energy stored in the field outside the sphere is $\mu_0 VM^2/9$, where V is the volume of the sphere. Show also that the energy (thus defined) stored in the field inside the sphere is $\mu_0 VM^2/18$. Thus the total energy is $\mu_0 VM^2/6$, and this is the magnetostatic energy of the sphere.

(*Hint.* Use the result of Problem 4.7. The question of the magnetostatic energy of a permanent magnet is discussed by Heine (1956).)

5.12. A bridge consists of a self-inductance L and resistance of $\sim 10\ \Omega$ in one arm and three non-inductive resistances in the remaining arms, so that an accurate steady balance is obtained. If, with the galvanometer in circuit, the battery key is depressed, a ballistic deflection of 10 cm of the galvanometer is recorded. If in a second experiment a resistance of $0{\cdot}02\ \Omega$ is connected in series with L, a steady deflection of 12 cm is obtained with the key depressed. The galvanometer is a moving-coil instrument with a period of 9 s. A switch in the galvanometer circuit is opened immediately after the flow of charge through it when it is used ballistically. Show that L is approximately 24 mH.

5.13. A ballistic galvanometer is calibrated by putting it in series with a 2-V battery and a resistance of $10^6\ \Omega$. A steady deflection of 17 cm is observed. The time of swing is 3·8 s. A capacitor is charged by a 4-V battery and when discharged through the galvanometer gives a throw of 24·2 cm. Find its capacitance.

(*Answer:* $C = 0{\cdot}43\ \mu$F.)

5.14. A small search coil with 8 turns of mean area 1·5 cm^2 is placed between the poles of an electromagnet with its plane normal to the magnetic field. It is connected to a ballistic galvanometer and the total resistance of the circuit is $1000\ \Omega$. When the coil is suddenly removed to a place where the field is negligible, the throw of the galvanometer is 23 divisions. When a capacitance of $1\ \mu$F charged to 1 V is discharged through the galvanometer, the throw is 25 divisions. Calculate B between the poles of the magnet.

(*Answer:* $B = 0{\cdot}77$ T.)

5.15. A ballistic galvanometer gives a throw of 10 cm when a charge of $3{\cdot}5\times10^{-7}$ C is passed. Its period is 2·2 s. Calculate the deflection when a steady current of $3\ \mu$A is passed.

(*Answer:* 30 cm.)

5.16. A capacitance of $10\ \mu$F is connected across a moving coil ballistic galvanometer. If the galvanometer coil turns through 0·05 rad for $1\ \mu$A of current, has a moment of inertia of 10^{-6} kg m^2, a period of 10 s, and negligible resistance, show that the fractional increase in the period is $\sim 2\times10^{-3}$.

5.17. In the circuit of Fig. 5.17, the moving coil (in series with R) may be connected either as shown or between the points A and C. Show that the fractional error in the reading of the power in the load (resistance Z) is less with the latter method of connection if $Z > (RR')^{\frac{1}{2}}$, and vice versa, where R' is the resistance of the fixed coil.

5.18. A 100-V dynamo is connected to a magnet whose resistance is $10\ \Omega$ and self-inductance 0·1 H. Show that the percentage increase in the heating of the magnet caused by the presence of a 100 Hz ripple voltage of 5 V amplitude in the output of the dynamo is 0·0031 per cent.

6. Magnetic materials and magnetic measurements

6.1. Origins of magnetism

THE fact that a substance placed in a magnetic field acquires a magnetic moment has been discussed in the theory of magnetostatics in § 4.3. The ratio of the magnetic moment per unit volume to the magnetic field strength H is known as the susceptibility χ, and substances are classed as diamagnetic, paramagnetic, or ferromagnetic according to the nature of their susceptibility. In the first two of these classes the induced magnetization is proportional to the applied field under ordinary conditions, so that the susceptibility is independent of the field strength. In diamagnetic substances the magnetization is in the opposite direction to the applied field, so that χ is negative, while in paramagnetic substances it is in the same direction, giving a positive value of χ. Ferromagnetic substances are distinguished by very large (positive) values of χ, which are not independent of the field strength; in addition they may possess a magnetic moment even in the absence of an applied field, as in a permanent magnet. In general magnetic properties are both non-linear and anisotropic; in a crystal of less than cubic symmetry the magnetic susceptibility is a tensor quantity, as in the case of the electric susceptibility (see § 1.5). Anisotropic effects are considered in Chapters 14–16; in this chapter we give a brief description of the origins of the magnetization, and describe methods of measuring magnetic properties.

After the discovery that a small coil carrying a current behaves like a magnet, Ampère suggested that the origin of all magnetism lay in small circulating currents associated with each atom. These 'amperean currents' would each possess a magnetic dipole moment, and the total magnetic moment of any substance would be just the vector sum of the magnetic dipole moments of its constituent atoms. This gave a natural explanation of the fact that no isolated magnetic pole had ever been observed, since even on the atomic scale only dipoles existed, and these were due to electric currents and did not consist of two actual magnetic poles of opposite sign separated by a small distance. Ampère's theory is essentially the same as that of modern atomic physics, the origin of his elementary current circuits being the motion of the negatively charged electrons in closed orbits round the positively charged atomic nucleus.

When a particle is moving in a closed orbit in a system where no external torque is acting, its angular momentum is constant. If the particle is charged, a magnetic moment will be associated with its motion, and there is a linear relation between the angular momentum and the magnetic moment (see Problem 4.8 for the simple case of a circular orbit). A general expression for the magnetic dipole moment associated with a distributed current has already been found in § 4.5. In eqn (4.59) we can replace $\mathbf{J}\,d\tau$ by $\mathbf{v}\,dq$ to find the equivalent expression for a moving charge, giving

$$\mathbf{m} = \int \tfrac{1}{2}(\mathbf{r} \wedge \mathbf{v})\, dq. \tag{6.1}$$

In the absence of an external force, the quantity $(\mathbf{r} \wedge \mathbf{v})$ is a constant, since $m(\mathbf{r} \wedge \mathbf{v})$ is equal to $\mathbf{G}$, the angular momentum for a particle of mass m. Hence

$$\mathbf{m} = \mathbf{G}\int (dq/2m) = (q/2m)\mathbf{G} = \gamma\mathbf{G}, \tag{6.2}$$

where $q = \int dq$ is the total charge circulating in the orbit. The quantity $\gamma = q/2m$ is called the magnetogyric ratio.

This close relation between magnetic moment and angular momentum is of great importance in atomic theories of magnetism, because on quantum theory the angular momentum of an electron in an atom, which is a constant of the motion, can only have discrete values. The electron possesses angular momentum not only in respect of its orbital motion round the nucleus, but also in respect of its intrinsic rotation (spin) about its own axis. The resultant angular momentum of an atom is the vector sum of the individual angular momenta of its electrons, and the resultant magnetic moment is a similar vector sum of the individual magnetic moments of the electrons. Because of the linear relation between the two, an atom, ion, or molecule will have no resultant permanent magnetic moment if the total angular momentum is zero. If the total angular momentum is not zero, the atom, ion, or molecule will have a permanent magnetic dipole moment. Most free atoms possess a permanent magnetic dipole moment because they have a resultant electronic angular momentum. Experiments such as the gyromagnetic effect (§ 15.4) in solids, and magnetic resonance (Chapter 24) for free particles and solids, give precise determinations of the ratio of the magnetic moment to the angular momentum for nuclei, and atoms and molecules, as well as magnetic ions in solids.

Such magnetic dipole moments are fundamentally different from electric dipole moments, where it has been shown (§ 2.3) that if parity is conserved in an atom or nucleus, no electric dipole moment can exist. The difference is clear if we write out the components of the magnetic

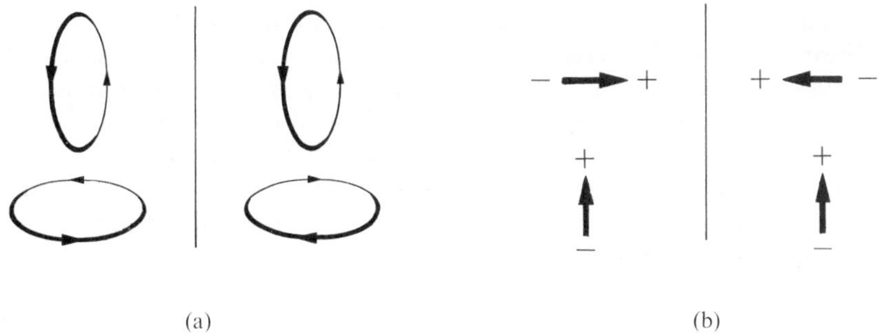

FIG. 6.1. (a) Reflection of circulating currents in a plane. (b) Reflection of dipoles in a plane. Note that the circulating currents behave differently from the equivalent dipoles on reflection: the magnetic dipole is an axial vector, the electric dipole a polar vector (see Appendix A.1).

dipole moment given by eqn (6.1); for example,

$$m_z = \int \tfrac{1}{2}\left(x\frac{\partial y}{\partial t} - y\frac{\partial x}{\partial t}\right) \mathrm{d}q.$$

If parity is conserved, we have inversion symmetry, that is, the value of $\mathrm{d}q$ is the same at the point $(-x, -y, -z)$ as at the point (x, y, z). So also is the quantity $\{x(\partial y/\partial t) - y(\partial x/\partial t)\}$, so that the integrand retains the same sign under the inversion operation, and the integral can have a finite value. This differs from the electric dipole moment in that the latter involves the first power of the coordinates (see eqn (2.25)), while the magnetic dipole moment involves the second power. The difference shows clearly under reflection in a plane, see Fig. 6.1; also, magnetic moments change sign under time reversal, $t \to -t$.

In general, however, atoms do not exist in the free state but are combined into molecules, and it so happens that the forces responsible for chemical binding strongly affect the arrangement of the individual magnetic moments of the various electrons in the molecule. As a result the stable state of the molecule is nearly always one in which the vector sum of these individual moments is just zero, so that the molecule as a whole has no resultant permanent magnetic moment. Similarly, in the majority of solids and liquids, the atomic constituents are ions with no permanent magnetic dipole moments. When such a substance is placed in a magnetic flux density $\mathbf{B}$, each individual electron, being a moving charge, experiences a force. Its orbital motion round the nucleus is altered in such a way that it acquires an angular momentum and hence a magnetic dipole moment which is proportional to the applied field, but in the opposite direction. This gives a negative susceptibility, whose value will now be calculated. It should be noted that this diamagnetic effect is present in all substances, but in paramagnetic substances there is a much larger positive contribution to the susceptibility which generally far outweighs the diamagnetic contribution. The positive contribution arises from molecules (or ions) where the vector sum of the individual electron moments is not zero, so that the molecule or ion possesses a permanent magnetic moment. Thus all substances show a diamagnetic effect, giving a susceptibility of the order of 10^{-5} in S.I. units, but in paramagnetic substances there is a positive susceptibility contribution which is much greater.

6.2. Diamagnetism

The fact that an atom placed in a magnetic flux density **B** acquires a magnetic moment parallel to **B** can be shown by finding the effect on the atom of establishing the field **B** from zero. In § 5.1 it was shown that this changes the momentum of a particle of charge q from $m\mathbf{v}_0$ to $m\mathbf{v}$, where

$$m\mathbf{v} = m\mathbf{v}_0 - q\mathbf{A}$$

and **A** is the vector potential associated with **B**. The corresponding current density is given by

$$\mathbf{J} = \rho_e\mathbf{v} = \rho_e\mathbf{v}_0 - \rho_e(q/m)\mathbf{A}. \tag{6.3}$$

If **A** is increased by $\delta\mathbf{A}$, the increase in potential energy is, by eqn (4.52),

$$\delta U_P = -\int(\mathbf{J}.\delta\mathbf{A})\,\mathrm{d}\tau = -\int\rho_e(\mathbf{v}_0.\delta\mathbf{A})\,\mathrm{d}\tau + \int\rho_e(q/m)(\mathbf{A}.\delta\mathbf{A})\,\mathrm{d}\tau. \tag{6.4}$$

For simplicity we take **B** to be uniform (which it will be over atomic dimensions), so that

$$\mathbf{A} = \tfrac{1}{2}(\mathbf{B}\wedge\mathbf{r}), \quad \text{and} \quad (\mathbf{A}.\delta\mathbf{A}) = \tfrac{1}{4}(B\,\delta B)r^2\sin^2\theta,$$

where θ is the angle between **B** and **r**. Hence

$$\begin{aligned}\delta U_P &= -\int\tfrac{1}{2}\rho_e\mathbf{v}_0.(\delta\mathbf{B}\wedge\mathbf{r})\,\mathrm{d}\tau + (q/4m)\int\rho_e B\,\delta B r^2\sin^2\theta\,\mathrm{d}\tau \\ &= -\delta\mathbf{B}.\int\tfrac{1}{2}\rho_e(\mathbf{r}\wedge\mathbf{v}_0)\,\mathrm{d}\tau + (q/4m)B\,\delta B\int\rho_e r^2\sin^2\theta\,\mathrm{d}\tau \\ &= -\delta\mathbf{B}.\mathbf{m}_0 - \delta\mathbf{B}.\mathbf{m}_i = -\delta\mathbf{B}.\mathbf{m}.\end{aligned} \tag{6.5}$$

This result shows that the magnetic moment **m** consists of two parts, of which the first, $\mathbf{m}_0$, is clearly the permanent magnetic dipole moment (see eqn (6.1)), while the second $\mathbf{m}_i$ is an induced dipole moment which is proportional to the flux density **B**. The value of $\mathbf{m}_i$ is

$$\mathbf{m}_i = -\mathbf{B}(q/4m)\int\rho_e r^2\sin^2\theta\,\mathrm{d}\tau = -\mathbf{B}(q^2/4m)\langle r^2\sin^2\theta\rangle,$$

where $q = \int \rho_e \, d\tau$ is the total charge. If ρ_e is independent of θ,

$$\langle r^2 \sin^2\theta \rangle = \langle r^2 \rangle \frac{1}{4\pi} \int_0^\pi \sin^2\theta . 2\pi \sin\theta \, d\theta = \tfrac{2}{3} \langle r^2 \rangle,$$

where $\langle r^2 \rangle$ is the mean-square radius of the orbit. On summing over all electrons in the atom or ion, $\mathbf{m}_0$ must be replaced by the resultant (if any) of the permanent dipole moments, while the induced moment per atom becomes

$$\mathbf{m}_i = -(e^2/6m_e)\mathbf{B} \sum \langle r^2 \rangle, \tag{6.6}$$

where the electronic charge $-e$ and mass m_e have been inserted. Hence the diamagnetic susceptibility of a sample containing n atoms (or ions) per unit volume will be

$$\chi = n\mathbf{m}_i/\mathbf{H} = -n\mu_0(e^2/6m_e) \sum \langle r^2 \rangle. \tag{6.7}$$

This result can be derived in another way using Larmor's theorem (Appendix A, § A.11). The result of establishing a field of flux density **B** is to set up a precessional motion of the electronic orbits with angular velocity $\boldsymbol{\omega} = -(q/2m)\mathbf{B}$, as a result of which each electron acquires an additional angular momentum

$$\mathbf{G}_i = m\langle a^2 \rangle \boldsymbol{\omega} = -m\langle a^2 \rangle (q/2m)\mathbf{B},$$

where $\langle a^2 \rangle = \langle r^2 \sin^2\theta \rangle$ is the mean-square distance of the electron from an axis parallel to **B** through the centre of gravity of the atom (the nucleus). Associated with $\mathbf{G}_i$ is an additional magnetic moment of magnitude (from eqn (6.2))

$$\mathbf{m} = (q/2m)\mathbf{G}_i = -(q^2/4m)\langle r^2 \sin^2\theta \rangle \mathbf{B}.$$

On writing $q = -e$, $m = m_e$ and summing over all electrons in the atom, this gives eqn (6.7) above.

This equation shows that the diamagnetic susceptibility is inherently negative in sign, and does not depend, for example, on the sign of the electronic charge. Fundamentally, the negative sign follows from Lenz's law. When the external magnetic flux density **B** is switched on, there is a change in the flux through the electron orbits which induces a momentary e.m.f. The change in the orbits which this causes gives an induced magnetic moment to the atom which is in such a direction as to oppose the change in flux through the orbit, that is, the magnetic moment is due to an induced current whose own lines of magnetic flux through the atom are in the opposite direction to those of the external flux. We have assumed that $\langle r^2 \rangle$ is unaltered by the presence of the magnetic field; this is justified because the magnetic forces at ordinary field strengths are negligible compared with the internal atomic forces (see Appendix § A.11).

As defined in Chapter 4, the susceptibility refers to unit volume of substance, and n is then the number of atoms in unit volume. Since χ is linearly proportional to n, it is permissible to take samples of different size, and refer to the 'susceptibility per unit mass', or 'susceptibility per gram atom'. These are simply equal to the susceptibility per unit volume (or 'volume susceptibility' for short) multiplied by the volume of the sample. Thus the 'mass susceptibility' $\chi_m = \chi/\rho$, where ρ is the density, since unit mass (1 kg) occupies a volume of $1/\rho$ m^3. Similarly, the molar susceptibility will be $\chi_M = 10^{-3} M\chi_m$, where M is the atomic or molecular weight in grams. Here the factor 10^{-3} occurs because our χ_m refers to a kilogram of substance. For a mole eqn (6.7) gives the numerical value

$$\chi_M = -3{\cdot}55 \times 10^9 \sum \langle r^2 \rangle. \tag{6.8}$$

In some reference tables the susceptibility is given in electromagnetic units, and for solids and liquids the diamagnetic volume susceptibility (per cm^3) is of the order -10^{-6} e.m.u. (for gases it is much smaller owing to the lower number of atoms per unit volume). In S.I. units, however, the volume susceptibility (per m^3) is greater by a factor 4π and so is of order -10^{-5}. As a rough rule, the molar susceptibility in S.I. is of order -10^{-11} Z, where Z is the atomic number, equal to the number of electrons in the atom. This indicates that the mean value of $\langle r^2 \rangle$ is about 10^{-21} m^2, as expected from atomic theory. The susceptibility of a diamagnetic substance is substantially independent of temperature, since $\sum \langle r^2 \rangle$ is practically unaltered by thermal motion.

In the first approximation $\sum \langle r^2 \rangle$ is constant for a particular type of atom or ion, and is not greatly altered by its surroundings. Thus aqueous solutions of alkali and alkaline earth halides obey quite accurately (and most substances approximately) an additivity rule known as Wiedemann's law. According to this rule the mass susceptibility χ_m of a solution containing a mass m_1 of a salt of mass susceptibility χ_1 in a mass m_2 of solvent of mass susceptibility χ_2 is

$$\chi_m = \frac{m_1\chi_1 + m_2\chi_2}{m_1 + m_2}. \tag{6.9}$$

Similar additivity rules are approximately valid for chemical compounds. Thus, for compounds which ionize in solution, the molar susceptibility of the compound is generally close to the sum of the ionic susceptibilities of its constituent ions in solution. By assuming a theoretical value for the susceptibility of one type of ion, approximate values for the susceptibilities of other ions can be obtained from measurements on compounds, and the validity of the additivity rules tested. (For a review of the diamagnetism of ions see Myers (1952).)

6.3. Paramagnetism

As already pointed out, paramagnetism occurs in those substances where the individual atoms, ions, or molecules possess a permanent magnetic dipole moment. In the absence of an external magnetic field, the atomic dipoles point in random directions and there is no resultant magnetization of the substance as a whole in any direction. This random orientation is the result of thermal agitation within the substance. When an external field is applied, the atomic dipoles tend to orient themselves parallel to the field, since this is a state of lower energy than the antiparallel position. This gives a net magnetization parallel to the field, and a positive contribution to the susceptibility. Since the thermal agitation, which tends to give a random orientation, is less at low temperatures, a bigger proportion of the dipoles are able to align themselves parallel to the field, and the magnetization is greater for a given field. It was discovered by Curie that, for ordinary fields and temperatures, the susceptibility of many substances follows the equation

$$\chi = M/H = C/T, \tag{6.10}$$

where C is a constant, and T is the absolute temperature; this is known as Curie's law. For large fields at low temperatures the magnetization produced is no longer proportional to the applied field, and tends to a constant value. This saturation effect is produced when all the atomic dipoles are aligned parallel to the field, so that the magnetization reaches a limiting maximum value.

The theoretical explanation of Curie's law was given by Langevin, using the classical statistics of Boltzmann. He assumed that each atom had a permanent magnetic moment $\mathfrak{m}$, and that the only force acting on it was that due to the external field **B**. Then, if a given atomic dipole is pointing in a direction making an angle θ with **B**, its magnetic potential energy is $W = -\mathfrak{m}B \cos\theta$. Now, on classical statistics, the number of atoms making an angle between θ and $\theta + \mathrm{d}\theta$ is

$$\mathrm{d}n = c\mathrm{e}^{-W/kT} \sin\theta \, \mathrm{d}\theta, \tag{6.11}$$

where k is Boltzmann's constant and T is the absolute temperature. c is a constant defined by the fact that integration of (6.11) over the whole possible range of energies must give just n, the total number of atoms in the system. Hence for our case

$$\mathrm{d}n = c\mathrm{e}^{\mathfrak{m}B\cos\theta/kT} \, . \sin\theta \, \mathrm{d}\theta$$

and n is equal to this integrated over all angles from 0 to π. The component of each dipole moment parallel to **B** is $\mathfrak{m}\cos\theta$, and hence the average component per atom is $\bar{\mathfrak{m}}$, where

$$n\bar{\mathfrak{m}} = \mathfrak{m} \int \cos\theta \, \mathrm{d}n.$$

Hence

$$\frac{\overline{\mathfrak{m}}}{\mathfrak{m}}=\frac{\int_0^{\pi}\cos\theta\,.\,c e^{\mathfrak{m}B\cos\theta/kT}\,.\sin\theta\,d\theta}{\int_0^{\pi} c e^{\mathfrak{m}B\cos\theta/kT}\,.\,\sin\theta\,d\theta}$$

On writing $\mathfrak{m}B/kT = y$, $\cos\theta = x$, this takes the form

$$\frac{\overline{\mathfrak{m}}}{\mathfrak{m}}=\frac{\int_{-1}^{+1} x e^{xy}\,dx}{\int_{-1}^{+1} e^{xy}\,dx}=\coth y-\frac{1}{y}=L(y), \qquad (6.12)$$

where $L(y)$ is known as the Langevin function. It is plotted as a function of y in Fig. 6.2. For large values of y the function tends to unity, saturation being reached when all the atomic dipoles are parallel to **B**. For small values of y the curve is linear, and $L(y) = y/3 = \mathfrak{m}B/3kT$. Then the susceptibility is

$$\chi = n\overline{\mathfrak{m}}/H = \mu_0 n\mathfrak{m}^2/3kT, \qquad (6.13)$$

where n is the number of atoms per unit volume. This is the same as Curie's law, eqn (6.10), if we identify the Curie constant C with $\mu_0 n\mathfrak{m}^2/3k$. The only unknown quantity in (6.13) is the atomic dipole moment $\mathfrak{m}$ so that by measuring the susceptibility as a function of the absolute temperature in a region where $\mathfrak{m}B/kT$ is small, the magnitude of the atomic dipole moments may be found. In general these are of the

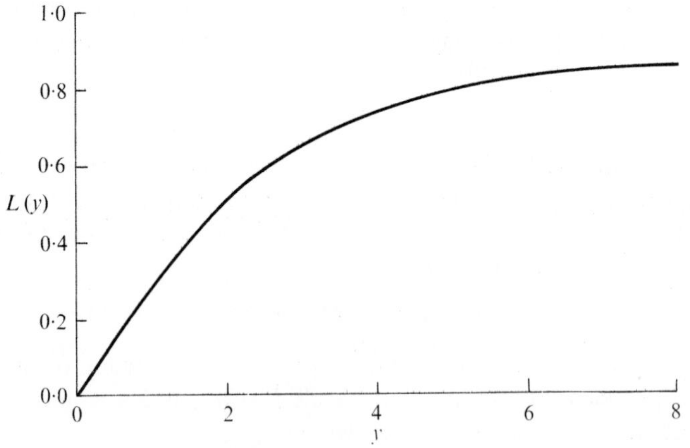

FIG. 6.2. The Langevin function $L(y) = \coth y - 1/y$.

order of 10^{-23} A m^2 or slightly greater, and the volume susceptibility (per m^3) at room temperature of solid paramagnetic substances which obey Curie's law $\approx +10^{-3}$. Thus the paramagnetism considerably outweighs the diamagnetism which is also present.

Langevin's theory applies strictly only to gases, where the molecules are sufficiently far apart for their mutual interactions to be negligible. In liquids and solids such interactions may be large, and many substances obey the modified Curie–Weiss law

$$\chi = C/(T-\theta). \tag{6.14}$$

θ is called the 'Weiss constant' and is characteristic of the substance; it may be either positive or negative (see Chapters 15 and 16). Eqn (6.14) holds only at temperatures where $T > |\theta|$, and for many substances no single equation represents the susceptibility variation adequately over a wide temperature range.

As already noted, the tendency in chemical combination is towards zero resultant angular momentum of the electrons and hence to zero permanent magnetic moment. Of the common gases, only oxygen (O_2) and nitric oxide (NO) are paramagnetic. In the solid state paramagnetism occurs in salts of the 'transition group' ions (see Chapter 14), and the magnetic moment is associated with the metallic ion itself. Thus in compounds such as $CrK(SO_4)_2.12H_2O$ ('chrome alum') or $CuSO_4.5H_2O$ (copper sulphate) only the Cr^{3+} ion and Cu^{2+} ions respectively have permanent magnetic moments, the other ions (K^+, SO_4^{2-}) and water molecules giving only a diamagnetic contribution to the susceptibility. In most metals the outer electrons are detached from the individual atoms, which are thus left as diamagnetic ions. The detached electrons are free to move through the metal and form the conduction electrons; these give rise to a diamagnetic and a paramagnetic effect, both of the same order of magnitude and both independent of temperature (see § 14.11). The outstanding exceptions are the ferromagnetic metals, iron, cobalt, nickel, the lanthanide metals, and many alloys.

6.4. Ferromagnetism

Ferromagnetic substances are all solids, and each is characterized by a certain temperature known as the Curie point at which its properties change abruptly. Above the Curie point the susceptibility is independent of field strength, and follows approximately a Curie–Weiss law (eqn (6.14)) with a Weiss constant θ whose value is close to that of the Curie point. Below the Curie temperature the behaviour is quite different; very large values of magnetization are produced by quite small fields, and the magnetization varies quite non-linearly with the field strength. This is shown by a characteristic plot of the flux density B as a function of the

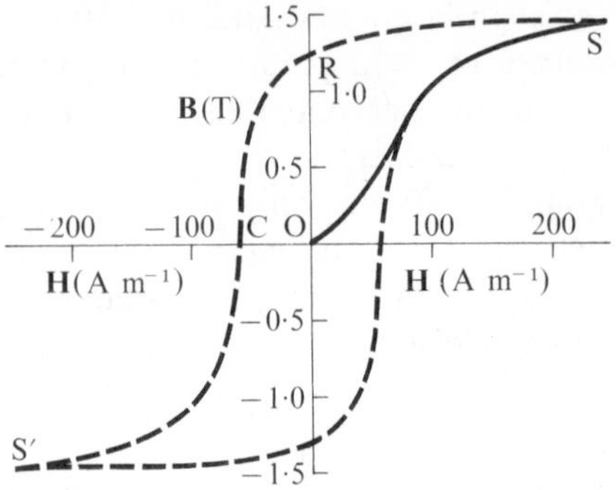

FIG. 6.3. Magnetization curve (full line OS) and hysteresis loop (broken line) for iron.

field strength H in a sample of iron (Fig. 6.3). If the iron is initially unmagnetized, and a field of slowly increasing magnitude is applied, B follows the full line in Fig. 6.3, known as the 'magnetization curve'. In a field of a few hundred amperes per metre the value of B becomes practically constant at about 1·5 T. If the magnetic field strength H is now reduced, B does not return along the magnetization curve but follows the broken line, and even at $H=0$, corresponding to the point R in the figure, B is still near the saturation value. The value of B at this point is known as the 'residual flux density', and the retention of magnetization in zero field is known as 'remanence'. On applying a reverse field the value of B falls and finally becomes zero (point C in Fig. 6.3); the value of the field strength at this point is called the 'coercive force'. As the magnitude of the reverse field is further increased, a reverse flux density is set up which quickly reaches the saturation value. Finally, if the reverse field is gradually removed and a positive field applied, the flux density traces out the lower broken curve in the direction S′S. The whole curve is called the 'hysteresis curve'. It shows that the change in the flux density always lags behind the change in the applied magnetic field strength.

The magnetization curve may also be represented in the form of a permeability curve, showing the variation of $\mu_r = B/(\mu_0 H)$ as a function of either B or H. Such a curve (μ_r against B) is shown in Fig. 6.4. When a field is applied μ_r goes through a maximum and then falls rapidly as the material becomes saturated. The values of μ_r, of the order of 10^4, are enormously greater than in paramagnetism (1·001 or so at ordinary temperatures).

To explain this behaviour, Weiss suggested that a ferromagnetic substance contains atoms with permanent magnetic moments, as in a paramagnetic substance, but that there are large forces acting between neighbouring atomic dipoles which cause groups of them all to point in the same direction. The substance would then be permanently magnetized

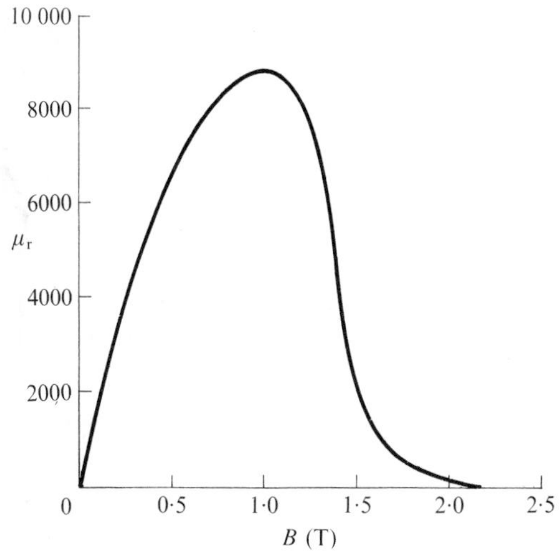

FIG. 6.4. Curve of relative permeability μ_r against flux density B for iron.

within each group; such groups are called 'domains' and their size is now known to vary from about $10^{-3}\,mm^3$ to $1\,mm^3$, or greater in single crystals. In an unmagnetized polycrystalline specimen the domains are oriented at random, so that there is no resultant magnetic moment in any direction. When a field is applied, domains where the magnetization is parallel or at a small angle with the field grow at the expense of those where the magnetization is antiparallel or nearly so, so that the boundary between domains is displaced. Initially, in the full curve of Fig. 6.3, the magnetization of the substance as a whole proceeds by small (reversible) boundary displacements, but the steeper part of the magnetization curve is due to larger (irreversible) displacements. Above the knee of the curve, magnetization proceeds by rotation of the direction of magnetization of whole domains; such a process is rather difficult and the increase in magnetization is relatively slow. These processes are shown schematically in Fig. 6.5. When the applied field is reduced, there is little change in the domain structure so that the magnetization remains quite high until reverse fields are applied, thus giving rise to the hysteresis described above.

Ferromagnetic substances may be broadly divided into two classes: (1) magnetically soft materials, which have high permeability, and are easily magnetized and demagnetized, and (2) magnetically hard materials, which have a relatively low permeability and are difficult to magnetize or demagnetize (high coercive force). The chief uses of the former are in electromagnetic machinery and transformers, and of the latter in permanent magnets. To obtain a soft magnetic material the domain walls must

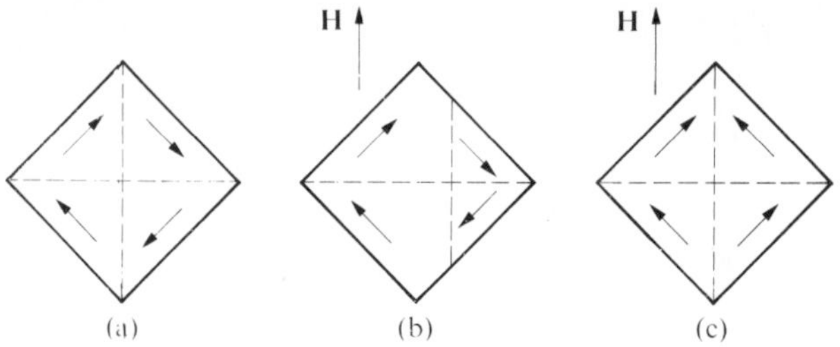

FIG. 6.5. Schematic representation of domains in a ferromagnetic substance: (a) unmagnetized; (b) magnetization through movement of domain boundary wall; domains oriented parallel to **H** grow at the expense of antiparallel domains; (c) magnetization by rotation of the magnetization of whole domains. The domains remain magnetized along a preferred direction in each crystallite; very large fields are required to swing the magnetization away from such a direction towards the applied field.

be able to move easily and reversibly, so that the magnetization changes by large amounts for small changes in the magnetizing field. This requires a material as free as possible from irregularities in the crystal structure due to strains or small particles of impurities. The main treatment of such materials consists therefore of heating to a temperature where sufficient movement of the atoms is possible for them to settle into an ordered lattice, followed by a slow cooling (annealing) so as not to disturb it. On the other hand, a hard magnetic material is one in which domain wall movement is difficult owing to lattice imperfections. These are produced by heating the material and then plunging it suddenly into cold oil (quenching), which sets up internal stresses. Some alloys are then reheated to a lower temperature to cause one of the constituents partially to separate out in small particles dispersed through the alloy. In an alternative process magnets are constructed from compressed powders of very fine particles. If the particles are below a certain size, each forms a single domain and there are no domain walls within the particle. Magnetization and demagnetization can only be accomplished by rotation of the direction of magnetization of each particle. With high anisotropy, which may arise either from the shape of the particles or from their crystal structure, the only possible rotation is reversal through 180°. This requires a higher field than wall movement, and so gives a higher coercive force. To produce a good permanent magnet, the particles must consist of a material with the highest possible saturation magnetization, and be aligned so that their 'easy' directions of magnetization are all parallel. Obviously a high Curie temperature is also required.

6.5. Production of magnetic fields

For many purposes it is necessary to maintain a large magnetic field which is constant over a certain volume. By 'large' is meant values of B

ranging from 0·1 T to 10 T, and the volume may vary from 10^{-4} m^3 up to many cubic metres in a nuclear accelerator or a plasma-containment apparatus. According to the particular application, either permanent magnets or electromagnets (with or without iron) may be employed, and the principles of their construction are outlined below

A modern permanent magnet of typical shape is shown in Fig. 6.6. The important quantities are the values of B in the air gap, and the volume of the air gap; from eqn (5.44) the energy density at any point is $\frac{1}{2}\mathbf{B}.\mathbf{H}$, and the total energy stored in the air gap is $(\frac{1}{2}BH)V_a$, where V_a is the volume of the gap, and the vectorial representation of B and H can be dropped since they are parallel to one another. The energy stored in the gap is simply related to the quantities involved in the magnet design, as follows. If we assume that there is no leakage, so that all the lines of magnetic flux within the magnet pass through the gap, we have

$$BA_a = B_m A_m, \tag{6.15}$$

where A_a, A_m are the cross-sections of the air gap and the magnet, respectively, and B_m is the value of the flux density in the magnet. Now, if we consider a circuit through the gap and the magnet as indicated by the broken line in Fig. 6.6, the total magnetomotive force (see § 4.2) is zero, since no electric currents are involved. Hence

$$\int \mathbf{H}.\mathrm{d}\mathbf{s} = Hd_a + H_m d_m = 0, \tag{6.16}$$

where d_a and d_m are the path lengths in the air gap and the magnet respectively, and H_m is the magnetic field strength in the magnet. On

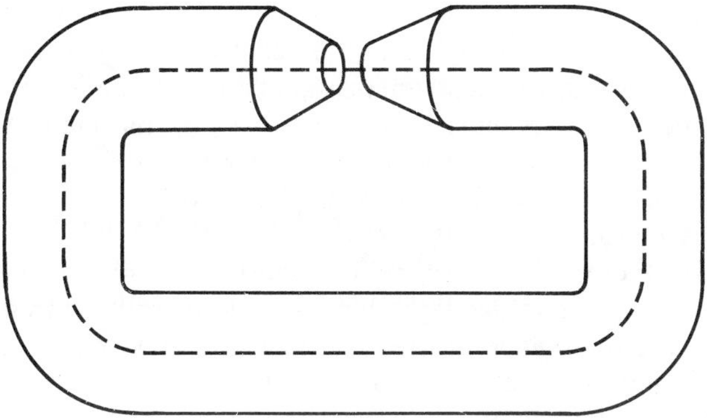

FIG. 6.6. Permanent magnet with coned pole pieces. The magnetomotive force is evaluated along the broken line.

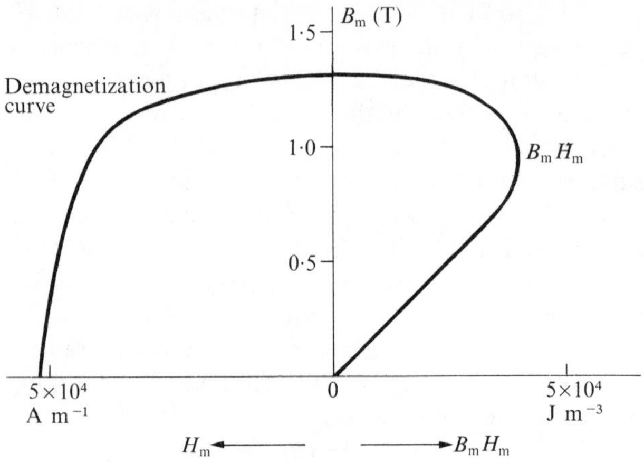

FIG. 6.7. Demagnetization curve (to the left) and plot of (B_mH_m) against B_m (to the right), for Alcomax III.

combining eqns (6.15) and (6.16) we have

$$(\tfrac{1}{2}BH)V_a = (\tfrac{1}{2}BH)A_a d_a$$
$$= -(\tfrac{1}{2}B_mH_m)A_m d_m = -(\tfrac{1}{2}B_mH_m)V_m. \tag{6.17}$$

This important relation shows that for a magnet of given volume V_m, the greatest amount of energy stored in the gap is obtained if the product B_mH_m has its maximum value.

The variation of the product B_mH_m with B_m for a typical material is shown in Fig. 6.7, together with the 'demagnetization curve', that is, the part of the hysteresis loop corresponding to the application of a reverse field. In a permanent magnet the field inside is the 'demagnetizing field' due to the 'free magnetic poles' near the ends of the magnet; this field is in the opposite direction to B_m, and the negative sign in eqn (6.17) arises from this. The size of the demagnetizing field is determined by the shape of the magnet, and this must be designed so that the material is at the point where B_mH_m is a maximum. A good working rule is that at this point the ratio of B_m to H_m is equal to the ratio B_r/H_c, where B_r is the residual flux density and H_c the coercive force.

In practice there is always some leakage of lines of magnetic field so that the energy stored in the gap is less than the theoretical value given by eqn (6.17). The best shape of the magnet is one in which B_mH_m at every point in the material is closest to its maximum value. Since the product B_mH_m is a characteristic of the material the volume of material required increases linearly with the energy stored in the gap. Many special materials are available with high values of the energy product, and their

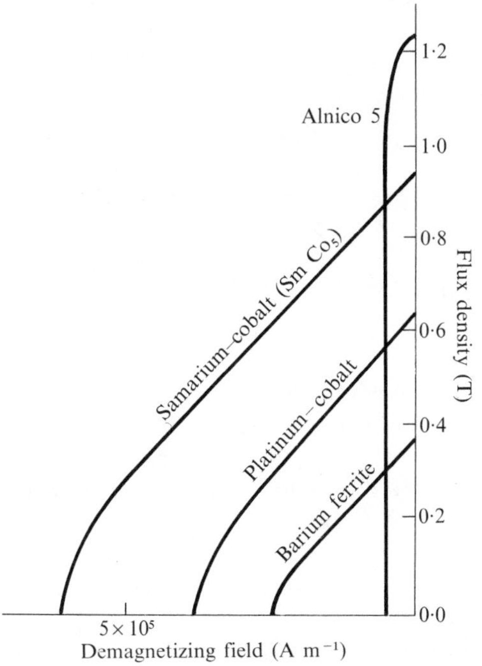

FIG. 6.8. Demagnetizing curves for some permanent magnet materials. Alnico 5 (alcomax III) is an alloy of 50% Fe, 25% Co, 13·5% Ni, 8% Al, 3% Cu, ½% Nb.

demagnetization curves are shown in Fig. 6.8. The intermetallic compound $SmCo_5$ has a value of B_mH_m of order 2×10^5 J m^{-3}, and a coercive force of over 5×10^5 A m^{-1}.

Iron-cored electromagnets

When fields of greater magnitude, or adjustable fields, are required, an electromagnet is used. For fields of flux density up to about 2 T the normal type of construction is as shown in Fig. 6.9. It consists of a yoke

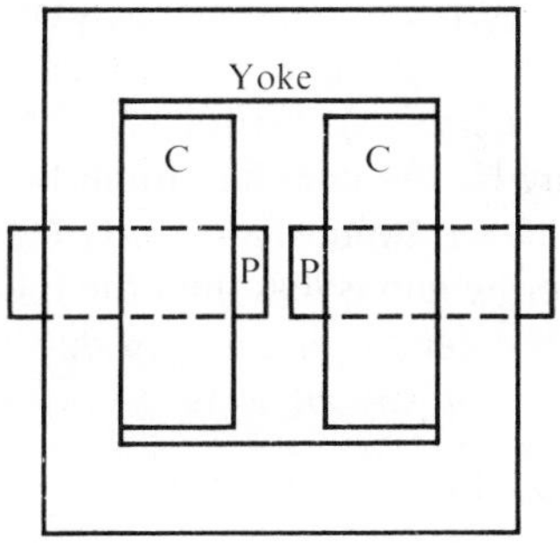

FIG. 6.9. Weiss type of electromagnet. C, C coils carrying electric current, P, P pole tips.

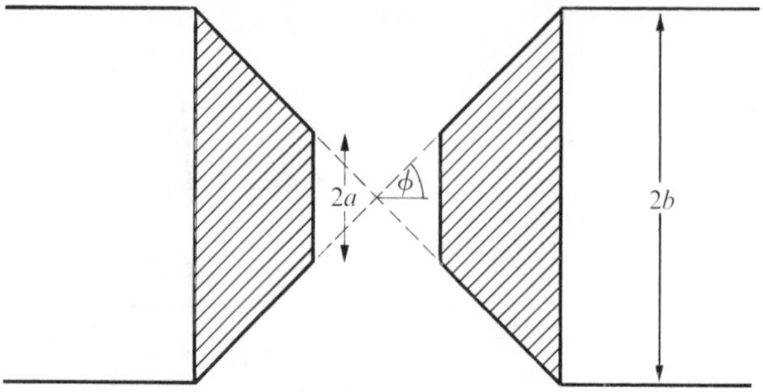

FIG. 6.10. Coned pole tips for magnet.

of soft iron or, more commonly, mild steel, which has a reasonably high magnetization for small values of the applied field. C and C are coils of copper wire, or copper tube through which cooling water may flow, carrying current from a d.c. generator. If the coils have a total of n turns each carrying a current I, the m.m.f. is

$$nI = \int \mathbf{H}.\mathrm{d}\mathbf{s} = Hd_{\mathrm{a}} + H_{\mathrm{m}}d_{\mathrm{m}} \tag{6.18}$$

round a circuit through the air gap and yoke similar to that shown in Fig. 6.6, with the same nomenclature as before. Since eqn (6.15) still holds, we have

$$nI = BA_{\mathrm{a}}\left(\frac{d_{\mathrm{a}}}{\mu_0 A_{\mathrm{a}}} + \frac{d_{\mathrm{m}}}{\mu A_{\mathrm{m}}}\right), \tag{6.19}$$

where the quantity in parenthesis, is known as the magnetic 'reluctance' of the system. In this analogy between the m.m.f. and the e.m.f. in a circuit, the flux BA_{a} is analogous to the electric current, and the flux passes through the two components—the air gap and the yoke—in series. The total reluctance of the 'magnetic circuit' is the sum of the two parts due to the gap and the iron. Since μ_{r} for iron is very large, the greater part of the reluctance is generally associated with the air gap.

In most iron-cored magnets the pole faces may be removed and replaced by others of different shape for special investigations, and the width of the air gap may be altered. To obtain higher values of B, coned pole pieces may be used. In the design of Fig. 6.10, if it is assumed that the magnetization is everywhere parallel to the axis and has the saturation value M_{s}, the field at the centre of the gap due to the conical portions indicated by shading may be shown to be (see Problem 6.7)

$$B = \mu_0 H = \mu_0 M_{\mathrm{s}} \sin^2\phi \cos\phi \ln(b/a). \tag{6.20}$$

This has a maximum value at $\phi = 54{\cdot}7°$; in practice the pole pieces are not everywhere completely saturated, and a value of about 60° gives the best results. The pole tips may be made of a special cobalt steel which has a higher saturation flux density than ordinary mild steel. A plot of the field in the gap against exciting current is usually fairly linear (apart from a small initial field due to the remanence) until the steel becomes saturated; the rate of increase then becomes much lower, any extra field being that due to the current in the coils themselves.

It was pointed out by Bitter (1936) that this arrangement, in which the iron is magnetized everywhere in the same direction, does not make the best use of the iron. In Fig. 6.11 the field at the origin O due to the dipole at A, oriented as shown, has a component parallel to the x-axis equal to

$$H_x = \frac{\mathfrak{m}}{4\pi r^3}(2\cos\theta\cos\phi + \sin\theta\sin\phi).$$

If the orientation of the dipole is variable, then H_x has a maximum value when $dH_x/d\theta = 0$, that is, $\tan\theta = \frac{1}{2}\tan\phi$, and this value is then

$$(H_x)_{\max} = \frac{\mathfrak{m}}{4\pi r^3}(1+3\cos^2\phi)^{\frac{1}{2}}. \tag{6.21}$$

If, on the other hand, the dipole points parallel to the x-axis, so that $\theta = -\phi$, the value of H_x is only

$$H_x = \frac{\mathfrak{m}}{4\pi r^3}(3\cos^2\phi - 1). \tag{6.22}$$

Both the angular functions in (6.21) and (6.22) have the value 2 at $\phi = 0$, but the latter falls to zero at $\phi = 54{\cdot}7°$ and then changes sign (the reason why the coned pole pieces discussed above have this as their optimum angle is that dipoles at a larger angle oriented parallel to the axis would give a reverse field and thus reduce the field in the gap). On the other hand, at this angle $(1+3\cos^2\phi)^{\frac{1}{2}}$ has fallen only to $\sqrt{2}$, and its smallest value is 1 at $\phi = 90°$.

At first sight it would seem impracticable to set up a magnetization in the iron with the angular distribution of the direction of magnetization

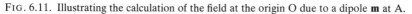

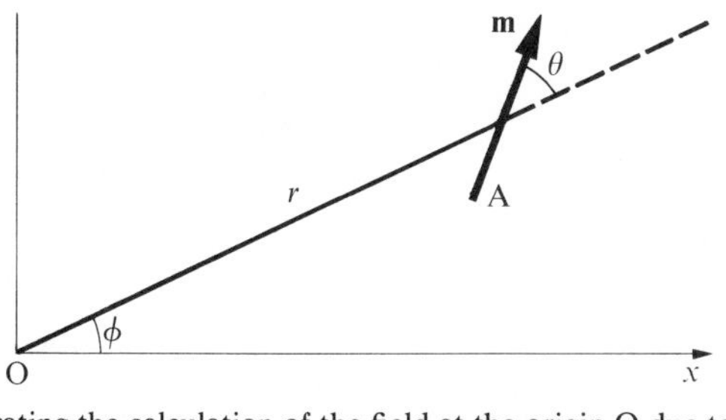

FIG. 6.11. Illustrating the calculation of the field at the origin O due to a dipole **m** at A.

required by the equation $\tan\theta = \frac{1}{2}\tan\phi$. The orientation of the magnetization at any point is, however, just the same as the direction of the lines of force set up by a point dipole at the origin pointing along the x-axis, and it follows that such a dipole would magnetize the iron in just the right direction. Bitter therefore designed a magnet of new type in which the magnetizing coil, whose field is approximately that of a point dipole, is at the centre, and surrounded by soft iron. This gave an appreciably better performance than the older type of design.

Air-cored electromagnets

Since iron and other magnetic materials saturate at 1–2 T, the contribution they make in magnets for fields of flux density above 3 T is not sufficient to justify the expense, and solenoids with no iron are used instead. Bitter has considered the question of the current distribution, and concluded that higher efficiency is obtained if the current density is not uniform, but falls off inversely with the radius. A non-uniform current distribution of this type can be obtained by using flat conductors of the shape shown in Fig. 6.12; current is led in along the edge AA and out along BB, so that the current density varies inversely with the radius because the resistance along a path of radius r increases with r. The solenoid is constructed of a number of such discs mounted one above the other, with holes drilled in them so that cooling water can be forced through. The low resistance of such a design ($\sim 0{\cdot}01\ \Omega$) reduces the chance of any breakdown in the insulation, and corrosion through electrolysis of the cooling water. In one such coil Bitter (1940) obtained a flux density of 10 T, uniform to 1 per cent over a volume of 25 cm^3 with a supply of 10 000 A at 170 V. It can be shown that the field obtained from an air-cored coil can be expressed as $(P\lambda/\rho r)^{\frac{1}{2}}$ times a factor depending only on the shape of the coil and the current distribution; here P is the

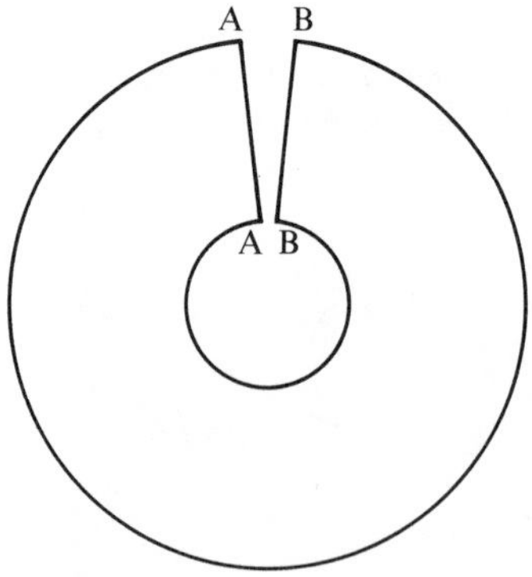

FIG. 6.12. Shape of copper disc known as a 'Bitter pancake'.

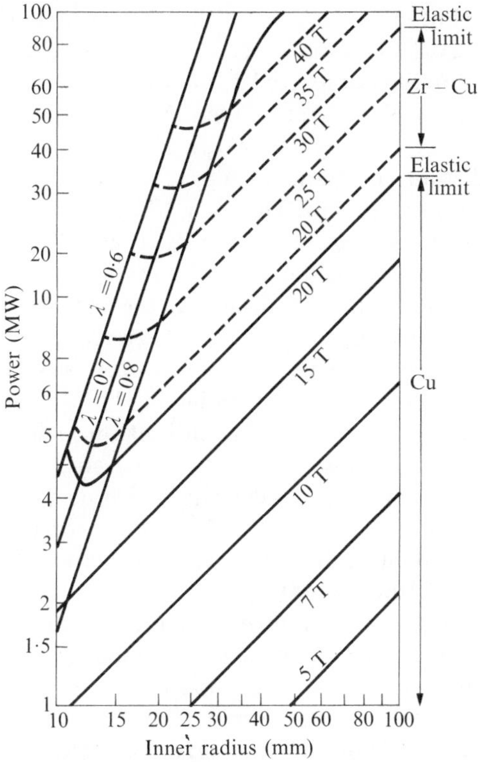

FIG. 6.13. Stress and cooling limitations for disc magnets with 1-mm-diameter cooling passages, and a surface-cooling rate in these passages of $1{\cdot}5\times10^7$ W m^{-2}. λ is the fraction of coil volume occupied by the conductor. (Bitter 1962.)

power dissipation, λ the fraction of the coil volume occupied by conductor of resistivity ρ (the remainder being insulation and coolant), and r a linear dimension such as the inner radius. Thus to double the field over the same volume requires four times the power.

The ultimate performance of an air-cored coil, whatever shape of conductor is used, is limited by the rate at which heat generated in the conductors can be removed and by the mechanical strength of the material, which must withstand enormous magnetic forces. These design limitations are illustrated in Fig. 6.13.

Superconductive magnets

Below its transition temperature a superconductive material (see Chapter 13) offers zero resistance to the flow of electric current, provided that any magnetic field present does not exceed the critical field above which normal resistivity is restored. This field is 0·1 T or less for most Type I superconductors, but in a number of Type II superconductors it lies between 5 T and 20 T. The discovery of materials such as niobium and the alloys niobium–titanium (NbTi), niobium–tin (Nb_3Sn), niobium–zirconium (NbZr), and vanadium–gallium (V_3Ga) has made it possible to construct superconductive magnets which, for fields above 2 T, are much cheaper to build than conventional magnets, and can be energized by small laboratory power supplies. The alloy NbTi can be used for fields up to 10 T at 2 K, while Nb_3Sn can reach 12 T at 4·2 K, but the critical current density falls rapidly as the field is increased. Immersion of the coil in liquid helium complicates the design of the magnet, but the cost of refrigeration is much less than that of the power supply for a conventional magnet. Once a superconductive magnet has been cooled and energized, the circuit can be closed by a superconducting link; a 'persistent current mode' is then established in which the current through the coil and link flows in a continuous superconducting circuit, and no external power supply is required to maintain the current, which decays at less than one part in 10^6 in an hour. To alter the field, the superconducting link is driven normal by a small heating strip while the external power supply is reconnected to adjust the current.

Many superconducting alloys have poor mechanical strength, and are fabricated in the form of a composite wire containing about 60 filaments of superconductor in a matrix of a good conductor such as copper or aluminium. The filaments are less than 0·05 mm in diameter to reduce the penetration of magnetic flux into the superconductor. Such penetration must be avoided, since it causes local heating, setting up an instability which can spread through the superconductor and heat it to the point at which it reverts to a normal conductor. The geometry of the coils is designed to maximize the field at the centre, but minimize the stray field in the region of superconducting current.

Hybrid magnets

In a coil where a number of turns are wound one on top of the other, the inner turns are subjected to the largest field. For a given available power, a higher field in the central experimental region can be achieved by replacing the inner turns of a superconductive magnet by a conventional water-cooled magnet. If the two sections of such a 'hybrid magnet' are designed to be used separately as well as in combination, the range of fields available from one set of equipment is greatly increased. Problems arise from the enormous mechanical forces between the two sections; these cannot be rigidly connected since the outer superconductive magnet is at liquid helium temperature and the inner at room temperature. These problems have been overcome and fields of 15–20 T are feasible with hybrid magnets. The optimum design of such equipment presents a complicated choice, where the capital cost of the superconductive alloy and of the power supply must be balanced together with the running costs.

6.6. Measurement of magnetic fields

The classical method of measuring the value of B at any point in air is by means of a 'flip coil' and an instrument for measuring flux. The flip coil consists of a number N of turns of wire wound on a small former of known area A; this is mounted on a handle and the leads to the coil are twisted and brought out through the handle. When the coil is placed with its axis parallel to a field of flux density $\mathbf{B}$, the flux through it is NAB; if the coil is then quickly removed to a point where $\mathbf{B}=0$, the flux change is just NAB and this can be measured either by a ballistic galvanometer or a fluxmeter (see § 5.7). For accurate results the ballistic galvanometer should be standardized as described in § 5.7 using a mutual inductance, with the secondary winding and the flip coil in series with the galvanometer throughout all the measurements so that the total resistance of the circuit remains constant. Flip coils can be made with different values of the product NA (turns×area), so that by choice of the right coil for the field to be measured a suitable deflection can be obtained on the ballistic galvanometer or fluxmeter. As noted in § 5.7, the accuracy obtainable with an average fluxmeter is about 1 per cent, and with a ballistic galvanometer about 10 times greater.

If the coil is rotated rapidly in the field, an alternating voltage is set up which is proportional to NAB times the angular velocity; measurement of this voltage requires a less sensitive instrument than the fluxmeter because the energy available is much greater than with a single throw. In a typical instrument, a coil of 3 mm outer diameter is rotated at 30 Hz, and gives full scale deflection in a field of about 0·05 T; with larger and smaller coils, fields ranging from the earth's field to 10 T can be measured quickly with an accuracy of about 1 per cent.

An absolute method which is capable of giving higher accuracy is the electromagnetic balance of Cotton, in which the force due to the flux density $\mathbf{B}$ on a length of wire carrying a known current is measured directly. This method can only be used for rather strong fields which are uniform over a fair volume. A long rectangular coil is suspended from an analytical balance with the lower end of the coil in the field to be measured, this field being directed horizontally. The long sides of the coil are vertical, so that no force is exerted on them in the vertical direction; they act as leads for the current in the horizontal lower edge of the coil, and the force measured is just that on this lower edge, assuming that the value of $\mathbf{B}$ along the upper edge is negligible. If the lower edge is directed perpendicular to the lines of $\mathbf{B}$, the net vertical force is $F=I\int B\,\mathrm{d}x$, where the integral is measured along the lower edge. Thus the integrated value of the flux density along this edge is determined, and to find the value at any point the flux density distribution must be known. The current is measured with a standard resistance and potentiometer, and the force by the change in the balance reading when the current through the rectangular coil is reversed in direction. With a balance of this type Thomas, Driscoll, and Hipple (1950) were able to measure a field of about 0·5 T with an accuracy of a few parts in 10^5.

For most purposes the most accurate way of measuring a magnetic flux density which is reasonably uniform is to determine the precession frequency $\omega/2\pi$ of a nuclear moment, such as that of the proton, by means of 'nuclear magnetic resonance' (see Chapter 24). This method is based on the equation

$$\omega = |\gamma|\, B,$$

where γ is the magnetogyric ratio (see eqn (6.2)). The frequency is linearly proportional to B, and may readily be determined with high accuracy (see Chapter 22); from the known value of γ this gives an accurate measure of B. For protons in a spherical sample of liquid H_2O, the value of γ is $2{\cdot}6751\times10^8\,\mathrm{s}^{-1}\,\mathrm{T}^{-1}$, corresponding to a frequency $\omega/2\pi$ of 42·576 MHz in a flux density of 1 T.

A convenient but less precise method, which has the advantage that it can be used in less uniform fields, is to measure the voltage in a 'Hall probe'. This contains a small sample of a semiconductor such as n-type indium antimonide or indium arsenide, in which the Hall voltage (see § 12.11) varies almost linearly with the magnetic flux density.

6.7. Measurement of magnetic moment and susceptibility

Most methods of determining the susceptibility of weakly magnetic substances depend on measuring the force on the substance in an inhomogeneous magnetic field. From eqn (4.9), the force on a magnetic dipole **m** has an x component

$$F_x = \mathfrak{m}_x\left(\frac{\partial B_x}{\partial x}\right)+\mathfrak{m}_y\left(\frac{\partial B_y}{\partial x}\right)+\mathfrak{m}_z\left(\frac{\partial B_z}{\partial x}\right). \tag{6.23}$$

Now, if instead of a permanent dipole we have a particle of magnetizable matter of susceptibility χ and volume v, its moment will be

$$\mathbf{m} = \chi v \mathbf{H}$$

and the x component of the force on it is

$$F_x = \chi\mu_0 v\left(H_x\frac{\partial H_x}{\partial x}+H_y\frac{\partial H_y}{\partial x}+H_z\frac{\partial H_z}{\partial x}\right) = \tfrac{1}{2}\chi\mu_0 v\left(\frac{\partial}{\partial x}H^2\right).$$

If the particle has a susceptibility χ_1 and is immersed in a medium (such as the atmosphere) with susceptibility χ_2, then the force on it is

$$F_x = \tfrac{1}{2}(\chi_1 - \chi_2)\mu_0 v \frac{\partial H^2}{\partial x}. \tag{6.24}$$

This can be seen from the fact that any displacement of the particle in the x direction requires an opposite displacement of an equal volume of the surrounding medium.

This equation may also be derived by considering the stored energy. The effect of the presence of the particle of volume v is to increase the stored energy by

$$\begin{aligned} U &= v(\tfrac{1}{2}\mathbf{B}_1 . \mathbf{H} - \tfrac{1}{2}\mathbf{B}_2 . \mathbf{H}) \\ &= \tfrac{1}{2}v\{(1+\chi_1)-(1+\chi_2)\}\mu_0 H^2 \\ &= \tfrac{1}{2}v(\chi_1 - \chi_2)\mu_0 H^2. \end{aligned}$$

From eqn (5.40) the force component is given by $F_x = +(\partial U/\partial x)$, giving the same formula as eqn (6.24). In both these derivations we have assumed that the magnetic field inside the specimen is the same as the value measured before the specimen was introduced. The difference due to the demagnetizing field, of order $M = \chi H$, is negligible in diamagnetic substances, and in many other cases only relative measurements, such as the change with temperature, rather than absolute measurements, are required. If the sample is in a vacuum, as in the apparatus described below, χ_2 is of course zero.

The force equation (6.23) or (6.24) is the basis of a number of methods of determining magnetic moment and susceptibility, but we describe only a modern version of a method due to Faraday. The sample is suspended by a quartz fibre from a vacuum microbalance which can support samples up to a weight of several grams, and measure force-changes from less than 1 μg up to 1 g weight. The sample hangs inside a cylindrical dewar vessel (Fig. 6.14) so that its temperature can be controlled and measured down to 1 K. Surrounding the dewar vessel is a superconductive solenoid giving a uniform vertical flux density of 5 T or more. The force on the sample is provided by a pair of coils carrying equal currents flowing in opposite senses; these give a controlled vertical gradient $\mathrm{d}B/\mathrm{d}x$ of up to $10\,\mathrm{T\,m^{-1}}$, so that $B(\mathrm{d}B/\mathrm{d}x)$ can have values up to about $50\,\mathrm{T^2\,m^{-1}}$. The geometry of the coils is designed to make the first three derivatives of the gradient zero (Problem 6.9), giving the largest possible volume in which the gradient is nearly uniform. A diamagnetic susceptibility of about $-10^{-9}\,\mathrm{kg^{-1}}$ can be measured to one part in 10^4 with a sample of 1 g. The paramagnetic susceptibility of very dilute samples (see Fig. 6.15) and the value of the magnetization in the cooperative magnetic state can be determined.

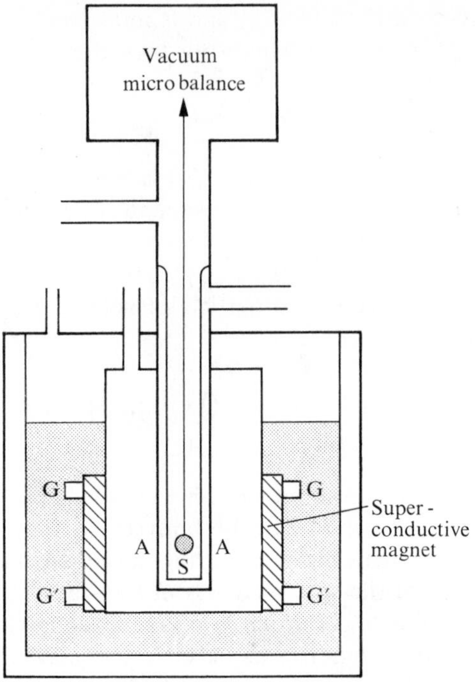

FIG. 6.14. Measurement of magnetic moment by Faraday's method. The force on the samples is measured by a vacuum microbalance. The superconductive magnet (giving a uniform field) and the gradient coils GG, G′G′ are immersed in liquid helium; a cryostat and heater (not shown) surround the sample at AA.

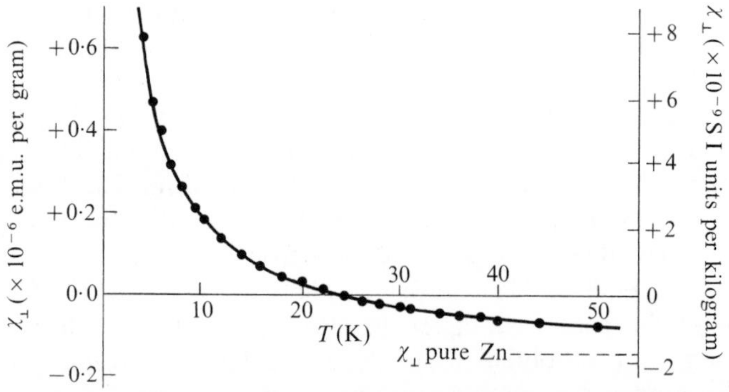

FIG. 6.15. Mass susceptibility (perpendicular to the hexagonal axis) of a single crystal of zinc containing 135 p.p.m. of chromium. At this low concentration the chromium ions are paramagnetic and give a positive susceptibility which follows Curie's law, but above 25 K this is outweighed by the diamagnetic contribution from the host metal. (See Appendix D for connection between e.m.u. and S.I. units) (P. L. Camwell and R. Dupree).

An alternative method of measuring a magnetic moment is to determine the flux change when a sample is moved in and out of a coil, or from one coil to another, in a known magnetic field. Movement of the sample is superior to movement of the coil, because the latter gives a flux change even in the absence of a sample due to residual inhomogeneity in the field. A null method is often used; one way of achieving this with a cylindrical sample is to wind a small coil round the sample and adjust the current through it until the magnetic moment of the coil just cancels that due to the sample, as shown by the zero flux change in a pick-up coil from which the (sample+coil) are suddenly removed. This is a good example of the equivalence of a current circuit and a magnetic shell, as discussed in § 4.2.

A sensitive magnetometer, due to Foner (1959), in which the sample is vibrated at 90 Hz, is shown in Fig. 6.16. The sample S, at the end of a long support which reaches down into a dewar vessel for work at low temperatures, is vibrated vertically by a loudspeaker transducer T fed by alternating current at 90 Hz. The sample is at the midpoint between two

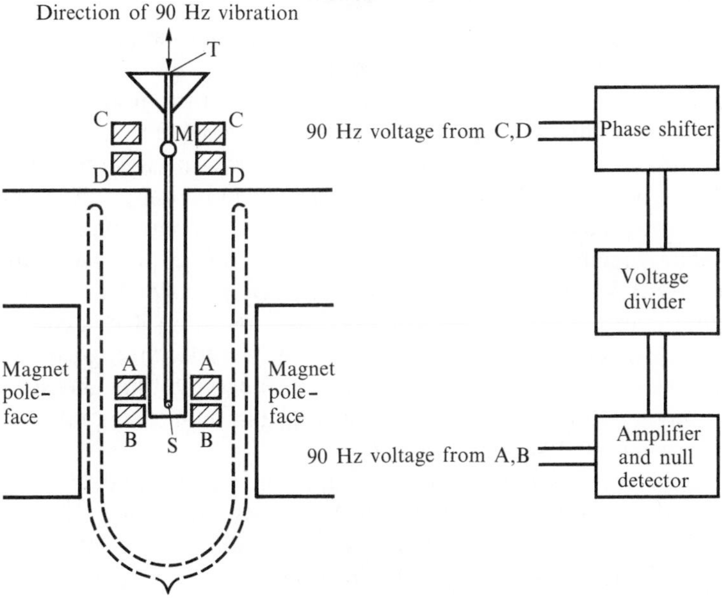

FIG. 6.16. Foner's vibrating magnetometer. The pair of coils A, B (and similarly C, D) are in series but opposing, thus reducing spurious voltages due to magnetic field instability or unwanted mechanical vibration.

S sample, producing 90 Hz voltage in coils A, B,
M permanent magnet producing comparison voltage in coils C, D,
T loudspeaker transducer driving sample S and magnet M in vertical vibration at 90 Hz.

coils AA, BB whose axes are vertical; the vertical component of the flux from the horizontal magnetic moment of the sample threads the top coil and the bottom coil in opposite sense, so that the net flux is zero if the sample is exactly opposite the midpoint. When the sample is displaced vertically during vibration, the net flux through the coils becomes finite and in first approximation is linearly proportional to the displacement of the sample. Thus an e.m.f. alternating at 90 Hz is induced in the coils, which can be balanced against a similar e.m.f. induced by a small permanent magnet M in the coils CC, DD. This gives a null method where the magnetic moment of the sample is read off on the calibrated voltage divider (the phase shifter is required because the two alternating voltages may have a small phase difference—see Chapter 7). The accuracy is about 1 per cent, and changes in magnetic moment of about 10^{-7} A m^2 can be detected in a field of flux density 1 T. Rotation of the whole assembly about a vertical axis enables the magnetic moment to be measured throughout a horizontal plane; permanent moments can be measured as well as induced moments.

6.8. Experimental investigation of the hysteresis curve

In order to determine the hysteresis curve of a substance it is necessary to know the values of B and H inside the substance. For this purpose the most satisfactory shape of the substance is in the form of an anchor ring, or toroid, since then there is no demagnetizing field due to the 'free poles' at the ends of the specimen. If the radius of the ring is large compared with the dimensions of its cross-section, a coil wound uniformly round the ring will produce a uniform field strength H everywhere within the ring. If n_1 is the number of turns per unit length in this (the 'primary' coil), and I the current flowing, then $H = n_1 I$. To measure B, a small secondary coil is wound over the primary at some point on the ring; if this coil has n_2 turns, and the cross-section of the ring is A, the flux through the coil is $n_2 AB$. Changes in the flux through the secondary coil are measured by connecting it to a ballistic galvanometer; the galvanometer is calibrated at the same time as described in § 5.7.

For many purposes sufficient accuracy can be obtained by the very convenient method of displaying the hysteresis curve on a cathode-ray oscilloscope. A small resistance is inserted in the primary of the circuit above, and the voltage across this resistance is amplified and used to deflect the electron beam horizontally by an amount proportional to the current in the primary circuit, and hence to H. A primary current alternating at a low frequency is used, and the changing flux in the secondary circuit produces an alternating voltage which is amplified and used to deflect the electron beam vertically by an amount proportional to B.

The area enclosed by the hysteresis curve is of importance because it represents the work done in taking the material once round the hysteresis curve. Thus in a transformer for alternating current of frequency f, the hysteresis curve is traversed f times per second and power is dissipated which appears as heat in the magnetic material of the core. From eqn (5.45) the work done per unit volume in moving from one point on the hysteresis curve to another is the integral, taken along the curve,

$$W = \int H \, dB. \tag{6.25}$$

It is readily seen that in a complete cycle the value of the integral is just given by the area enclosed by the hysteresis curve. Since this curve is traversed once per cycle, the energy dissipated rises linearly with the frequency of the alternating current.

To minimize losses, a soft magnetic material is required, whose hysteresis loop is as thin as possible; ideally the permeability should be high and constant, with the area of the hysteresis loop zero, so that the B–H curve is a straight line of high slope through the origin. Alloys of iron with a few per cent of silicon approach this ideal more closely than pure iron, and are used in power transformers for supply frequencies. For small transformers and other uses at higher frequencies, more expensive alloys requiring lengthy heat treatment give improved performance. For example, an alloy of about 78 % Ni and 22 % Fe (permalloy), if slowly cooled from 1150 K and then rapidly cooled from 850 K, gives an initial permeability of nearly 10^4 and a maximum permeability of nearly 10^5. Supermalloy, with the composition 80 % Ni, 15 % Fe, 5 % Mo (molybdenum), after heating in very pure hydrogen at over 1500 K, gives initial and maximum permeabilities some 10 times higher. These high permeabilities are accompanied by very low values of the coercive force, $\sim 4\ \mathrm{A\,m^{-1}}$ for permalloy and $0{\cdot}3\ \mathrm{A\,m^{-1}}$ for supermalloy.

Another effect of the alternating magnetic flux in the core is to set up induced voltages and produce power loss through eddy currents. The core is therefore constructed of very thin laminations, insulated from each other, and oriented so that the insulation lies across the path of the eddy current. This is therefore constrained to flow within the lamination, and the losses are reduced (see Problem 8.12). In this respect the special magnetic alloys mentioned above have the further advantage of a high electrical resistivity. At radio-frequencies (10^5–10^8 Hz) the use of magnetic cores also reduces the size of inductors and improves the coupling between the coils of transformers; such cores are made from magnetic oxides ('ferrites') whose resistivity may be as high as $10^8\ \Omega\,\mathrm{m}$ at room temperature.

6.9. Terrestrial magnetism

To a good approximation the magnetic flux density at the surface of and outside the earth is the same as that of a magnetic dipole at the centre of the earth. More detailed measurements show that the magnetic potential must be analysed using a series of spherical harmonics, of which the dipole term is much the largest. Leaving aside localized distortions due to iron-bearing minerals in the earth's crust, one finds a dipole term which has decreased in magnitude by about 5 per cent in the last hundred years, together with quadrupolar and higher terms which have strong and fairly rapid secular variations with lifetimes less than a hundred years. The field at any point also contains daily variations which are irregular and unpredictable; these are caused by currents in the ionosphere due to solar and lunar perturbations. Charged particles approaching the earth, including the stream of charged particles from the sun known as the 'solar wind', are deviated by the earth's magnetic field, and give rise to many fluctuating phenomena in the upper atmosphere, including such dramatic effects as the aurorae.

Archaeomagnetic data have revealed the unexpected result that the equivalent magnetic dipole moment of the earth had a maximum, about 50 per cent greater than the present value, around the beginning of the Christian era, with a minimum at about 4000 B.C. preceded by an earlier maximum at about 6000 B.C. Much more dramatic changes, including many reversals of the earth's field, are revealed by palaeomagnetic studies (see, for example, Aitken 1970; McElhinney 1973; Stacey and Banerjee 1974).

These 'sudden' changes, in which the geomagnetic field is reversed in a time of 1000–10 000 years, are difficult to understand except on a 'dynamo theory', in which electric currents flow round the core of the earth. This core, with a radius of (3473±4) km, is assumed to be liquid, since no transverse seismic waves are transmitted through it, and to consist mainly of liquid silicates of iron, magnesium, and calcium at high temperatures and pressures. The electric currents are associated with convective heat currents caused by radioactive or chemical sources; the mathematics of the process (energy source → kinetic energy of fluid → electrical energy) has been studied by Elsasser, Bullard, and others, and it seems probable that electric currents can be maintained in this way. For detailed accounts reference should be made to Chapman and Bartels (1940) and Elsasser (1950, 1955–6).

References

AITKEN, M. J. (1970). *Rep. Prog. Phys.* **33,** 941.
BITTER, F. (1936). *Rev. scient. Instrum.* **7,** 479; **7,** 482.
— (1940). *Rev. scient. Instrum.* **11,** 373.

BITTER, F. (1962). *Proceedings of the International Conference on high magnetic fields*, p. 85, M.I.T. Press and Wiley, New York.
CHAPMAN, S. and BARTELS, J. (1940). *Geomagnetism*. Clarendon Press, Oxford.
ELSASSER, W. M. (1950). *Rev. mod. Phys.* **22,** 1.
— (1955–6). *Am. J. Phys.* **23,** 590; **24,** 85.
FONER, S. (1959). *Rev. mod. Phys.* **30,** 548.
MCELHINNEY, M. W. (1973). *Palaeomagnetism and plate tectonics*. Cambridge University Press.
MYERS, W. R. (1952). *Rev. mod. Phys.* **24,** 15.
STACEY, F. D. and BANERJEE, S. K. (1974). *The physical principles of rock magnetism*. Elsevier, Amsterdam.
THOMAS, H. A., DRISCOLL, R. L., and HIPPLE, J. A. (1950). *Phys. Rev.* **78,** 787.

Problems

6.1. The susceptibility of a mole of helium gas is $-2{\cdot}4\times10^{-11}$. Show that this corresponds to a value for the mean-square radius of each electronic orbit in the helium atom of $1{\cdot}22a_0^2$, where $a_0=0{\cdot}528\times10^{-10}$ m is the radius of the first Bohr orbit in the hydrogen atom.

6.2. If the mutual repulsion of the two electrons in the helium atom is neglected, the wavefunction of each electron in the ground state is

$$\psi=(Z^3/\pi a_0^3)^{\frac{1}{2}}\exp(-Zr/a_0),$$

where $-e\psi^2$ is the density of electronic charge at a distance r from the nucleus, and Ze is the effective charge of the helium nucleus. Show that this wavefunction leads to a value for the mean-square radius of each electronic orbit of

$$\overline{r^2}=3a_0^2/Z^2.$$

Verify that agreement with the value of $1{\cdot}22a_0^2$ in the previous problem is obtained if we take Z as about 1·6 (we should expect it to be less than 2 because each electron partially shields the other from the field of the nucleus).

6.3. When 23·15 g of $NiCl_2$ are dissolved in 100 g water, the density of the solution is 1255 kg m^{-3}. Show that the maximum height of a column of the solution which can be supported by magnetic force when one surface of the column is in a uniform flux density of 1 T is 3·0 mm. (Susceptibilities of 1 kg of $NiCl_2$ and water are $0{\cdot}438\times10^{-6}$ and $-0{\cdot}0090\times10^{-6}$ respectively; volume susceptibility of air $=+0{\cdot}4\times10^{-6}$.)

6.4. The susceptibility of a mole of $NiK_2(SO_4)_2.6H_2O$ is found to be $1{\cdot}6\times10^{-5}\,T^{-1}$. Assuming that the diamagnetic contribution is negligible, and that the only paramagnetic contribution comes from the Ni^{2+} ion, calculate the size of the permanent dipole moment on each Ni^{2+} ion.

(*Answer:* $3{\cdot}0\times10^{-23}$ A m^2.)

6.5. An iron anchor ring of large mean radius R and uniform cross-section has a gap of thickness d cut in it. It is wound with a single-layer coil. Over the range under consideration the permeability of the iron is $(1+a/H)$, where H is the field strength in the iron. Show that, if $d \ll R$, 4 times as much power is required to maintain a field strength $3a$ in the gap as is required for a field strength $2a$.

6.6. It was shown by Rayleigh that at low values of the magnetic flux density the hysteresis loop with tips at B_0, H_0 and $-B_0$, $-H_0$ is described by the equations

$$B = \mu H + \tfrac{1}{2}a(H_0^2 - H^2) \quad \text{(upper half of loop)}$$

and

$$B = \mu H - \tfrac{1}{2}a(H_0^2 - H^2) \quad \text{(lower half of loop)},$$

where $\mu = B_0/H_0$. Show that the energy loss per cycle represented by the area of the loop is

$$W = \tfrac{4}{3}aH_0^3$$

in each unit volume of the substance.

This relation is valid only for low values of B_0 (in iron below about 0·05 T). At high values W varies approximately as $B_0^{1\cdot 6}$, an empirical law due to Steinmetz.

6.7. Deduce eqn (6.20) by the use of equivalent magnetic shells and integration of eqn (4.66), or from integration of the magnetic flux density due to the 'free poles' (polarization charges) on the surfaces of the cones.

6.8. An air-cored solenoid of length $2l$ is constructed from n equally spaced Bitter pancakes each of thickness t, inner radius a, outer radius b, and resistivity ρ. If n is large, and P is the power supplied, show that the field strength H at the centre is approximately

$$H = \left(\frac{Ptn}{8\pi l^2 \rho \ln(b/a)}\right)^{\frac{1}{2}} \ln \frac{b\{l + (l^2 + a^2)^{\frac{1}{2}}\}}{a\{l + (l^2 + b^2)^{\frac{1}{2}}\}}.$$

Note that this is of the form $(P\lambda/\rho r)^{\frac{1}{2}}$ times a factor depending on the shape of the solenoid (as given in § 6.5), since $\lambda = nt/2l$.

6.9. Two circular coaxial coils, each carrying N turns of radius a, carry equal currents but in opposite senses. Show that at the midpoint on the axis between the coils, the value of H_z and the derivatives $\mathrm{d}^2H_z/\mathrm{d}z^2$, $\mathrm{d}^4H_z/\mathrm{d}z^4$ along the axis are zero, and that $\mathrm{d}^3H_z/\mathrm{d}z^3 = 0$ provided that the distance between the coils is $\sqrt{3}a$.

6.10. The iron core of a transformer has laminations of thickness a and resistivity ρ; it is subject to a sinusoidally varying flux density with a maximum value B and frequency f. Show that the power loss per unit volume (neglecting skin effects, see Problem 8.12) due to eddy currents is approximately $\pi^2B^2a^2f^2/6\rho$. If $a = 0{\cdot}1$ mm, $\rho = 4\times10^{-7}\ \Omega$ m, $B = 0{\cdot}5$ T, and $f = 50$ Hz, show that the power loss is approximately 26 W m^{-3}. Compare this with the power loss through hysteresis, if the energy dissipated per cycle for permalloy at this flux density is 20 J m^{-3}.

(*Answer:* Hysteresis loss = 10^3 W m^{-3}.)

7. Alternating current theory

7.1. Forced oscillations

IN § 5.3 we considered the transient currents which flow when a capacitor, initially charged, is allowed to discharge through a circuit containing both inductance and resistance, and found that an oscillatory current of decaying amplitude flowed through the circuit provided that the resistance in the circuit was not too high. The theory of such transients is due to Lord Kelvin, and its correctness was verified by early experimenters. With the invention of the dynamo and, later, the electronic vacuum tube, it became possible to produce continuous alternating currents whose frequency of oscillation may be anything up to about 10^{11} Hz. In the simplest case the form of the current is that of a simple sine wave, and may be written as

$$I = I_0 \cos \omega t,$$

where I is the value of the current at time t. The maximum value of I is I_0, known as the 'amplitude' of the current, and the frequency of alternation is $f = \omega/2\pi$ Hz. The current generated by a dynamo or other device may or may not have a simple sinusoidal waveform, but whatever the actual waveform it may be resolved by Fourier analysis into a sum of sine and cosine terms whose frequencies are integral multiples of the fundamental frequency. This frequency is given by the inverse of the period between instants at which the whole waveform is repeated. Since the behaviour of a circuit is in general different at different frequencies, it is necessary to consider each component of such a Fourier series separately, and in the theory that follows we shall assume that the waveform is sinusoidal, varying at one frequency only. Except in non-linear circuits where the behaviour of a circuit element depends on the size of the current or voltage applied to it (that is, elements in which the amplitudes of current and voltage are not linearly proportional to one another) any non-sinusoidal fluctuations may be resolved into their Fourier components, and the required solution is simply a sum of such components.

In the circuit of Fig. 7.1, a voltage $V_0 \cos \omega t$ is applied to an inductance, a resistance, and a capacitance in series. If I is the current flowing at any instant, the e.m.f. set up in the inductance is $-L(\mathrm{d}I/\mathrm{d}t)$, and the voltage drop across the capacitance is Q/C, where Q is the charge on the

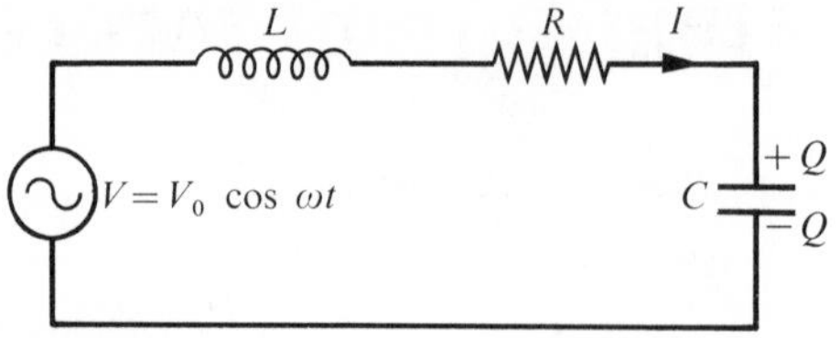

FIG. 7.1. Forced oscillations in a circuit containing L, C, R.

capacitor. We have therefore for the circuit

$$V_0 \cos \omega t - L\frac{dI}{dt} - \frac{Q}{C} = RI$$

or

$$L\frac{dI}{dt} + RI + \frac{Q}{C} = V_0 \cos \omega t. \tag{7.1}$$

Now the rate of increase of the charge on the capacitor is $dQ/dt = I$, the current flowing, and hence by differentiation we have

$$L\frac{d^2I}{dt^2} + R\frac{dI}{dt} + \frac{I}{C} = \frac{dV}{dt} = -\omega V_0 \sin \omega t. \tag{7.2}$$

This is a differential equation whose solution consists of two parts. The first of these, known as the 'complementary function', is found by solving the equation obtained by setting the right-hand side equal to zero; that is, it is a solution of eqn (5.27), and hence is of the form given by eqn (5.28). This solution represents a transient flow of current produced by the act of applying the e.m.f. $V_0 \cos \omega t$, it being assumed that this starts to act at the instant $t = 0$. In all practical applications this transient current decays rapidly in amplitude, owing to the exponential term $e^{-tR/2L}$ and becomes negligible within a few seconds or less of the circuit being closed. If conditions are such that the transient current is oscillatory, its frequency is the natural frequency determined by the values of L, C, and R, and not that of the applied e.m.f.

The second part of the solution is known as the 'particular integral', and for eqn (7.2) it may be written as

$$I = (V_0/Z)\cos(\omega t - \phi), \tag{7.3}$$

where

$$Z = \{R^2 + (\omega L - 1/\omega C)^2\}^{\frac{1}{2}} \tag{7.4}$$

and the phase angle ϕ is given by

$$\tan \phi = \frac{(\omega L - 1/\omega C)}{R}. \tag{7.5}$$

This is also known as the 'steady-state' solution, since it gives the current flow at any time after the transient current has become negligible. The frequency of the current is the same as that of the applied e.m.f., so that the circuit is in 'forced' oscillation. In general the phase of the current is different from that of the applied voltage, except when $\omega L - 1/\omega C = 0$. This occurs when

$$\omega = \omega_0 = 1/(LC)^{\frac{1}{2}}, \tag{7.6}$$

and hence is the same as the angular frequency of natural oscillation of the circuit in the absence of any damping resistance. The circuit is then said to be in 'resonance', and the amplitude of the current is, by eqns (7.3) and (7.4), a maximum.

The quantity Z in eqn (7.4) is called the impedance of the circuit, and at resonance the value of Z is just equal to R, the total resistance in the circuit. At other frequencies the value of Z is related to the quantities R, L, and C, but is not given by the simple additive relation that holds (§ 3.4) for resistances in series. The reason for this is that the voltages across the different elements are not in phase, and the total voltage amplitude is therefore not just the sum of the individual amplitudes. In the following sections we shall see how this difficulty can be overcome by the introduction of complex numbers to represent the impedances. The use of such complex impedances enables us to apply Kirchhoff's laws (§ 3.4) to alternating current (a.c.) networks, and we can find the steady-state values of the current and voltage in any branch without having to solve a differential equation.

An important consideration in a.c. circuits is the rate of doing work. At any instant the rate at which work is done by the generator in Fig. 7.1 is

$$\begin{aligned} \mathrm{d}W/\mathrm{d}t &= VI = (V_0 \cos \omega t)\times(V_0/Z)\cos(\omega t - \phi) \\ &= (V_0^2/Z)(\cos^2\omega t \cos\phi + \cos\omega t \sin\omega t \sin\phi). \end{aligned}$$

To find the mean rate $\mathrm{d}\bar{W}/\mathrm{d}t$ of doing work, this expression must be averaged over one or more periods of oscillation. Now the mean value of $\cos^2\omega t$ averaged in this way is just $\frac{1}{2}$, while that of $\cos\omega t \sin\omega t$ is 0. Hence the mean power drawn from the generator is

$$\mathrm{d}\bar{W}/\mathrm{d}t = \tfrac{1}{2}(V_0^2/Z)\cos\phi = \tfrac{1}{2}V_0 I_0 \cos\phi = \tfrac{1}{2}I_0^2 Z \cos\phi, \tag{7.7}$$

where $I_0 = V_0/Z$ is the amplitude of the current given by eqn (7.3). From eqn (7.5) it is readily shown that $\cos\phi = R/Z$, and the expression for the mean power can therefore be written

$$\mathrm{d}\bar{W}/\mathrm{d}t = \tfrac{1}{2}I_0^2 R. \tag{7.8}$$

Now the mean rate at which power is dissipated in the resistance of the circuit is the average value of $RI_0^2\cos^2(\omega t - \phi)$ taken over a complete period, and this is just equal to $\frac{1}{2}RI_0^2$. Hence all the power delivered by the generator, averaged over a period, is dissipated in the resistance of the circuit. The term in $\cos \omega t \sin \omega t$ in the expression for VI represents work done by the generator in increasing the energy stored in the inductance and capacitance; since the product $\cos \omega t \sin \omega t$ is as often negative as positive, this work is returned to the generator in other parts of the cycle and no mean power is drawn from the generator for this purpose in the steady state. Of course, power was drawn from the generator initially to provide the stored energy, and this is represented by the transient current; when this has decayed, the mean stored energy remains constant and no further work is done by the generator on the average except to supply that dissipated in the resistance.

Since the rate at which power is dissipated is proportional to the square of the current, it is convenient to specify the root-mean-square (r.m.s.) value $\tilde{I}$ of the current, defined by the fact that

$$(\tilde{I})^2 = \langle I^2 \rangle,$$

where the average of the square of the current is taken over a whole period. The r.m.s. value of the voltage $\tilde{V}$ may be defined in a similar way, $(\tilde{V})^2 = \langle V^2 \rangle$. If the waveform is sinusoidal, the r.m.s. value is just $(1/\sqrt{2})$ times the amplitude, and we can write eqn (7.7) as

$$\mathrm{d}\bar{W}/\mathrm{d}t = (\tilde{V}^2/Z)\cos\phi = \tilde{V}\tilde{I}\cos\phi = \tilde{I}^2 Z\cos\phi \tag{7.9}$$

and the ratio $(\mathrm{d}\bar{W}/\mathrm{d}t)/\tilde{V}\tilde{I} = \cos\phi$ is called the 'power factor' of the circuit. It represents the fraction of the product $\tilde{V}\tilde{I}$ which is dissipated as Joule heat. If the circuit behaves as a pure resistance, as occurs when the resonance condition (7.6) is fulfilled in the circuit of Fig. 7.1, the power factor is unity, while if the circuit contains no resistance the power factor is zero.

The general practice in a.c. circuits is to specify the r.m.s. values of the current and voltage, and it should be understood that any values quoted are r.m.s. values unless the contrary is specifically stated.

7.2. Use of vectors and complex numbers

The values of the voltage across the individual components of the circuit in Fig. 7.1 will now be considered in more detail. For this purpose we assume that a current $I = I_0 \cos \omega t$ flows through them all in series, as in Fig. 7.2. The voltages across the three circuit elements are:

$$\text{across } R: \quad V = IR = RI_0 \cos \omega t$$

$$\text{across } L: \quad V = L(\mathrm{d}I/\mathrm{d}t) = -\omega L I_0 \sin \omega t = \omega L I_0 \cos(\omega t + \tfrac{1}{2}\pi)$$

$$\text{across } C: \quad V = Q/C = (1/\omega C) I_0 \sin \omega t = (1/\omega C) I_0 \cos(\omega t - \tfrac{1}{2}\pi)$$

where in the last case we have used the relation $Q = \int I\,\mathrm{d}t$. It will be seen

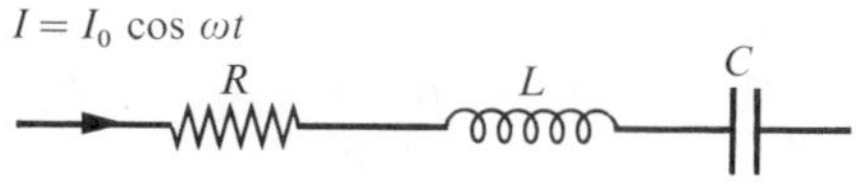

FIG. 7.2. Current I flowing through R, L, C in series.

that the voltage across the resistance is in phase with the current, the voltage across the inductance leads the current in phase by 90°, while that across the capacitance lags behind by 90°. We may represent these voltages by vectors such that, if the voltage across the resistance is represented by a vector drawn parallel to the x-axis, that across the inductance is represented by a vector parallel to the y-axis, and that across the capacitance by a vector parallel to the latter but in the opposite sense. The lengths of the vectors are proportional to R, ωL, and $1/\omega C$, respectively.

If we require the voltage across two of the elements, say the resistance and inductance, it may be found by adding the two individual voltage vectors together, as in Fig. 7.3(a). For the total voltage will be

$$V = IR + L(\mathrm{d}I/\mathrm{d}t) = RI_0 \cos \omega t - \omega L I_0 \sin \omega t$$
$$= I_0(R^2+\omega^2L^2)^{\frac{1}{2}} \cos(\omega t+\phi) = I_0 Z \cos(\omega t+\phi), \qquad (7.10)$$

where $\tan \phi = \omega L/R$. From Fig. 7.3(a) it will be seen that $(R^2+\omega^2L^2)^{\frac{1}{2}}$ is just the length of the hypotenuse of the triangle, while the phase angle ϕ is just the angle between the vectors representing R and $(R^2+\omega^2L^2)^{\frac{1}{2}}$. Thus the magnitude of the total voltage is represented in amplitude by the hypotenuse, and its phase relative to the current is given by the angle through which this vector is rotated with respect to vector representing R. Similarly it can be shown that for a resistance and capacitance in series the total voltage may be found by adding the vectors for R and $-(1/\omega C)$ as in Fig. 7.3(b), while the case of a resistance R, inductance L, and capacitance C all in series is represented by Fig. 7.3(c). Here the amplitude of the resultant vector is $\{R^2+(\omega L-1/\omega C)^2\}^{\frac{1}{2}}$, which is just the value of Z (eqn (7.4)) and the voltage leads the current by the phase

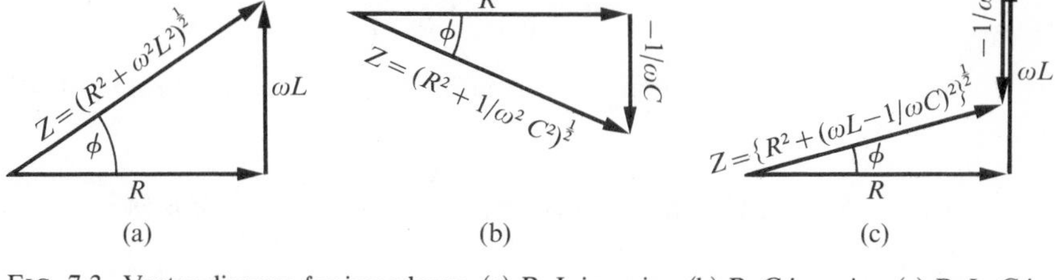

FIG. 7.3. Vector diagram for impedance. (a) R, L in series; (b) R, C in series; (c) R, L, C in series. The voltage vectors are the same as the impedance vectors if the current is represented by a unit vector parallel to the direction of R.

angle ϕ, where

$$\tan\phi = \frac{(\omega L - 1/\omega C)}{R},$$

as in eqn (7.5). The term 'impedance' has already been introduced for Z, which represents the ratio of the amplitude of voltage to current for the whole circuit. The quantities ωL and $1/\omega C$ associated with inductance and capacitance respectively are known as reactances, and are usually denoted by the symbol X. Thus the total reactance of the circuit of Fig. 7.2 is $X = \omega L - 1/\omega C$, and the impedance is given by the vector sum of R and X, represented by mutually perpendicular vectors. Thus $Z = (R^2 + X^2)^{\frac{1}{2}}$, and the phase angle is given by $\tan\phi = X/R$.

The vector representation of the impedance is similar to the representation of a complex number on the Argand diagram. Using the relation

$$\exp \mathrm{j}\omega t = \cos\omega t + \mathrm{j}\sin\omega t,$$

where $\mathrm{j}^2 = -1$, we may replace our cosine and sine functions by complex exponentials on the understanding that we are interested only in the real or imaginary parts respectively. Then we may write the current $I_0\cos\omega t$ as $I_0\,\mathrm{Re}(\exp \mathrm{j}\omega t)$, and quantities such as $\mathrm{d}I/\mathrm{d}t$ or $Q = \int I\,\mathrm{d}t$ become

$$\frac{\mathrm{d}I}{\mathrm{d}t} = I_0\left(\frac{\mathrm{d}}{\mathrm{d}t}\right)\cos\omega t = I_0\,\mathrm{Re}\left(\frac{\mathrm{d}}{\mathrm{d}t}\exp \mathrm{j}\omega t\right) = I_0\,\mathrm{Re}(\mathrm{j}\omega\exp \mathrm{j}\omega t)$$

and

$$Q = I_0\int\cos\omega t\,\mathrm{d}t = I_0\,\mathrm{Re}\left(\int\exp \mathrm{j}\omega t\,\mathrm{d}t\right) = I_0\,\mathrm{Re}\left(\frac{1}{\mathrm{j}\omega}\exp \mathrm{j}\omega t\right)$$

respectively. Hence the total voltage across the circuit of Fig. 7.2 is

$$\begin{aligned} V &= RI + L(\mathrm{d}I/\mathrm{d}t) + (1/C)\int I\,\mathrm{d}t \\ &= \left(R\exp \mathrm{j}\omega t + \mathrm{j}\omega L\exp \mathrm{j}\omega t + \frac{1}{\mathrm{j}\omega C}\exp \mathrm{j}\omega t\right)I_0 \\ &= \left(R + \mathrm{j}\omega L + \frac{1}{\mathrm{j}\omega C}\right)I_0\exp \mathrm{j}\omega t \\ &= \left(R + \mathrm{j}\omega L + \frac{1}{\mathrm{j}\omega C}\right)I. \end{aligned} \qquad (7.11)$$

If the current is $I_0\cos\omega t$, then to find the voltage we take the real part of (7.11). Thus

$$V = I_0\,\mathrm{Re}\left(R\cos\omega t + \mathrm{j}R\sin\omega t + \mathrm{j}\omega L\cos\omega t - \omega L\sin\omega t - \frac{\mathrm{j}}{\omega C}\cos\omega t + \right.$$
$$\left. + \frac{1}{\omega C}\sin\omega t\right)$$

$$= I_0\left(R\cos\omega t - \omega L\sin\omega t + \frac{1}{\omega C}\sin\omega t\right),$$

where the three component terms are just the voltages across the three circuit elements derived at the beginning of this section. Similarly, if the current had been $I_0 \sin \omega t = I_0 \operatorname{Im}(\exp j\omega t)$, the voltage would be found by taking the imaginary part of (7.11), giving

$$V = I_0\left(R \sin \omega t + \omega L \cos \omega t - \frac{1}{\omega C} \cos \omega t\right).$$

The importance of eqn (7.11) lies in the fact that it shows we may represent the inductance and capacitance by impedance operators $j\omega L$ and $1/j\omega C$ respectively, and these operators may be added to one another when the elements are in series in a similar way to resistances. The phase of the voltage, which leads the current by $\frac{1}{2}\pi$ in the inductance and lags by $\frac{1}{2}\pi$ in the capacitance, is taken care of by the presence of j. If, as is usual in the Argand diagram, real quantities are represented by vectors drawn parallel to the x-axis, and imaginary quantities by vectors parallel to the y-axis, then the complex impedance operator is as shown in our vector diagram (Fig. 7.3(c)). The circuit impedance is given by the modulus of the complex impedance operator, and the phase angle by its argument. If we write the impedance operator as

$$\mathbf{Z} = R + j\omega L + \frac{1}{j\omega C} = R + jX = Z \exp j\phi$$

then $Z = (R^2 + X^2)^{\frac{1}{2}}$, and $\tan \phi = X/R$, as before. Then

$$V = \mathbf{Z}I = Z \exp j\phi \times I_0 \exp j\omega t = ZI_0 \exp j(\omega t + \phi).$$

If $I = I_0 \cos \omega t$, the real part of this gives $V = ZI_0 \cos(\omega t + \phi)$, as in eqn (7.10). If, on the other hand, the voltage is given as $V_0 \cos \omega t$, as in § 7.1, the current is found by taking the real part of

$$I = V/\mathbf{Z} = V_0 \exp j\omega t/(Z \exp j\phi) = (V_0/Z) \exp j(\omega t - \phi),$$

which gives

$$I = (V_0/Z) \cos(\omega t - \phi),$$

as in eqn (7.3).

From an extension of the treatment given above it may readily be shown that the impedance operator for a circuit consisting of a number of impedances $\mathbf{Z}_1, \mathbf{Z}_2, \ldots, \mathbf{Z}_n$ in series is

$$\mathbf{Z} = \mathbf{Z}_1 + \mathbf{Z}_2 + \ldots + \mathbf{Z}_n. \tag{7.12}$$

The corresponding formula for a number of impedances in parallel is found by noting that the voltage across each is the same. The current through the impedance $\mathbf{Z}_k$ is $V/\mathbf{Z}_k$, and the total current is the sum of a number of similar terms. Hence the net impedance is given by

$$\frac{I}{V} = \frac{1}{\mathbf{Z}} = \frac{1}{\mathbf{Z}_1} + \frac{1}{\mathbf{Z}_2} + \ldots + \frac{1}{\mathbf{Z}_n}. \tag{7.13}$$

When a number of elements are in parallel it is generally convenient to work in terms of the reciprocal of the impedance, known as the admittance $\mathbf{Y}$. Thus $I = V/\mathbf{Z} = \mathbf{Y}V$, and eqn (7.13) may be written as

$$\mathbf{Y} = \mathbf{Y}_1 + \mathbf{Y}_2 + \ldots + \mathbf{Y}_n. \tag{7.14}$$

In general $\mathbf{Y}$ is complex; thus $\mathbf{Y} = G + \mathrm{j}S$, where G is called the conductance and S the susceptance. For the three simple circuit elements we have conductance $G = 1/R$; susceptance of an inductance is $S = (\mathrm{j}\omega L)^{-1} = -\mathrm{j}/\omega L$; susceptance of a capacitance is $S = \mathrm{j}\omega C$; but note that for a circuit containing both resistance and reactance these simple reciprocal relations do not hold. For

$$\mathbf{Y} = 1/\mathbf{Z} = 1/(R + \mathrm{j}X) = (R - \mathrm{j}X)/(R^2 + X^2),$$

so that $Y = |\mathbf{Y}| = |\mathbf{Z}^{-1}| = Z^{-1}$, but

$$\left.\begin{aligned} G &= \frac{R}{(R^2+X^2)} = Y\cos\phi, \qquad S = -\frac{X}{(R^2+X^2)} = -Y\sin\phi \\ &\text{and similarly} \\ R &= \frac{G}{(G^2+S^2)} = Z\cos\phi, \qquad X = -\frac{S}{(G^2+S^2)} = Z\sin\phi \end{aligned}\right\}. \tag{7.15}$$

In order to calculate the power consumed in a circuit we must use the relation $dW/dt = \mathrm{Re}(V) \times \mathrm{Re}(I)$; note that this is not the same as $\mathrm{Re}(VI)$, for $\mathrm{Re}\{V_0 \exp \mathrm{j}\omega t\} \times \mathrm{Re}\{I_0 \exp \mathrm{j}(\omega t - \phi)\} = V_0 \cos \omega t \times I_0 \cos(\omega t - \phi)$, which does not equal $\mathrm{Re}\{V_0 I_0 \exp \mathrm{j}(2\omega t - \phi)\} = V_0 I_0 \cos(2\omega t - \phi)$. However, the mean energy $\mathrm{d}\bar{W}/\mathrm{d}t$ dissipated is given by any of the relations (where $\tilde{V}$ and $\tilde{I}$ are the r.m.s. values)

$$\begin{aligned} \mathrm{d}\bar{W}/\mathrm{d}t &= \tilde{V}^2\,\mathrm{Re}(\mathbf{Y}) = \tilde{V}^2 G = \tilde{V}^2 R/(R^2+X^2) \\ &= \tilde{I}^2\,\mathrm{Re}(\mathbf{Z}) = \tilde{I}^2 R = \tilde{I}^2 G/(G^2+S^2), \end{aligned} \tag{7.16}$$

as can be verified by comparison with eqn (7.9). An alternative relation is given in Problem 7.13.

The unit of reactance and impedance is the same as that of resistance, the ohm; when ω is expressed in radians per second, L in henries, and C in farads, the corresponding reactances are in ohms. Similarly, the unit of admittance and susceptance is the same as that of conductance, the reciprocal ohm (or siemen).

We may illustrate the use of admittance by considering the case of a lossy capacitor, represented by a pure capacitance C shunted by a resistance R, as in Fig. 7.4(a). The admittance operator is

$$\mathbf{Y} = (1/R) + \mathrm{j}\omega C = G + \mathrm{j}\omega C.$$

This may be represented by the admittance diagram shown in Fig. 7.4(b). The phase angle of the admittance is given by $\tan\psi = \omega C/G = \omega CR$; by

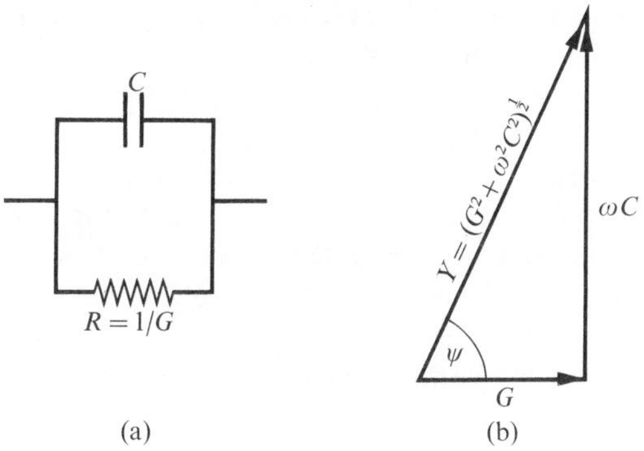

FIG. 7.4. (a) Imperfect capacitor with conductance G, represented by capacitance C shunted by resistance $R = 1/G$; (b) corresponding admittance diagram.

use of eqn (7.15) or by plotting the corresponding impedance diagram it is readily shown that $\tan\psi = -\tan\phi$, so that $\psi = -\phi$. Hence the power factor $\cos\phi = \cos\psi = (1+\omega^2C^2R^2)^{-\frac{1}{2}}$. For a perfect capacitance, as for a perfect inductance, the power factor is zero; in practice this is not quite true, since at radio-frequencies power is lost through eddy currents in the plates and imperfections in the dielectric. The latter are caused by hysteresis effects which cause the vector **D** to lag behind the vector **E** in the dielectric as in the corresponding case of **B** and **H** for a ferromagnetic substance. The quality of a dielectric is expressed in terms of its loss tangent' $\tan\delta$, where

$$\delta = \tfrac{1}{2}\pi - \psi = \tfrac{1}{2}\pi + \phi;$$

hence

$$\tan\delta = \cot\psi = -\cot\phi = G/\omega C = (\omega CR)^{-1}.$$

In an oscillating electric field the hysteresis loop is traversed once per cycle, so that the loss conductance G is proportional to the frequency. The value of $\tan\delta$ should therefore be independent of frequency, and this is generally true except for strongly polar dielectrics (see Chapter 10); for a good dielectric, such as quartz, the value of $\tan\delta$ at room temperature is about 10^{-4}.

7.3. Tuned circuits

The circuit of Fig. 7.1 consisting of an inductance, resistance, and capacitance in series is known as a 'series resonant' circuit. The impedance operator is

$$\mathbf{Z} = R + \mathrm{j}(\omega L - 1/\omega C), \tag{7.17}$$

and, as pointed out in § 7.1, the modulus of this has a minimum value if the frequency is adjusted to make $\omega L = 1/\omega C$. This is the resonant frequency of the circuit, given by eqn (7.6)

$$\omega_0 = 1/(LC)^{\frac{1}{2}}.$$

At this frequency the current through the circuit is a maximum and is in phase with the applied voltage, since $\mathbf{Z}$ is real; the magnitude of the current is V/R. The voltage across the resistance R is thus equal to the voltage across the whole circuit, and the voltages across the inductance and capacitance are therefore equal and exactly 180° out of phase with one another, making the voltage across the two zero. The voltage across the capacitance alone at resonance is

$$\mathrm{V_C} = \mathrm{I}/\mathrm{j}\omega_0 C = \mathrm{V}/\mathrm{j}\omega_0 CR,$$

and hence, using (7.6),

$$\left|\frac{\mathrm{V_C}}{\mathrm{V}}\right| = (\omega_0 CR)^{-1} = \frac{1}{R}\left(\frac{L}{C}\right)^{\frac{1}{2}} = \frac{\omega_0 L}{R} = Q_F \tag{7.18}$$

where Q_F is the 'quality factor' of the circuit as defined in § 5.3. It is also known as the 'circuit magnification factor', since from eqn (7.18) we see that it equals the ratio of the voltage across the capacitor to the voltage across the whole circuit. Thus the tuned circuit acts as a transformer; since the current through the capacitor is a maximum at resonance, the ratio of the voltages $\mathrm{V_C}/V$ is also a maximum at this point. The voltage across the inductance L is equal to that across the inductance at resonance, but in general R is associated with L and only the voltage across the combination $R + \mathrm{j}\omega L$ can be measured in practice.

The quality factor Q_F is very useful both in theory and in practice. Eqn (7.17) can be written as

$$\mathbf{Z} = R\left\{1 + \mathrm{j}Q_F\left(\frac{\omega}{\omega_0} - \frac{\omega_0}{\omega}\right)\right\}, \tag{7.19}$$

and near resonance, on making the substitution $\omega = \omega_0 + \delta\omega$, we have

$$\mathbf{Z} = R\{1 + 2\mathrm{j}Q_F(\delta\omega/\omega_0)\}, \tag{7.20}$$

which can also be written as

$$\mathbf{Z} = R + 2\mathrm{j}\,\delta\omega L. \tag{7.21}$$

The amplitude of the impedance is

$$Z = |\mathbf{Z}| = R\{1 + 4Q_F^2(\delta\omega/\omega_0)^2\}^{\frac{1}{2}} = \{R^2 + 4(\delta\omega L)^2\}^{\frac{1}{2}}. \tag{7.22}$$

The reactance passes through zero at $\delta\omega = 0$ or $\omega = \omega_0$, but it is varying rapidly if Q_F is large. Eqn (7.22) shows that if a given current flows through the circuit, the voltage across the circuit rises by a factor $\sqrt{2}$ when the frequency deviates from the resonant frequency by a fraction $\delta\omega/\omega_0 = \pm 1/(2Q_F)$. Similarly, if a given voltage is applied to the circuit, the current through it falls to $1/\sqrt{2}$ of its maximum value when the frequency deviates by this amount; the power dissipated in the circuit falls to one-half of the maximum, and these points are therefore often referred to as the 'half-power' points. The form of eqn (7.21) shows that the variation of impedance with frequency gives a curve of universal shape but whose spread in frequency is determined by the value of Q_F. Since in general the ratio of $|V_C|$ to $|V|$ is $(\omega CZ)^{-1}$, the voltage step-up obtained falls sharply away on either side of the resonance point. Thus the circuit is selective in its response to signals of different frequency; its 'selectivity' is determined by the value of Q_F since Q_F is the ratio of the resonant frequency to the difference of frequency between the two half-power points:

$$Q_F = \omega_0/2\delta\omega = f_0/2\delta f.$$

A plot of the amplitude of the current in a series resonant circuit when a signal of given voltage but varying frequency is applied to it is shown in Fig. 7.5. Since the voltage across the capacitance $V_C = I/\omega C$, and near resonance the variation in I is very much more rapid (if Q_F is large) than that of ω, the voltage step-up obtained is also given to a good approximation by a curve of the same shape as in Fig. 7.5.

The approximation used in the deduction of eqn (7.20) is that $\delta\omega/\omega_0 \ll 1$. At frequencies where this approximation is not valid, the variation of **Z** still follows a universal curve, given by eqn (7.19). Since the reactive part of the impedance varies as $2Q_F(\delta\omega/\omega_0)$ near resonance, the current falls to quite small values before the approximation $\delta\omega/\omega_0 \ll 1$ becomes invalid, provided that Q_F is fairly high. In an ordinary tuned circuit the resistance

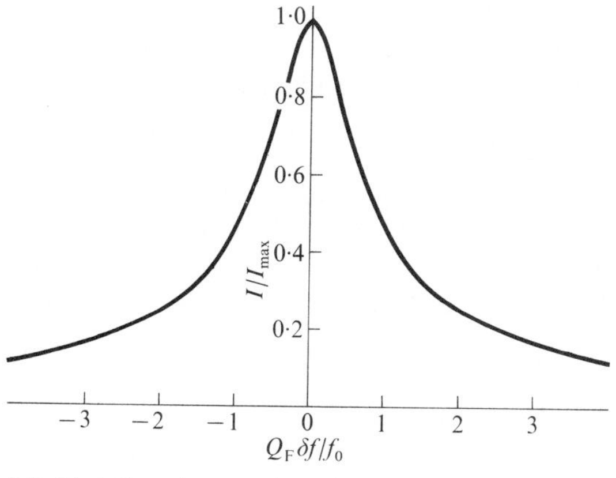

FIG. 7.5. Variation of current in series tuned circuit near resonance.

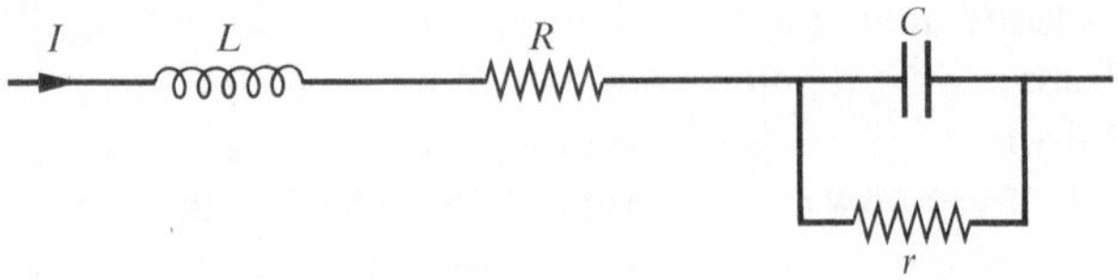

FIG. 7.6. Series resonant circuit with lossy capacitor.

is that of the wire used in winding the inductance coil; if we use a larger inductance in order to increase the value of Q_F at a given frequency, the resistance goes up and so does the self-capacitance between the different parts of the coil. The latter sets up an upper limit to the size of the coil since at some point the coil will resonate at the desired frequency without any external capacitance, because of its self-capacitance. As a rough guide one may take the Q_F of a coil designed to resonate at an audiofrequency as about 20; at frequencies of the order of 1 MHz a Q_F of 100–200 may be obtained; at frequencies of 10^3 to 10^5 MHz, where tuned transmission lines and waveguide cavities (see Chapters 9, 21, and 22) are used instead of lumped circuits, values of 1000–10 000 are obtained for Q_F. Thus in most radio work with tuned circuits eqn (7.20) is a good approximation.

Hitherto we have assumed no loss in the capacitor; if we include some loss the circuit becomes that shown in Fig. 7.6. The impedance operator is now

$$\mathbf{Z} = R + \mathrm{j}\omega L + \frac{1}{\mathrm{j}\omega C + 1/r} = R + \mathrm{j}\omega L + \frac{(r - \mathrm{j}\omega Cr^2)}{(1 + \omega^2 C^2 r^2)}.$$

If the power factor of the capacitor is small, $\omega Cr \gg 1$ and the impedance operator is approximately (cf. Problem 7.3)

$$\mathbf{Z} = R + \mathrm{j}\omega L + 1/\omega^2 C^2 r - \mathrm{j}/\omega C. \tag{7.23}$$

Thus the resonance frequency is unaltered in the first approximation, but the resistance of the circuit is increased. The value of Q_F is now given by

$$\frac{1}{Q_F} = \frac{R + 1/\omega_0^2 C^2 r}{\omega_0 L} = \frac{R}{\omega_0 L} + \frac{1}{\omega_0 Cr} = \frac{1}{Q_L} + \tan\delta,$$

where $Q_L = \omega_0 L/R$ is the Q_F of the circuit due to the loss in the inductance alone. Since for a good capacitor $\tan\delta < 10^{-3}$, while for lumped circuits with inductive coils (as distinct from transmission line circuits) $1/Q_L \approx 10^{-2}$, loss in the capacitor can generally be neglected.

Parallel resonant circuits

A parallel resonant circuit consists of a capacitance shunted across an inductance+resistance, as in Fig. 7.7. The admittance operator for the circuit is

$$\mathbf{Y} = \mathrm{j}\omega C + \frac{1}{R+\mathrm{j}\omega L} = \mathrm{j}\omega C + \frac{R-\mathrm{j}\omega L}{R^2+\omega^2 L^2}. \tag{7.24}$$

We define the resonance point for this circuit as the point at which the admittance is real. Then, equating the imaginary terms to zero gives

$$R^2+\omega_0^2 L^2 = L/C \tag{7.25a}$$

and hence

$$\omega_0 = \left(\frac{1}{LC}\right)^{\frac{1}{2}}\left(1-\frac{R^2C}{L}\right)^{\frac{1}{2}} = \left(\frac{1}{LC}\right)^{\frac{1}{2}}\left(1-\frac{1}{Q_F^2}\right)^{\frac{1}{2}}, \tag{7.25b}$$

while the admittance at this point is a pure conductance

$$G = R/(R^2+\omega_0^2 L^2) = RC/L = 1/(Q_F^2 R). \tag{7.26}$$

We see that the resonance frequency is not quite the same as in the series tuned circuit, but the difference is small and can usually be neglected if Q_F is large (if $Q_F = 100$, the fractional difference in ω_0 is $0{\cdot}5\times10^{-4}$). Because of this difference, the alternative formulae $\omega_0 L/R$ and $1/\omega_0 CR$ for Q_F are not exactly equivalent to $(L/C)^{\frac{1}{2}}/R$, but they are a good approximation if Q_F is large. The reciprocal of the admittance at resonance is called the 'parallel resistance' of the circuit, and is equal to $Q_F^2 R$ without any approximation.

The definition of resonance as the point at which the admittance is real is to a certain extent arbitrary, but is very convenient to use; it has the further advantage of being unique. The admittance of the circuit at this point is very small, but is not necessarily a minimum. If resonance is defined as the point at which Y is a minimum, then the resonance condition depends on what is being adjusted to make Y a minimum. If the capacitance is altered, and other quantities are kept fixed, then, by differentiating $Y = |\mathbf{Y}|$ with respect to C, it can readily be shown that Y is a minimum when the value of the capacitance satisfies eqn (7.25a). On the other hand, if L or the applied frequency is altered, slightly different resonance conditions are obtained; the fractional difference is, however, only of order $1/Q_F^2$, and so can be neglected if Q_F is large. The calculation involved in finding these resonance conditions by differentiation is very

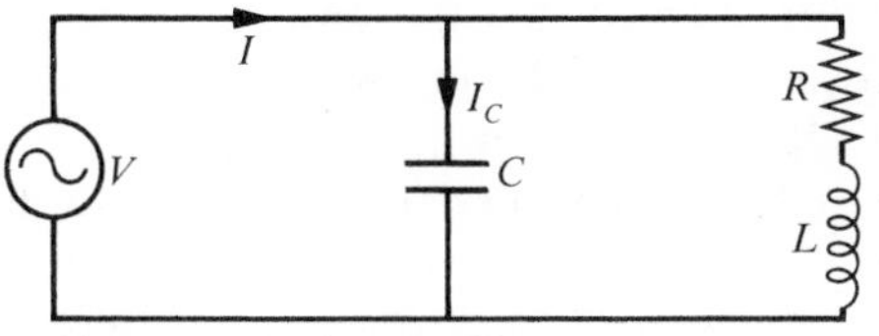

FIG. 7.7. Parallel resonant circuit.

tedious in comparison with the simple device of setting the imaginary part of $\mathbf{Y}$ equal to zero, and the simple definition of resonance as the point at which the circuit has unity power factor (that is, the admittance is real) has many advantages.

It is useful to have an approximate expression for $\mathbf{Y}$ near resonance similar to eqns (7.19)–(7.21) for the series resonance circuit. Writing $\omega=\omega_0+\delta\omega$, we have from eqn (7.24)

$$\mathbf{Y}=\frac{R}{R^2+\omega_0^2L^2}\left(1-\frac{2\omega_0\,\delta\omega L^2}{R^2+\omega_0^2L^2}\right)+\mathrm{j}\left\{\delta\omega C-\frac{\delta\omega L}{R^2+\omega_0^2L^2}\left(1-\frac{2\omega_0^2L^2}{R^2+\omega_0^2L^2}\right)\right\}$$

$$=\frac{RC}{L}\{1-2\delta\omega(\omega_0 CL)\}+2\mathrm{j}\;\delta\omega C^2L\omega_0^2,$$

where we have used eqn (7.25a). In the term of order $\delta\omega$ we may put $\omega_0^2LC=1$, giving

$$\mathbf{Y}=\frac{RC}{L}\left(1-\frac{2\delta\omega}{\omega_0}\right)+2\mathrm{j}\;\delta\omega C.$$

In general the small change in the conductance can be neglected, so that

$$\mathbf{Y}=\frac{RC}{L}+2\mathrm{j}\;\delta\omega C=\frac{RC}{L}\left\{1+2\mathrm{j}Q_{\mathrm{F}}\frac{\delta\omega}{\omega_0}\right\}, \tag{7.27}$$

showing that the behaviour of the admittance near resonance for the parallel resonant circuit is similar to that of the impedance of the series tuned circuit. The selectivity is the same in that Y rises by a factor $\sqrt{2}$ when the fractional deviation of the frequency is

$$\delta f/f_0=\delta\omega/\omega_0=\pm 1/2Q_{\mathrm{F}}.$$

Thus the current drawn from the voltage generator is a minimum at resonance (apart from the small corrections which can be neglected when Q_{F} is large). The current in each of the arms is larger than that drawn from the source by a factor nearly equal to Q_{F}. Thus the ratio of the current through the capacitance to that drawn from the generator is

$$\left|\frac{I_{\mathrm{C}}}{I}\right|=\frac{\omega_0 CV}{RCV/L}=\frac{\omega_0 L}{R}=Q_{\mathrm{F}} \tag{7.28}$$

in our approximation; this relation is complementary to that given by eqn (7.18). That equation showed that the series resonant circuit can be used as a tuned transformer where the output voltage across the condenser is larger by a factor Q_{F} than that injected in series with the inductance. The impedance measured at the output (for example, by finding the current drawn from a generator applied across the capacitance as in Fig. 7.7) is Q_{F}^2R at resonance (by eqn (7.26)) whereas the impedance of the series tuned circuit is R. Thus the ratio of the impedances is Q_{F}^2, the square of the voltage ratio. This, which is a characteristic of all transformers, can be seen as follows: if we apply a generator of voltage V to the series tuned circuit, the current drawn from it is I; to have the same current flowing through the capacitance in the parallel resonant circuit formed from the same elements, we must apply a generator of voltage $Q_F V$ across the capacitance, and the current drawn from it will be I/Q_F. The power dissipated in the circuit is VI, the same in each case.

The importance of the quality factor Q_F in determining the properties of a resonant circuit can be readily appreciated from the following summary. For a simple resonant circuit as considered hitherto:

(1) $Q_F = (L/C)^{\frac{1}{2}}/R \approx \omega_0 L/R \approx (\omega_0 CR)^{-1}$;

(2) Q_F is equal to the voltage step-up obtained by using the circuit as a tuned transformer;

(3) the parallel resistance is Q_F^2 times the series resistance;

(4) the fractional frequency difference $2\delta f/f_0$ between the points at which the impedance or admittance changes by a factor $\sqrt{2}$ is $1/Q_F$;

(5) $Q_F = \pi/\Lambda$, where Λ is the logarithmic decrement of free oscillations in the circuit (see § 5.3);

(6) in forced resonance,

$$Q_F = \frac{\omega_0 \times \text{stored energy}}{\text{rate of energy dissipation}}.$$

This last relation can be used to define Q_F in more complicated resonant circuits, and in other resonant systems such as waveguide cavities (see Chapter 9) where the values of L, C, and R cannot be specified. For the series resonant circuit of Fig. 7.2 it is readily shown that this definition agrees with that in (1) above. For the stored energy at any instant is

$$\begin{aligned}\tfrac{1}{2}LI^2 + \tfrac{1}{2}Q^2/C &= \tfrac{1}{2}LI_0^2\cos^2\omega_0 t + \tfrac{1}{2}(I_0^2/\omega_0^2 C)\sin^2\omega_0 t \\ &= \tfrac{1}{2}LI_0^2 \quad \text{at resonance,}\end{aligned}$$

while the mean rate of energy dissipation is $\frac{1}{2}RI_0^2$. Hence

$$Q_F = \frac{\omega_0 \times \frac{1}{2}LI_0^2}{\frac{1}{2}RI_0^2} = \frac{\omega_0 L}{R}.$$

7.4. Coupled resonant circuits

Two circuits are said to be coupled together if they have a common impedance. The impedance may be a resistance, inductance, or capacitance, and may be a part of each circuit, as in the example shown in Fig.

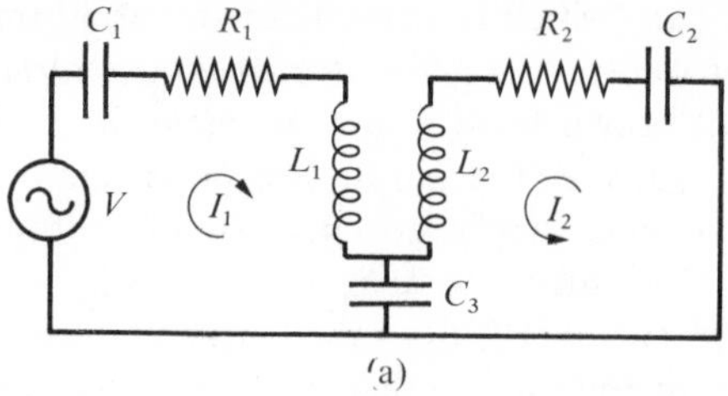

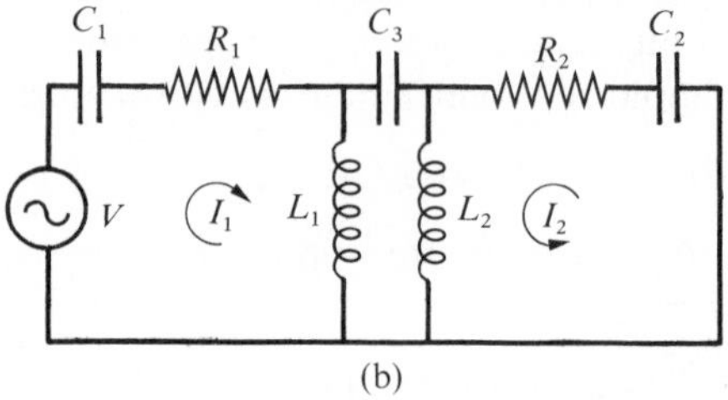

FIG. 7.8. Types of coupled circuits: (a) with common impedance C_3; (b) with impedance C_3 connected between them.

7.8(a), or connected between the two circuits as in Fig. 7.8(b). Two circuits may also be coupled together if one of them is in an electric or magnetic field set up by the other; for example, the oscillating magnetic flux due to a coil in one circuit may induce a voltage in a coil in the second circuit, so that there is a mutual inductance between the two circuits.

In Fig. 7.9 two resonant circuits 1 and 2 are coupled together by a mutual inductance M between the coils L_1 and L_2. The voltage V_1 applied to the first circuit produces a current I_1, and this induces a voltage $M(\mathrm{d}I_1/\mathrm{d}t)$ in the secondary circuit. For a simple sinusoidal waveform this voltage may be written as $\mathrm{j}\omega MI_1$, where $\mathrm{j}\omega M$ is the impedance operator for the mutual inductance. Similarly, if I_2 is the current in the secondary, a voltage $\mathrm{j}\omega MI_2$ will be induced in the primary. Thus for the two circuits we have

$$\begin{aligned} V_1 &= I_1\mathbf{Z}_1+\mathrm{j}\omega MI_2, \\ 0 &= I_2\mathbf{Z}_2+\mathrm{j}\omega MI_1, \end{aligned} \tag{7.29}$$

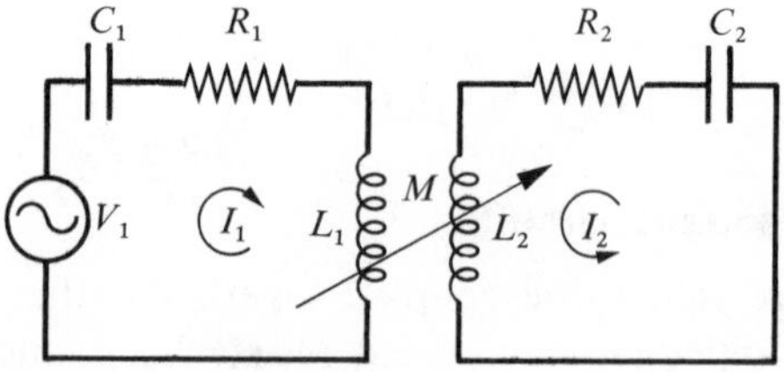

FIG. 7.9. Two resonant circuits coupled by mutual inductance M.

where $\mathbf{Z}_1$, $\mathbf{Z}_2$ are the impedance operators for the primary and secondary circuits in the absence of the mutual inductance. On eliminating I_2 we have

$$V_1 = I_1\left(\mathbf{Z}_1 + \frac{\omega^2 M^2}{\mathbf{Z}_2}\right). \tag{7.30}$$

The quantity $\omega^2 M^2/\mathbf{Z}_2$ is called the impedance 'reflected into the primary circuit'. For the secondary circuit, elimination of I_1 gives

$$-\frac{\mathrm{j}\omega M}{\mathbf{Z}_1} V_1 = I_2\left(\mathbf{Z}_2 + \frac{\omega^2 M^2}{\mathbf{Z}_1}\right), \tag{7.31}$$

showing that the current flow is that produced by an apparent voltage $-(\mathrm{j}\omega M/\mathbf{Z}_1)V_1$ working into $\mathbf{Z}_2$ plus the impedance 'reflected into the secondary circuit', $\omega^2 M^2/\mathbf{Z}_1$.

To investigate the behaviour of the primary circuit impedance in more detail we write $\mathbf{Z}_1 = R_1 + \mathrm{j}X_1$, $\mathbf{Z}_2 = R_2 + \mathrm{j}X_2$. Then

$$\begin{aligned} \frac{V_1}{I_1} &= R_1 + \mathrm{j}X_1 + \frac{\omega^2 M^2}{(R_2 + \mathrm{j}X_2)} \\ &= R_1 + \frac{R_2 \omega^2 M^2}{R_2^2 + X_2^2} + \mathrm{j}\left(X_1 - \frac{X_2 \omega^2 M^2}{R_2^2 + X_2^2}\right) \\ &= R_\mathrm{p} + \mathrm{j}X_\mathrm{p}. \end{aligned} \tag{7.32}$$

The current and voltage in the primary are in phase if X_p is zero. This requirement is satisfied if $X_1 = X_2 = 0$, and in the following analysis we shall assume that the elements of the primary and secondary circuits are the same, so that $R_1 = R_2$, $X_1 = X_2$ at all frequencies (the latter implies $L_1 = L_2$, $C_1 = C_2$). Then X_p is zero when either

$$X = 0 \quad \text{or} \quad X^2 = \omega^2 M^2 - R^2. \tag{7.33}$$

The second of these conditions can be fulfilled only if $\omega M > R$. If $\omega M = R$, the three roots are identical, and if $\omega M < R$ only one real root exists.

The significance of these three roots becomes apparent when we examine the behaviour of the secondary current. When

$$X_1 = X_2 = X = 0,$$

from eqn (7.31) we have

$$I_2 = \frac{-\mathrm{j}\omega_0 M V_1}{R^2 + \omega_0^2 M^2}, \tag{7.34}$$

where $\omega_0 = (LC)^{-\frac{1}{2}}$ is the resonance frequency of either circuit by itself. If we can vary the mutual inductance M, then the value of I_2 at this frequency rises as M is increased from zero, passes through a maximum value of $-jV_1/2R$ when $\omega_0 M = R$, and then falls again. At the maximum the impedance reflected into the primary from the secondary is just equal to R, so that the secondary circuit is 'matched' to the primary circuit, and the power dissipated in the secondary circuit is a maximum at this point. From eqns (7.30) and (7.34) it can be seen that the currents in the primary and secondary circuits are both equal to $V_1/2R$ at this point, but they differ in phase by $\frac{1}{2}\pi$.

At the second two points given by eqn (7.33) where the effective primary impedance is real, the secondary current is

$$I_2 = \frac{-j\omega MV_1}{\mathbf{Z}^2+\omega^2M^2} = \frac{-j\omega MV_1}{R^2-X^2+\omega^2M^2+2jRX}$$
$$= \frac{-j\omega MV_1}{2R^2+2jRX} = \frac{-j\omega MV_1}{2R(R+jX)},$$

where we have used the condition given by eqn (7.33). Hence

$$|I_2| = \frac{\omega MV_1}{2R(R^2+X^2)^{\frac{1}{2}}} = \frac{V_1}{2R}.$$

These results show that when the circuits are 'over-coupled' ($\omega M > R$) the secondary current rises to $V_1/2R$ at the second two points given by eqn (7.33), while it has fallen below this value at the point $X=0$. The primary-circuit impedance is purely resistive at all three points, but only at the second two points is it equal to $2R$, as can be seen by substituting in eqn (7.32). Thus the secondary current is a maximum at these two points because the secondary circuit is again 'matched' to the primary circuit.

The general behaviour of the secondary current is shown in Fig. 7.10, where the degree of coupling is specified in terms of the 'coefficient of coupling' defined by eqn (5.14) as

$$k = M/(L_1L_2)^{\frac{1}{2}} = M/L \tag{7.35}$$

when the two circuits are identical.

The critical condition $\omega_0 M = R$ then occurs when the coefficient of coupling has the value

$$k_0 = R/\omega_0 L = 1/Q_F. \tag{7.36}$$

The curves in Fig. 7.10 are drawn for $Q_F = 100$, and coefficients of coupling of $k = \frac{1}{2}k_0$, k_0, and $2k_0$ respectively. The first curve is similar to an ordinary resonance curve, but the current is always less than the maximum possible value $V_1/2R$. The second curve has a single peak in the centre, but the peak is decidedly flattened because of the three coincident roots. As soon as $k > k_0$ we have two side maxima in the current with a central minimum. As k/k_0 increases these peaks move

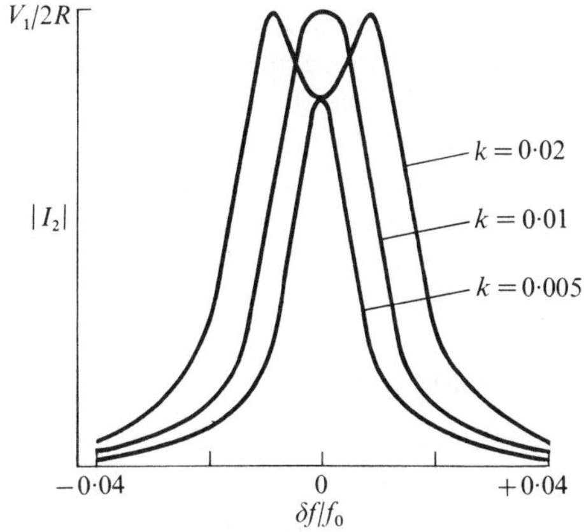

FIG. 7.10. Secondary current in the circuit of Fig. 7.9 for two identical circuits each with $Q_F = 100$. Critical coupling occurs at $k = 0{\cdot}01$.

outwards and the trough in the middle deepens. When Q_F is large and the frequency is close to resonance we may write $X = 2\delta\omega L$, and eqn (7.33) gives for the separation $2\delta f$ of the peaks

$$\frac{2\delta f}{f_0} = \frac{2\delta\omega}{\omega_0} \approx \left(\frac{M^2}{L^2} - \frac{R^2}{\omega_0^2 L^2}\right)^{\frac{1}{2}} = (k^2 - 1/Q_F^2)^{\frac{1}{2}} = (k^2 - k_0^2)^{\frac{1}{2}}. \qquad (7.37)$$

Although identical circuits have been assumed in this analysis, the behaviour of any pair of coupled circuits whose natural resonant frequencies are the same is similar. Thus for two circuits with the same X but different R_1, R_2, optimum coupling occurs (for $X = 0$) when

$$\omega_0 M = (R_1 R_2)^{\frac{1}{2}} \qquad (k = 1/(Q_1 Q_2)^{\frac{1}{2}}),$$

and the secondary current is then $V_1/2(R_1R_2)^{\frac{1}{2}}$. Side peaks occur when the coupling is greater than this value, and the secondary current at these peaks is $V_1/(R_1+R_2)$, which is less than the optimum. The values of the effective primary impedance are $2R_1$ when the coupling is optimum and $X = 0$, and R_1+R_2 at the side peaks when the circuits are over-coupled.

Coupled circuits have an important application in the reception of radio signals when it is desired to accept a narrow band of frequencies and reject frequencies outside this band. Clearly in radio reception the ideal receiver should have a uniform response over this band, combined with total rejection outside, so that the ideal response curve would be rectangular in shape. By the use of tuned circuits with rather more than critical coupling a response curve with steep sides is obtained, and the slight dip in the middle can be compensated by the use of somewhat under-coupled circuits (with a central peak) elsewhere in the receiver. In a tunable receiver of this kind a constant bandwidth is required independent of the central frequency. This cannot be achieved by mutual-inductance coupling alone, for eqn (7.37) shows that the bandwidth is then proportional to the central frequency. A combination of mutual inductance coupling together with capacitance coupling as in Fig. 7.8(a) may be used to give a more-or-less constant bandwidth, for the coupling through the capacitance decreases with frequency, because of the fall in the impedance common to the two circuits.

7.5. Low-frequency transformers

The coupled resonant circuits discussed in the last section may be regarded as a tuned transformer; they are used as such at radio-frequencies where the effects of stray capacitance can be reduced by making it part of the tuning capacitance. At low frequencies (such as those of power supplies) effects of stray capacitance are small, and transfer of power can be made very efficiently by use of the transformer whose principle was mentioned in § 5.2. It consists of two coils, the primary or input coil of n_1 turns, and the secondary or output coil of n_2 turns, which are closely wound together on an iron core so that all the flux due to one coil passes through the other. In practice there is always a small leakage of flux, so that not all the flux of one circuit passes through the other. Then the transformer may be represented by the circuit of Fig. 7.11, where L_1, L_2 are the self-inductances of the primary and secondary windings, and M is the mutual inductance between them. An alternating voltage V_1 of angular frequency ω is applied to the primary in series with an impedance $\mathbf{Z}_1$, and the secondary is connected to a load impedance $\mathbf{Z}_2$. Then, if the primary and secondary currents are I_1, I_2, the equations for the primary and secondary circuits are

$$\begin{aligned} V_1 &= (\mathbf{Z}_1+\mathrm{j}\omega L_1)I_1+\mathrm{j}\omega M I_2, \\ 0 &= (\mathbf{Z}_2+\mathrm{j}\omega L_2)I_2+\mathrm{j}\omega M I_1. \end{aligned} \tag{7.38}$$

These equations differ from (7.29) in that the self-inductances L_1, L_2 are not included in the impedances $\mathbf{Z}_1$, $\mathbf{Z}_2$ because their reactances are normally very much larger than any other impedances in the circuits.

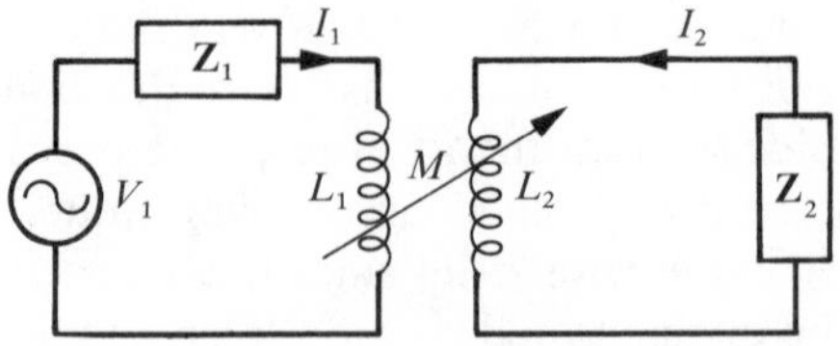

FIG. 7.11. Circuit diagram of transformer with load $\mathbf{Z}_2$.

Alternate elimination of I_2 and I_1 between these equations gives the following formulae for the primary and secondary circuits

$$V_1/I_1 = \mathbf{Z}_1 + \mathrm{j}\omega L_1 + \omega^2 M^2/(\mathbf{Z}_2 + \mathrm{j}\omega L_2) \quad (7.39)$$

and

$$-(M/L_1)V_1/I_2 = \mathbf{Z}_2 + \mathbf{Z}_1(L_2/L_1) + \mathrm{j}\omega(L_2 - M^2/L_1), \quad (7.40)$$

where in the second equation a term $\mathbf{Z}_1\mathbf{Z}_2/\mathrm{j}\omega L_1$ has been omitted since it is an order of magnitude smaller than $\mathbf{Z}_2$.

For a perfect transformer, there is complete coupling between the two coils, so that $M^2 = L_1L_2$. Since the self-inductances are proportional to the squares of the number of turns, the turns ratio $n = (L_2/L_1)^{\frac{1}{2}}$, and with complete coupling we have also $n = M/L_1 = L_2/M$. Hence, assuming $\mathbf{Z}_2 \ll \omega L_2$, from eqn (7.39) the primary impedance $\mathbf{Z}_\mathrm{p}$ may be written

$$V_1/I_1 = \mathbf{Z}_\mathrm{p} = \mathbf{Z}_1 + \mathrm{j}\omega L_1 + \omega^2 M^2(\mathbf{Z}_2 - \mathrm{j}\omega L_2)/(\mathbf{Z}_2^2 + \omega^2 L_2^2)$$

$$\approx \mathbf{Z}_1 + \mathbf{Z}_2(M^2/L_2^2) + \mathrm{j}\omega(L_1 - M^2/L_2) = \mathbf{Z}_1 + \mathbf{Z}_2/n^2. \quad (7.41)$$

The current in the secondary circuit is the same as that due to an e.m.f. $-(M/L_1)V_1 = -nV_1$ working into an impedance $\mathbf{Z}_\mathrm{s}$, where

$$\mathbf{Z}_\mathrm{s} = \mathbf{Z}_2 + \mathbf{Z}_1 n^2. \quad (7.42)$$

These equations show that for a perfect transformer on load ($\mathbf{Z}_2 \ll \omega L_2$) the inductive terms such as $\mathrm{j}\omega(L_1 - M^2/L_2)$ are exactly zero when there is complete coupling between the two coils. The effect of the secondary circuit on the current in the primary is represented by the additional impedance $\mathbf{Z}_2/n^2$, known as the 'reflected impedance'. In the secondary circuit the effective e.m.f. is $-nV_1$, with an apparent internal impedance $\mathbf{Z}_1 n^2$. Hence the impedances are transformed by n^2 while the voltages are transformed by n; the current transformation ratio is $1/n$, so that the power on the two sides is the same ($V_1I_1 = V_2I_2$), as we should expect for a perfect transformer with no losses.

In practice M^2 is slightly less than L_1L_2 because not all the flux from one circuit passes through the other, and we write $M = k(L_1L_2)^{\frac{1}{2}}$, where k is the coupling coefficient defined by eqn (5.14). It is useful to derive an equivalent circuit for the transformer, and in order to include the case where $\mathbf{Z}_2$ is of the same order of magnitude as ωL_2, we rewrite the expression for the primary impedance (eqn (7.39)) in the following way

$$\mathbf{Z}_\mathrm{p} = \mathbf{Z}_1 + \mathrm{j}\omega L_1(1-k) + \frac{\omega^2 k^2 L_1 L_2 + \mathrm{j}\omega L_1 k(\mathbf{Z}_2 + \mathrm{j}\omega L_2)}{\mathbf{Z}_2 + \mathrm{j}\omega L_2}. \quad (7.43)$$

The last term may be expressed in the form

$$\frac{\mathrm{j}\omega k L_1\{\mathbf{Z}_2 + \mathrm{j}\omega(1-k)L_2\}}{\{\mathbf{Z}_2 + \mathrm{j}\omega(1-k)L_2\} + \mathrm{j}\omega k L_1 n^2},$$

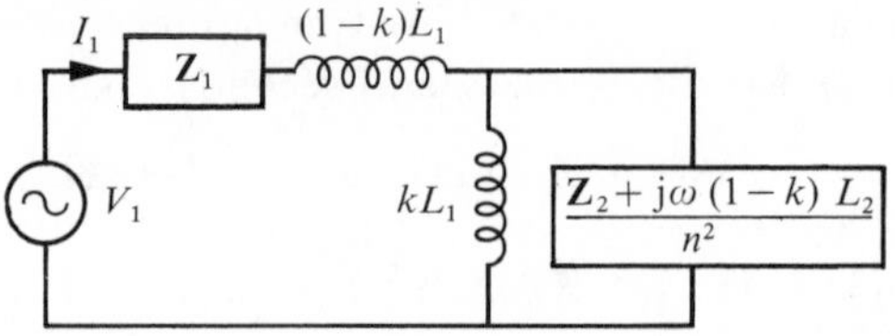

FIG. 7.12. Equivalent circuit of primary of imperfect transformer. $n^2 = L_2/L_1$; $k = M/(L_1L_2)^{\frac{1}{2}}$.

since in the denominator $\mathrm{j}\omega kL_1n^2 - \mathrm{j}\omega kL_2 = 0$. The last term is now

$$\left\{\frac{1}{\mathrm{j}\omega kL_1} + \frac{n^2}{\mathbf{Z}_2+\mathrm{j}\omega(1-k)L_2}\right\}^{-1},$$

which is equivalent to two impedances in parallel. Hence the primary circuit may be represented by Fig. 7.12. Here $(1-k)L_1$ is the 'leakage inductance' due to the imperfect coupling, and the impedance on the extreme right is the reflected impedance of the secondary, which is in parallel with the remainder of the primary inductance.

For the secondary circuit, since $M/L_1 = k(L_2/L_1)^{\frac{1}{2}} = kn$, we have, from eqn (7.40),

$$\begin{aligned} -knV_1/I_2 &= \mathbf{Z}_2 + n^2\mathbf{Z}_1 + \mathrm{j}\omega L_2(1-k) + \mathrm{j}\omega(kL_2 - M^2/L_1) \\ &= \mathbf{Z}_2 + n^2\{\mathbf{Z}_1 + \mathrm{j}\omega L_1(1-k)\} + \mathrm{j}\omega L_2k(1-k), \end{aligned} \tag{7.44}$$

which is represented by Fig. 7.13. The reflected impedance is the transformed value of $\mathbf{Z}_1$ plus the primary leakage inductance $L_1(1-k)$, while the secondary leakage inductance is, apart from a factor k, just that which appears in series with $\mathbf{Z}_2$ in the impedance reflected in the primary. If k is close to unity, this approximation is a good one, for the leakage inductance will be small compared with $\mathbf{Z}_2$ except for very small values of $\mathbf{Z}_2$. The equivalent circuit of the transformer can be drawn as in Fig. 7.14, where the centre portion enclosed in the dotted rectangle is regarded as a perfect transformer. The resistances r_1 and r_2 are the resistances of the windings, which previously we have regarded as part of $\mathbf{Z}_1$ and $\mathbf{Z}_2$. The resistance R in parallel with kL_1 allows for the dissipation of energy through hysteresis and eddy currents in the iron core.

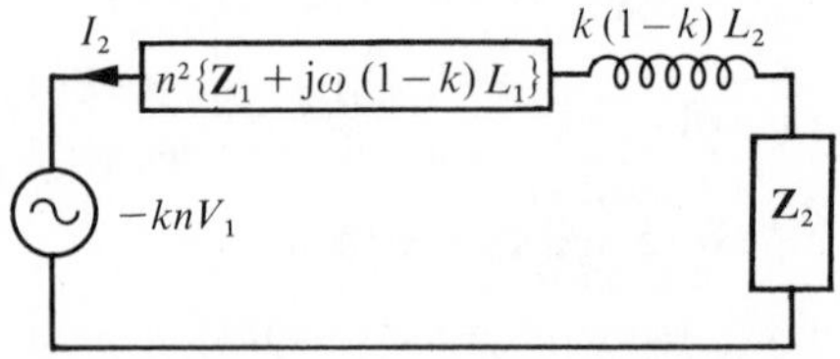

FIG. 7.13. Equivalent circuit for secondary of imperfect transformer.

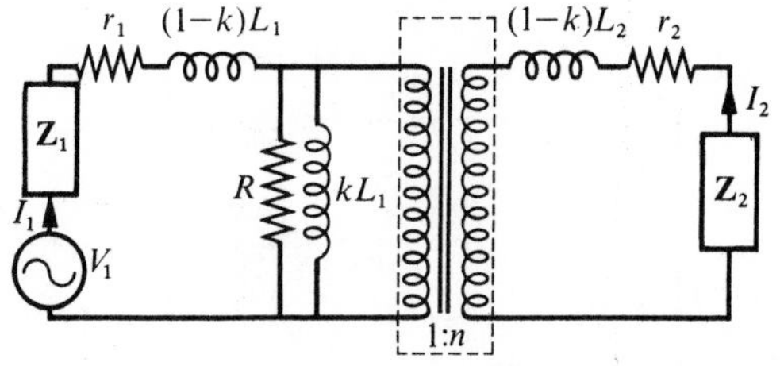

FIG. 7.14. Approximate equivalent circuit of imperfect transformer. r_1, r_2 resistance of primary and secondary windings, R resistance equivalent to hysteresis and eddy-current losses in iron core. The portion within the dotted rectangle is regarded as a perfect transformer of turns ratio n.

The prime requirements in a transformer are therefore a high primary inductance, to keep the 'magnetizing current' through kL_1 in Fig. 7.14 small, and the smallest possible leakage of flux between primary and secondary. These requirements are fulfilled by winding the primary and secondary round an iron core whose magnetic circuit is completed by a yoke. If the two coils are interwound the leakage is reduced to a minimum, but if good insulation between primary and secondary is required they may be wound side by side.

7.6. Linear circuit analysis

It is apparent from the examples considered in the previous sections that the behaviour of a network of impedances at a given frequency can be analysed in the same way as the network of resistances considered in § 3.4. The analysis is based on the two laws of Kirchhoff:

$$\sum_k I_k = 0, \qquad (7.45)$$

$$\sum_j V_j = \sum_k I_k \mathbf{Z}_k = \sum_k I_k/\mathbf{Y}_k. \qquad (7.46)$$

These equations are formally similar to the equations in § 3.4 with the substitution of $\mathbf{Z}_k$ for R_k, and $\mathbf{Y}_k$ for G_k, but the fact that the impedance $\mathbf{Z}$ and the admittance $\mathbf{Y}$ are complex numbers reminds us that each equation must be satisfied independently for the real and imaginary parts. That is, both amplitude and phase must be considered in the addition of voltages and currents. Allowing for this complication, a number of important theorems from Chapter 3 can be applied at once to alternating current circuits. These include the reciprocity theorem, Thévenin's theorem, and Norton's theorem (see § 3.4). The maximum power theorem (§ 3.3) needs one modification. In Fig. 7.15 we have an a.c. voltage generator with internal impedance $\mathbf{Z}_s = R_s + jX_s$ working into a load $\mathbf{Z}_L = R_L + jX_L$. The current through the load is $I = V_S/(\mathbf{Z}_S + \mathbf{Z}_L)$, and

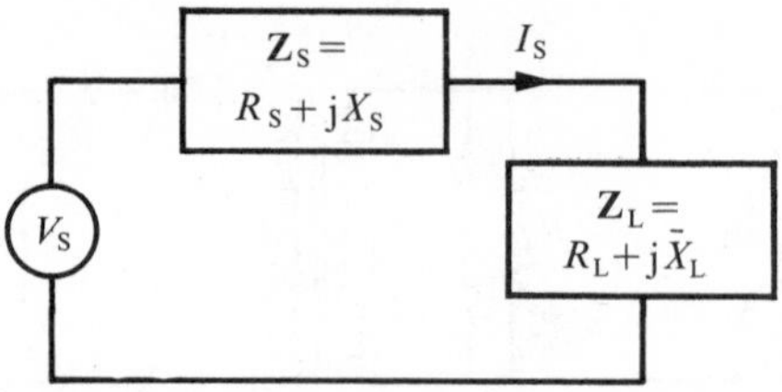

FIG. 7.15. The maximum power theorem for an a.c. circuit.

the power dissipated in it is

$$I^2R_L = V_S^2 \frac{R_L}{|\mathbf{Z}_S+\mathbf{Z}_L|^2} = V_S^2 \frac{R_L}{(R_S+R_L)^2+(X_S+X_L)^2}.$$

Since the reactances can be either positive or negative, maximum power is obtained by making first

$$X_S + X_L = 0, \tag{7.47}$$

and the problem is then the same as in § 3.3, that is, we need

$$R_S = R_L \tag{7.48}$$

provided that R_L is the variable quantity.

The four-terminal network

If in a network, however complicated, there is only one voltage generator v_S, with internal impedance $\mathbf{z}_S$, and we are interested in the output at one pair of terminals into a load $\mathbf{z}_L$, we may formally represent this as in Fig. 7.16. Here the rectangular box which replaces the network has just two input terminals, through which a current i_1 flows and between which the voltage is v_1, and two output terminals, through which a current i_2 flows and between which the voltage is v_2. The box is known as a 'four-terminal network', or 'quadrupole', whose behaviour can be represented by two simple equations, relating the voltages v_1, v_2 and the currents i_1, i_2. The analysis in this section is applicable to non-linear devices provided it is

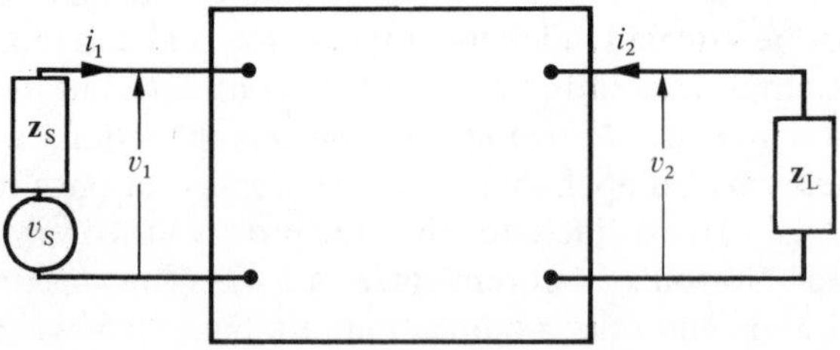

FIG. 7.16. The four-terminal network, with a voltage generator v_S (internal impedance $\mathbf{z}_S$) applied to the input terminal and a load $\mathbf{z}_L$ to the output terminals.

restricted to small changes, including small alternating components, of voltage and current. Such changes are denoted here by lower-case symbols v, i; similarly, lower-case symbols are used for circuit parameters to indicate that their values must be those appropriate to such small changes.

The low-frequency transformer is a simple example of a 'four-terminal network', as can be seen by redrawing Fig. 7.14 in the form illustrated by Fig. 7.16. Here v_S, $\mathbf{z}_S$ are the open-circuit voltage and internal impedance of an external generator applied to the terminals of the primary coil, while $\mathbf{z}_L$ is a load connected across the terminals of the secondary coil. The current which flows in the primary coil depends on the impedance in the secondary circuit, and vice versa; this is shown by the presence of $\mathbf{z}_2$ in eqn (7.43) and $\mathbf{z}_1$ in eqn (7.44), or, for the ideal transformer, in eqns (7.41) and (7.42) respectively. These results followed from the fundamental eqns (7.38), which can be written in terms of the voltages at the input and output terminals in Fig. 7.16, and the currents to these terminals, as follows:

$$v_1 = \mathbf{z}_{11} i_1 + \mathbf{z}_{12} i_2, \tag{7.49a}$$

$$v_2 = \mathbf{z}_{21} i_1 + \mathbf{z}_{22} i_2. \tag{7.49b}$$

By a simple rearrangement these equations can be written instead as

$$v_1 = \mathbf{h}_{11} i_1 + \mathbf{h}_{12} v_2, \tag{7.50a}$$

$$i_2 = \mathbf{h}_{21} i_1 + \mathbf{h}_{22} v_2. \tag{7.50b}$$

Here the symbols $\mathbf{h}$ stand for 'hybrid parameters'; obviously $\mathbf{h}_{11}$ has the dimensions of impedance, $\mathbf{h}_{22}$ those of admittance, while $\mathbf{h}_{12}$ and $\mathbf{h}_{21}$ are dimensionless. It is clear also that we can write

$$i_1 = \mathbf{y}_{11} v_1 + \mathbf{y}_{12} v_2, \tag{7.51a}$$

$$i_2 = \mathbf{y}_{21} v_1 + \mathbf{y}_{22} v_2. \tag{7.51b}$$

Eqns (7.49) are known as the 'impedance equations' and (7.51) as the 'admittance equations'. The three forms are all equivalent, and the parameters in one set may be related to those in another by simple algebraic processes. In matrix language any of these relations may be written in the form (taking (7.49) as an example)

$$\begin{bmatrix} v_1 \\ v_2 \end{bmatrix} = \begin{bmatrix} \mathbf{z}_{11} & \mathbf{z}_{12} \\ \mathbf{z}_{21} & \mathbf{z}_{22} \end{bmatrix} \begin{bmatrix} i_1 \\ i_2 \end{bmatrix}. \tag{7.52}$$

This is called the **Z**-matrix; $\mathbf{z}_{11}$, $\mathbf{z}_{22}$ are its 'diagonal elements', and $\mathbf{z}_{12}$, $\mathbf{z}_{21}$ its 'off-diagonal elements'. The matrix equivalents of (7.50) and (7.51) are known as the **H**- and **Y**-matrices. It is often convenient to analyse combinations of networks using matrix algebra, though we shall not use such methods here.

The input impedance is v_1/i_1, which from eqn (7.50a) is

$$v_1/i_1 = \mathbf{h}_{11} + \mathbf{h}_{12}(v_2/i_1), \tag{7.53}$$

where the presence of v_2 means that the input impedance depends on the impedance $\mathbf{z}_L$ connected to the output terminals as the load. Conversely, the output impedance ($=v_2/i_2$) of the network depends on the impedance $\mathbf{z}_S$ of the input source. However, in limiting cases all the parameters in the various relations between v_1, v_2, i_1, and i_2 have simple interpretations, of which we mention only the following which arise from eqns (7.50).

(1) When the output is short-circuited, $v_2 = 0$, and

$$\text{the input impedance} = \mathbf{h}_{11}; \tag{7.54}$$

$$\text{the current gain } A_i = \mathbf{h}_{21}. \tag{7.55}$$

(2) When the input is on open-circuit, $i_1 = 0$, and

$$\text{the output impedance} = \mathbf{h}_{22}^{-1}; \tag{7.56}$$

$$\text{the voltage gain } A_v = \mathbf{h}_{12}^{-1}. \tag{7.57}$$

The passive four-terminal network

If a network contains no voltage or current generator, it is known as a 'passive network', and for a passive four-terminal network we can derive some simple relations for the 'off-diagonal' elements of the matrices.

From eqns (7.49) we have

$$i_2/v_1 = -\mathbf{z}_{21}/(\det \mathbf{z}) \quad \text{when} \quad v_2 = 0,$$

and

$$i_1/v_2 = -\mathbf{z}_{12}/(\det \mathbf{z}) \quad \text{when} \quad v_1 = 0.$$

By the reciprocity theorem (cf. § 3.4) these quantities must be equal, so that $\mathbf{z}_{12} = \mathbf{z}_{21}$. Applying similar methods to the **H**- and **Y**-matrices, we have the results

$$\mathbf{z}_{12} = \mathbf{z}_{21}, \tag{7.58}$$

$$\mathbf{y}_{12} = \mathbf{y}_{21}, \tag{7.59}$$

$$\mathbf{h}_{12} + \mathbf{h}_{21} = 0. \tag{7.60}$$

In a network containing resistance, any current must decay with time because of the dissipation of energy in the resistance, unless there is some mechanism in the circuit for creating energy. Such a mechanism is equivalent to a 'negative resistance'; if more energy is created than is dissipated, the net resistance is negative, and oscillations will increase in amplitude instead of decreasing, as is obvious from eqn (5.28). The condition under which oscillations can just be sustained is that the net resistance is zero, and under such conditions it is often useful to have a method of determining the natural frequency of oscillation which is simpler than solving a differential equation such as (5.27). This is particularly true for coupled circuits, where calculations such as those in § 7.4 are clearly rather tedious. In Fig. 7.9, if we put $R_1 = R_2 = 0$, and $V_1 = 0$, we have

$$\begin{aligned} 0 &= \left(\mathrm{j}\omega L_1 - \frac{1}{\mathrm{j}\omega C_1}\right) i_1 + \mathrm{j}\omega M i_2, \\ 0 &= \mathrm{j}\omega M i_1 + \left(\mathrm{j}\omega L_2 - \frac{1}{\mathrm{j}\omega C_2}\right) i_2. \end{aligned} \tag{7.61}$$

These equations are equivalent to (7.49) with $v_1 = v_2 = 0$, and the condition for a non-trivial solution is $\det \mathbf{z} = 0$. This gives, writing $\omega_1^2 L_1 C_1 = 1$, $\omega_2^2 L_2 C_2 = 1$ for the natural resonant frequencies of the uncoupled circuits,

$$\left(1 - \frac{\omega_1^2}{\omega^2}\right)\left(1 - \frac{\omega_2^2}{\omega^2}\right) = \frac{M^2}{L_1 L_2} = k^2, \tag{7.62}$$

where k is the coefficient of coupling defined by eqns (5.14) or (7.35). The two natural resonant frequencies are found to be given by

$$\frac{1}{\omega^2} = \frac{1}{2}\left(\frac{1}{\omega_1^2} + \frac{1}{\omega_2^2}\right) \pm \frac{1}{2}\left\{\left(\frac{1}{\omega_1^2} - \frac{1}{\omega_2^2}\right) + \frac{4k^2}{\omega_1^2 \omega_2^2}\right\}^{\frac{1}{2}}. \tag{7.63}$$

This method can be used in dealing with the natural resonant frequencies of other coupled systems.

The active four-terminal network

A network containing a voltage or a current generator is known as an 'active network'. In general such a generator is an electronic device, and we choose as an example the network shown in Fig. 7.17, which is applicable to thermionic triodes and to various solid-state devices.

In Fig. 7.17 the input circuit has a source of voltage v_S and internal impedance $\mathbf{z}_S$, giving a voltage v_1 at the input to the network. The magnitude of v_1 depends on i_1 and is determined by the values of $\mathbf{h}_{11}$ and $\mathbf{h}_{12}$ as in eqn (7.50a). The output circuit contains a current generator $g_m v_1$ shunted by a resistance r_o, producing a voltage v_2 at the output terminals

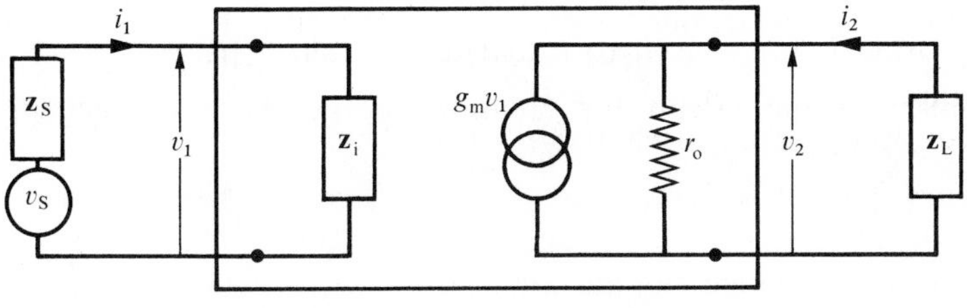

FIG. 7.17. An active four-terminal network.

across a load $\mathbf{z}_L$. This representation is chosen because the constants g_m and r_0 are the parameters of an electronic device which are most readily determined in practice, whereas the **h** parameters cannot be directly measured. The combination of $\mathbf{h}_{11}$ and the voltage generator $\mathbf{h}_{12}v_2$ can be represented by an impedance $\mathbf{z}_i$ which is the input impedance of the network, but it must be remembered that the magnitude of $\mathbf{z}_i$ is thus dependent on v_2 and hence also on the load $\mathbf{z}_L$. We have then the equations

$$v_S = v_1 + \mathbf{z}_S i_1 = (\mathbf{z}_S + \mathbf{z}_i) i_1, \tag{7.64a}$$

$$i_2 = g_m v_1 + v_2 / r_0. \tag{7.64b}$$

In the first equation $\mathbf{z}_i = \mathbf{h}_{11}$ if $\mathbf{h}_{12} = 0$; in the second, $g_m = \mathbf{y}_{21}$ and $r_0^{-1} = \mathbf{y}_{22}$ in eqn (7.51b). If $\mathbf{h}_{12} = 0$, $\mathbf{y}_{22} = \mathbf{h}_{22}$, and r_0 is the output impedance of the network in the limit when $i_1 = 0$. This is a good approximation for thermionic triodes and field-effect transistors where i_1 is zero or very small, but is less true in junction transistors.

When a load $\mathbf{z}_L$ is connected to the output terminals, the current generator $g_m v_1$ works into an admittance $(r_0 + \mathbf{z}_L)/r_0\mathbf{z}_L$, since r_0 and $\mathbf{z}_L$ are in parallel. The voltage across $\mathbf{z}_L$ is

$$-g_m v_1 \frac{r_0 \mathbf{z}_L}{(r_0 + \mathbf{z}_L)}, \tag{7.65}$$

and the voltage amplification is

$$A_v = \frac{v_2}{v_1} = -g_m \frac{r_0 \mathbf{z}_L}{(r_0 + \mathbf{z}_L)}. \tag{7.66a}$$

When $|\mathbf{z}_L| \ll r_0$ this reduces to

$$A_v = -g_m \mathbf{z}_L, \tag{7.66b}$$

and under the same condition, since $v_1 = i_1 \mathbf{z}_i$,

$$A_i = \frac{i_2}{i_1} = \frac{g_m v_1}{i_1} = g_m \mathbf{z}_i, \tag{7.67}$$

so that

$$-A_v = A_i (\mathbf{z}_L / \mathbf{z}_i). \tag{7.68}$$

This relation is often used to calculate the fourth parameter when the other three are known.

In the other extreme, where $|\mathbf{z}_L| \gg r_0$, we have

$$-A_v = g_m r_0 = \mu, \tag{7.69}$$

where μ is the voltage amplification factor of the device. The parameter g_m is known as the 'mutual conductance' or 'transconductance'.

Devices such as the thermionic triode and transistors are usually three-terminal devices, but they can obviously be represented by a four-terminal network in which two terminals are joined together. A more serious difficulty is that they are inherently non-linear devices in which the currents are not simply proportional to the voltages, and the parameters in the equations above are not constants independent of the working conditions. Nevertheless, for small changes in current and voltage it is still possible to use linear circuit analysis, provided that its limitation to small signals is recognized. In Fig. 7.17, if the output current $I_2 = f(V_1, V_2)$, then for small changes v_1, v_2 in V_1, V_2 we can use a Taylor series expansion, giving

$$I_2 = (I_2)_0 + v_1\left(\frac{\partial I_2}{\partial V_1}\right)_{V_2} + v_2\left(\frac{\partial I_2}{\partial V_2}\right)_{V_1} + $$
$$+\tfrac{1}{2}v_1^2\left(\frac{\partial^2 I_2}{\partial V_1^2}\right)_{V_2} + v_1 v_2\left(\frac{\partial^2 I_2}{\partial V_1\,\partial V_2}\right)_{V_2,V_1} + \tfrac{1}{2}v_2^2\left(\frac{\partial^2 I_2}{\partial V_2^2}\right)_{V_1} + \ldots \quad (7.70)$$

The presence of second- and higher-order terms means that the change $i_2 = I_2 - (I_2)_0$ in output current is not linear with the voltage changes; this is generally undesirable because it represents distortion in the output signal. For sufficiently small changes v_1, v_2 this effect becomes negligible, and (7.70) can be written as

$$i_2 = g_m v_1 + (1/r_0)v_2, \quad (7.71)$$

where

$$g_m = (\partial I_2/\partial V_1)_{V_2}, \qquad 1/r_0 = (\partial I_2/\partial V_2)_{V_1}. \quad (7.72)$$

From the relation

$$\left(\frac{\partial V_2}{\partial I_2}\right)_{V_1}\left(\frac{\partial I_2}{\partial V_1}\right)_{V_2}\left(\frac{\partial V_1}{\partial V_2}\right)_{I_2} = -1,$$

we have

$$-\left(\frac{\partial V_2}{\partial V_1}\right)_{I_2} = g_m r_0 = \mu, \quad (7.73)$$

a result which can also be obtained directly from eqn (7.71) by putting $i_2 = 0$. The differentials in which V_2, V_1, and I_2 are kept constant are equivalent to equations in which v_2, v_1, and i_2 are respectively zero. Comparison with eqn (7.51b) shows that

$$\left(\frac{\partial I_2}{\partial V_1}\right)_{V_2} = y_{21}, \qquad \left(\frac{\partial I_2}{\partial V_2}\right)_{V_1} = y_{22} \quad \text{and} \quad \left(\frac{\partial V_2}{\partial V_1}\right)_{I_2} = \frac{y_{21}}{y_{22}}.$$

Problems

7.1. A resistance R, inductance L, and capacitance C are all connected in parallel. Show that the admittance of the circuit at frequencies near resonance is

$$\mathbf{Y} = 1/R + 2\mathrm{j}\ \delta\omega C.$$

If $R = 3\times10^5\ \Omega$, $L = 10^{-3}$ H, $C = 100$ pF, calculate the current in each arm when a voltage of 10 V r.m.s. at a frequency of 0·5 MHz is applied, and the phase of the total current drawn from the generator.

(*Answers:* 0·033 mA, 3·18 mA, and 3·14 mA; 51°.)

7.2. A circuit is required to accept a signal of frequency 1·1 MHz and to reject a signal of frequency 1·2 MHz. A coil of self-inductance 200 μH and resistance 10 Ω is tuned to parallel resonance at 1·2 MHz by a capacitance C_1. A capacitance C_2 is then placed in series with the combination, so that the whole is in series resonance at 1·1 MHz. Find the values of C_1 and C_2.

(*Answers:* 88 pF and 16 pF.)

7.3. Show that a capacitance C shunted by a resistance r is equivalent to a capacitance C' in series with a resistance R at any given frequency. If $\omega Cr \gg 1$, show that approximately $R = (\omega^2 C^2 r)^{-1}$, and $C' = C$.

7.4. Four impedances $\mathbf{Z}_1$, $\mathbf{Z}_2$, $\mathbf{Z}_3$, and $\mathbf{Z}_4$, in that order, are placed in the arms of a 'generalized' Wheatstone's bridge using alternating current. Show that the balance condition is

$$\mathbf{Z}_1/\mathbf{Z}_2 = \mathbf{Z}_4/\mathbf{Z}_3.$$

Note that this is really a double balance condition, since the real and imaginary parts of this equation must be separately satisfied. This is because a null reading is obtained on the detector only if the voltages at each of its terminals are equal both in amplitude and phase.

7.5. The four arms of a Wheatstone's bridge, taken in cyclic order round the bridge, are a, b, c, d. a and b are equal resistances R; c is a resistance R in series with a capacitance C; d is a resistance R shunted by a capacitance C. Show that such a bridge will not be balanced at any frequency, but that if the resistance in arm b is doubled, a balance will be obtained at a frequency

$$f = (2\pi RC)^{-1}.$$

7.6. A capacitor of capacitance C, a resistance R_C, and a coil whose inductance is L and resistance is R_L are connected all three in parallel. An e.m.f. of variable frequency is applied across the capacitor. Show that the frequency of parallel resonance is independent of the value of R_C, and prove that at resonance the conductance of the combination is

$$\frac{1}{R_C} + \left(\frac{R_L C}{L}\right).$$

If the Q_F of the circuit is high, show that it is given approximately by

$$\frac{1}{Q_F} = \frac{1}{R_C}\left(\frac{L}{C}\right)^{\frac{1}{2}} + R_L\left(\frac{C}{L}\right)^{\frac{1}{2}}.$$

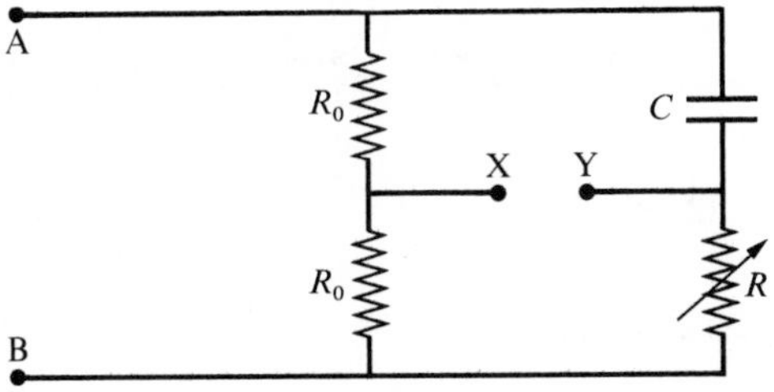

FIG. 7.18. Phase-shift network (see Problem 7.8).

7.7. A series resonant circuit is connected across a constant voltage generator operating at 6·50 MHz. As the capacitance is varied, the current is observed to fall to $1/\sqrt{2}$ of its maximum value when the capacitance is 12·47 pF, and again when the capacitance is 12·64 pF. Find the values of Q_F, L, R for the circuit.

(*Answer:* ≈150; 48 μH; 13 Ω.)

7.8. An alternating voltage is applied to the terminals A, B of the network shown in Fig. 7.18. Show that, as R is varied, the amplitude of the potential difference between the terminals X, Y remains constant, but its phase is shifted by π rad. Explain your results by means of a vector diagram. (This network is commonly used for producing a variable phase shift without change of the output amplitude.)

7.9. A high-frequency transformer has primary inductance 100 μH and primary resistance 5 Ω, secondary inductance $2{\cdot}5\times10^3$ μH and secondary resistance 100 Ω. If the coefficient of coupling $M/(L_1L_2)^{\frac{1}{2}}$ is 0·9, show that at high frequencies the voltage across the secondary is approximately 4·5 times that applied to the primary.

If primary and secondary are each separately tuned by capacitors to resonance at 50 kHz, show that the high-frequency power required to produce a r.m.s. voltage of 2000 V across the capacitance in the secondary circuit is approximately 650 W.

7.10. A wire-wound resistance has a small inductance and self-capacitance which may be represented by placing an inductance L in series with the resistance R, and shunting the combination by a capacitance C. Show that the reactance at low frequencies is zero in the first approximation if the wire is wound so that $L/C = R^2$. Show also that under these conditions the apparent resistance is (to the second approximation)

$$R(1+\omega^2LC)$$

7.11. An inductance L with small resistance r has self-capacitance which can be represented approximately by a capacitance C shunted across the series combination of L and r. Show that as the frequency increases the apparent self-inductance of the coil is increased by the factor $(1+\omega^2LC)$ while the apparent series resistance increases by the factor $(1+2\omega^2LC)$, in the region where $\omega^2LC \ll 1$.

7.12. A parallel-plate capacitor is filled with a medium which has permittivity ϵ and conductivity σ. Show that at a frequency $f = \omega/2\pi$ the power factor of the capacitor is $\cos\phi = \sin\delta$, where the loss tangent of the dielectric medium is given by the relation $\tan\delta = \sigma/\omega\epsilon$.

Note that this is independent of the shape of the capacitor, as would be expected from eqn (3.18).

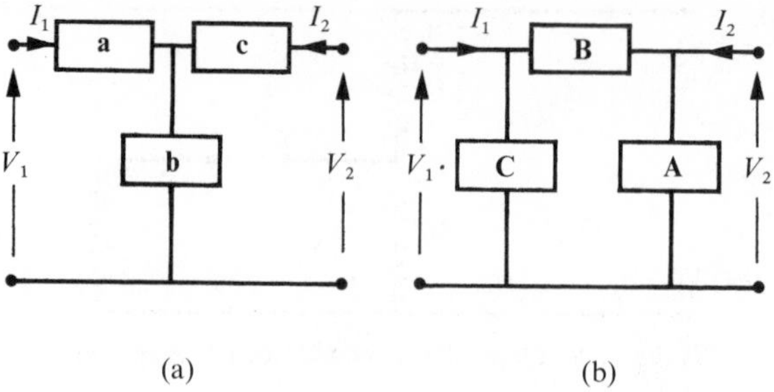

FIG. 7.19. (a) T-network with impedances **a, b, c.** (b) π-network with impedances **A, B, C.**

7.13. The voltage and current in a circuit are given by $V = Ae^{j\omega t}$ and $I = A'e^{j\omega t}$, where $A = a+jb$ and $A' = a'+jb'$ are each complex quantities. Show that the mean power averaged over a cycle is

$$\tfrac{1}{2}(aa'+bb') = \overline{\mathrm{Re}(V)\times\mathrm{Re}(I)} = \tfrac{1}{2}\mathrm{Re}(VI^*),$$

where I^* is the complex conjugate of I.

7.14. A 100-V dynamo is connected to a magnet whose resistance is 10 Ω and self-inductance is 0·1 H. Show that the percentage increase in the heating of the magnet caused by the presence of a 100 Hz ripple voltage of amplitude 5 V in the output of the dynamo is 0·0031 per cent.

7.15. Two networks such as those shown in Fig. 7.19 are equivalent provided that they obey any of the sets of eqns (7.49)–(7.51) with identical sets of parameters. Show that this is true provided that $\mathbf{aA} = \mathbf{bB} = \mathbf{cC} = \mathbf{ab}+\mathbf{bc}+\mathbf{ca}$.

This formula for converting a T-network into a π-network (see § 9.1) often simplifies circuit calculations, as in the transformer ratio bridge (§ 22.2).

8. Electromagnetic waves

8.1. Maxwell's equations of the electromagnetic field

So far we have considered the propagation of electrical currents in material conductors. The possibility of the propagation of an electromagnetic wave through space was first suggested by Faraday, and this suggestion was confirmed by the work of Maxwell. Maxwell was able to show that the laws of electromagnetism could be expressed in the form of some fundamental equations which, with an important modification, lead to a differential equation whose solutions represent transverse waves travelling through free space with the velocity of light. Further work showed that the properties of these waves—reflection, refraction, diffraction—are the same as those established experimentally for light waves, and we are therefore justified in assuming that they are identical, and that light waves are a form of electromagnetic radiation.

The theory of Maxwell deals entirely with macroscopic phenomena, making the assumption that matter is continuous and has no atomistic structure. This assumption places certain limitations on the theory: thus it offers no explanation of the phenomenon of dispersion—the change of refractive index with frequency. This phenomenon will be discussed in Chapter 10; it arises from the change in the electric permittivity and magnetic permeability of a medium with frequency. These changes can be related to the effect of electromagnetic waves on the individual electrons in an atom, but for the present we shall regard the electric permittivity and magnetic permeability as macroscopic quantities whose values are obtained by experiment.

The fundamental laws of electromagnetism which have already been derived may be summarized as follows:

(1) the theorem of Gauss applied to electrostatics (eqn (1.20)):

$$\operatorname{div} \mathbf{D} = \rho_e; \tag{8.1}$$

(2) the corresponding result for magnetic fields (eqn (4.6)):

$$\operatorname{div} \mathbf{B} = 0; \tag{8.2}$$

(3) Faraday's and Lenz's law of electromagnetic induction (eqn (5.3)):

$$\operatorname{curl} \mathbf{E} = -\partial \mathbf{B}/\partial t; \tag{8.3}$$

(4) Ampère's law for magnetomotive force (eqn (4.26)):

$$\operatorname{curl} \mathbf{H} = \mathbf{J}'. \tag{8.4}$$

The reason for writing $\mathbf{J}'$ rather than the ordinary current density $\mathbf{J}$ in this last equation is as follows. Since (div curl) of any vector is identically zero, it follows that eqn (8.4) implies that div $\mathbf{J}'$ is zero. If we had written $\mathbf{J}$ instead of $\mathbf{J}'$, we should have had div $\mathbf{J}=0$, and this conflicts with the equation of continuity (3.3) which gives

$$\operatorname{div}\mathbf{J} = -\partial\rho_e/\partial t. \tag{8.5}$$

This equation represents the law of conservation of charge, which is confirmed by all experiments. Maxwell realized that the difficulty arose from an incomplete definition of the total current density in eqn (8.4), which is not entirely given by the current flow due to the motion of electric charges. By using eqn (8.1) we may write (8.5) in the form

$$\operatorname{div}\mathbf{J} = -\frac{\partial}{\partial t}(\operatorname{div}\mathbf{D}) = \operatorname{div}\left(-\frac{\partial\mathbf{D}}{\partial t}\right),$$

or

$$\operatorname{div}\left(\mathbf{J}+\frac{\partial\mathbf{D}}{\partial t}\right)=0$$

Hence if we define $\mathbf{J}'$ as

$$\mathbf{J}' = \mathbf{J}+\frac{\partial\mathbf{D}}{\partial t}, \tag{8.6}$$

then div $\mathbf{J}'=0$, and Ampère's law takes the form

$$\operatorname{curl}\mathbf{H} = \mathbf{J}+\frac{\partial\mathbf{D}}{\partial t} = \sigma\mathbf{E}+\frac{\partial\mathbf{D}}{\partial t}. \tag{8.7}$$

The term $\mathbf{J}$ is generally called the 'conduction current' and the second term ($\partial\mathbf{D}/\partial t$) the 'displacement current', since it arises when the electric displacement $\mathbf{D}$ is changing with time. We may obtain some physical picture of what is implied by the displacement current by considering a simple circuit such as in Fig. 8.1, where a current I is flowing from a battery to charge a capacitance C. If we apply Ampère's law to a closed circuit such as LMNL which encircles the wire we find that $\int\mathbf{H}.\mathrm{d}\mathbf{s}$ round this circuit is just equal to I. To define the current which threads the circuit we must take some surface bounded by the circuit, and integrate the normal component of the current density crossing this surface. If we take a surface intersecting the wire, this clearly gives just I, the current in the wire. But if we take a surface which passes between the two plates of the capacitor, then $\int\mathbf{J}.\mathrm{d}\mathbf{S}=0$, since no conduction current flows through the surface. The value of this integral should be independent of what surface we choose, since $\int\mathbf{H}.\mathrm{d}\mathbf{s}$ depends only on the circuit bounding this surface; thus it is clear that we have omitted some contribution. If for simplicity we take a parallel-plate capacitor with plates of area A,

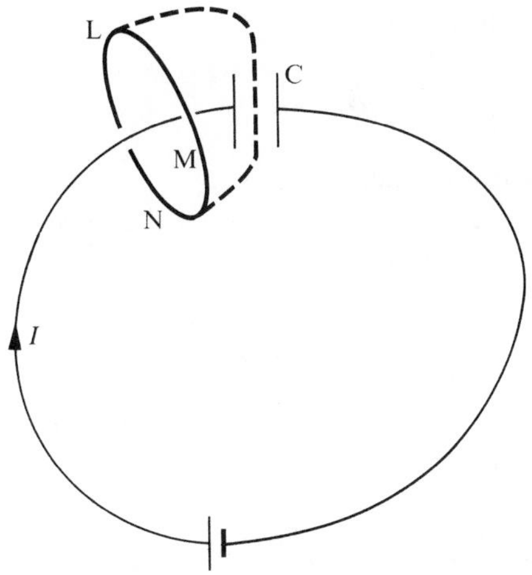

FIG. 8.1. Application of Ampère's law to calculate the magnetomotive force $\int \mathbf{H}.d\mathbf{s}$ round the circuit LMNL for the case of a current I charging a capacitor C.

surrounded by a guard ring, the field in between the plates is uniform, and the displacement $\mathbf{D}$ has the value $D = Q/A$, where Q is the total charge on the positive plate. Then the total displacement current between the plates is

$$A(\partial D/\partial t) = \partial Q/\partial t = I,$$

and our difficulties with the surface integral of the current density disappear if we include the displacement current.

For the case of an infinite homogeneous medium of electric permittivity ϵ and magnetic permeability μ, containing no free charges ($\rho_e = 0$) and having zero conductivity ($\sigma = 0$), our equations become

$$\operatorname{div} \mathbf{D} = \operatorname{div}(\epsilon \mathbf{E}) = \epsilon \operatorname{div} \mathbf{E} = 0, \tag{8.8}$$

$$\operatorname{div} \mathbf{B} = \operatorname{div}(\mu \mathbf{H}) = \mu \operatorname{div} \mathbf{H} = 0, \tag{8.9}$$

$$\operatorname{curl} \mathbf{E} = -\partial \mathbf{B}/\partial t = -\mu(\partial \mathbf{H}/\partial t), \tag{8.10}$$

$$\operatorname{curl} \mathbf{H} = \partial \mathbf{D}/\partial t = \epsilon(\partial \mathbf{E}/\partial t). \tag{8.11}$$

These form a set of simultaneous partial differential equations whose solutions can be found by eliminating one of the dependent variables, $\mathbf{E}$ or $\mathbf{H}$. This can be done by means of the vector identity

$$\operatorname{curl}(\operatorname{curl} \mathbf{E}) = \operatorname{grad}(\operatorname{div} \mathbf{E}) - \nabla^2 \mathbf{E} = -\nabla^2 \mathbf{E} \quad \text{(using eqn (8.8))}.$$

Then

$$\nabla^2 \mathbf{E} = -\operatorname{curl}(\operatorname{curl} \mathbf{E}) = \operatorname{curl}(\mu \partial \mathbf{H}/\partial t)$$

$$= \mu \frac{\partial}{\partial t}(\operatorname{curl} \mathbf{H}) = \mu\epsilon\left(\frac{\partial^2 \mathbf{E}}{\partial t^2}\right). \tag{8.12}$$

Similarly it may be shown that

$$\nabla^2\mathbf{H} = \mu\epsilon\left(\frac{\partial^2\mathbf{H}}{\partial t^2}\right). \tag{8.13}$$

These two equations are each of the general form for a wave motion in three dimensions; if the velocity with which the waves are propagated is v, the general wave equation is

$$\nabla^2\mathbf{X} = \frac{1}{v^2}\frac{\partial^2\mathbf{X}}{\partial t^2},$$

where $\mathbf{X}$ is some scalar or vector quantity. Comparison with our case shows that the wave velocity must be

$$v = (\mu\epsilon)^{-\frac{1}{2}}, \tag{8.14}$$

with the particular value in a vacuum ($\mu_r = 1$, $\epsilon_r = 1$) of

$$c = (\mu_0\epsilon_0)^{-\frac{1}{2}}. \tag{8.15}$$

Now the value of μ_0 has been defined as $4\pi \times 10^{-7}$, while the best direct measurement of ϵ_0 is that of Rosa and Dorsey (see § 5.8): this leads to a wave velocity of our electromagnetic waves in a vacuum of

$$(299\,784 \pm 10)\ \mathrm{km\,s^{-1}}.$$

Within the experimental error this is the same as the velocity of light determined by direct measurement, and it is accepted that light is a form of electromagnetic radiation, of the same form as radio waves, heat waves, X-rays, and γ-rays, which differ from light waves and from each other only in frequency and wavelength. The frequency range is from 10^4–10^{11} Hz for radio waves to 10^{20} Hz and over for γ-rays. Maxwell's theory does not give an adequate account of the interaction of electromagnetic radiation with atoms, and it has been found necessary (for example, in the photoelectric effect) to consider the electromagnetic energy as travelling about in 'packets' or 'quanta', which are indivisible. The size of a quantum for radiation of frequency ν is $h\nu$, where h is a universal constant known as Planck's constant. For radio waves such quantum effects are too small to affect the interaction with matter seriously, and we shall not consider them further at present.

If we accept the identification of our electromagnetic waves with light and radio waves, then we may use one of the accurate methods of determining their velocity (see § 22.7) to infer the value of the constant ϵ_0, since this gives a more accurate value than the direct comparison of a capacitance and resistance. The value obtained is given in Appendix C.

Eqn (8.14) shows that the velocity of the waves in a material medium is less than in a vacuum, since

$$v = c/(\mu_r\epsilon_r)^{\frac{1}{2}}.$$

For a light wave the velocity in a non-dispersive medium is $v = c/n$, where n is the refractive index of the medium. Hence we have

$$n = (\mu_r\epsilon_r)^{\frac{1}{2}}. \tag{8.16}$$

In using this equation we must remember that it applies only if we determine the values of n, μ_r, and ϵ_r at the same frequency, and nonsensical results may be obtained if we use values determined at different frequencies. Thus for water the optical refractive index is 1·33, but the static values of ϵ_r and μ_r are 81 and 1, leading to a refractive index (for very low frequencies only) of 9! Similarly, μ_r may be very high for a soft magnetic material at radiofrequencies, but close to unity at optical frequencies. The changes in ϵ_r, μ_r with frequency give rise to 'dispersion' in the refractive index (see Chapter 10).

8.2. Plane waves in isotropic dielectrics

In order to examine the behaviour of the electric and magnetic fields in more detail we shall consider the case of a plane wave. For simplicity we assume that this is moving in the direction of the z-axis of a set of right-handed Cartesian axes x, y, z. Then the definition of a plane wave is one in which the quantities have the same value over any plane normal to the direction of propagation; mathematically this is expressed by setting all partial differentials with respect to the x and y coordinates equal to zero (that is, $\partial/\partial x = \partial/\partial y = 0$). If these conditions are put into eqns (8.8) and (8.9) we find that

$$\partial E_z/\partial z = \partial H_z/\partial z = 0,$$

and from the z components of the curl equations ((8.10) and (8.11)) we have

$$\partial E_z/\partial t = \partial H_z/\partial t = 0.$$

This shows that apart from a uniform steady field in the z direction, which is not part of any wave motion, both E_z and H_z must be zero. The wave is therefore purely transverse in that no components of the electric or magnetic fields exist in the direction of propagation. The remaining components of the curl equations are, taking the z components of **E** and **H** to be zero, and $\partial/\partial y = \partial/\partial x = 0$,

$$\partial E_x/\partial z = -\mu\,\partial H_y/\partial t, \qquad -\partial H_y/\partial z = \epsilon\,\partial E_x/\partial t;$$

$$-\partial E_y/\partial z = -\mu\,\partial H_x/\partial t, \qquad \partial H_x/\partial z = \epsilon\,\partial E_y/\partial t.$$

These equations show that the x component of **E** is associated with the y component of **H**, while the y component of **E** is associated with the x component of **H**. The two components of **E** or **H** correspond to waves which are plane-polarized in directions normal to one another and normal to the direction of propagation. Thus we have two linearly independent solutions, in each of which the magnetic field is normal to the electric field. They differ only in the plane of polarization, which we shall take to be that of the electric vector; this differs from the convention accepted in light before the electromagnetic nature of the radiation was understood, which adopted the plane now known to be that of the magnetic vector. The electric vector is more important in the theory of dispersion (Chapter 10) as it determines the force on an electron in an atom.

Since we need consider only one state of polarization, we shall take the solution where the electric vector is parallel to the x-axis; it then follows that the magnetic vector has only a component parallel to the y-axis. The wave equation for E_x (eqn (8.12)) becomes

$$\frac{\partial^2 E_x}{\partial z^2} - \mu\epsilon\frac{\partial^2 E_x}{\partial t^2} = 0, \qquad (8.17)$$

which has a general solution of the form

$$E_x = F_1(z-vt)+F_2(z+vt), \qquad (8.18)$$

where F_1, F_2 may be functions of any form. The two functions represent waves travelling with velocity v, whose value is given by eqn (8.14). F_1 represents a wave travelling in the direction of z increasing, since any given point in the wave moves according to the equation

$$z = vt+\text{constant},$$

while F_2 is a wave travelling in the opposite direction, a given point moving as $z = -vt+\text{constant}$.

We may find the value of H_y by using one of the remaining components of our curl equations. For

$$\partial H_y/\partial z = -\epsilon\, \partial E_x/\partial t = \epsilon v\{F_1'(z-vt)-F_2'(z+vt)\},$$

where F' is the differential of F. Hence, since $v = (\epsilon\mu)^{-\frac{1}{2}}$, we find on integration

$$H_y = (\epsilon/\mu)^{\frac{1}{2}}\{F_1(z-vt)-F_2(z+vt)\}. \qquad (8.19)$$

This shows that the value of E_x bears a constant ratio to H_y in each of the travelling waves, but has the opposite sign for a wave travelling to the left. The ratio of E_x to H_y in a plane wave is often useful, and it has been called the 'intrinsic impedance' Z_0 of the medium. In our case of a non-conducting medium, where the wave velocity is v,

$$Z_0 = E_x/H_y = (\mu/\epsilon)^{\frac{1}{2}} = \mu v. \tag{8.20}$$

For a plane wave in free space $Z_0 = (\mu_0/\epsilon_0)^{\frac{1}{2}} = \mu_0 c = (\epsilon_0 c)^{-1}$, and its value is approximately 376·7 Ω. The fact that the dimensions of Z_0 are the same as those of the impedance discussed in earlier chapters is readily seen from the relations

$$\int \mathbf{H}.\mathrm{d}\mathbf{s} = I, \qquad \int \mathbf{E}.\mathrm{d}\mathbf{s} = V,$$

where V is the potential drop in a conductor; it appears also from the fact that E is measured in volts per metre and H in amperes per metre. The fact that the intrinsic impedance is a real quantity independent of z or t shows that the waveform of H_y is everywhere the same as that of E_x without any phase difference, in a travelling wave. This is true only of a non-conducting medium. Another important restriction is that the wave velocity must be independent of frequency if the waveform is to remain the same as the wave progresses. If it is not (that is, if we have a dispersive medium) we must perform a Fourier analysis of the initial waveform and treat each component of a given frequency separately. In the remainder of this chapter we shall assume that we are dealing with waves of one frequency only, and our field components are the real or imaginary parts of the exponential functions of the form

$$\exp\{\mathrm{j}\omega(t \pm z/v)\}.$$

Then the general solution for waves travelling both to the right and to the left will be

$$\begin{aligned} E_x &= A \exp\{\mathrm{j}\omega(t - z/v)\} + A' \exp\{\mathrm{j}\omega(t + z/v)\}, \\ Z_0 H_z &= A \exp\{\mathrm{j}\omega(t - z/v)\} - A' \exp\{\mathrm{j}\omega(t + z/v)\}. \end{aligned} \tag{8.21}$$

The exponentials are sometimes written in the forms $\exp\{\mathrm{j}\omega(t \pm nz/c)\}$ or $\exp \mathrm{j}(\omega t \pm \beta z)$; β is called the 'phase constant', and is equal to $2\pi/\lambda$, where λ is the wavelength in the medium. The phase velocity v is given by the relation

$$v = \omega/\beta. \tag{8.22}$$

It might be thought that a phase constant δ should be included in one of the waves in eqn (8.21), since otherwise some particular choice of zero is implied for either z or t which makes $\delta = 0$. We shall, however, allow A' (or A) to be complex, and of the form $A'' \exp(\mathrm{j}\delta)$, and thus include the phase constant in A' rather than in the complex exponential.

8.3. The Poynting vector of energy flow

Since an electromagnetic wave consists of electric and magnetic fields, we may expect that stored energy is associated with these fields. If the energy density, for which we now use the symbol U, is given by the equations derived for static fields, then for a plane wave in an isotropic dielectric it will be

$$U = \tfrac{1}{2}(\mathbf{D}.\mathbf{E}+\mathbf{B}.\mathbf{H}) = \tfrac{1}{2}(\epsilon\mathbf{E}^2+\mu\mathbf{H}^2).$$

For a wave propagated in the positive z-direction, the energy crossing unit area per second will just be the velocity times the energy density. This is

$$\begin{aligned} vU &= \frac{1}{2}\left\{\left(\frac{\epsilon}{\mu}\right)^{\frac{1}{2}}E_x^2+\left(\frac{\mu}{\epsilon}\right)^{\frac{1}{2}}H_y^2\right\} \\ &= \tfrac{1}{2}(E_x^2/Z_0+Z_0H_y^2) = E_xH_y = E_x^2/Z_0 = Z_0H_y^2. \end{aligned}$$

These various relations show that the energy stored in the magnetic field is just equal to that in the electric field. Since the direction of energy flow is normal to both **E** and **H**, and its magnitude is E_xH_y, we can represent the rate of energy flow per unit area by the vector (see Fig. 8.2)

$$\mathbf{E}\wedge\mathbf{H}. \tag{8.23a}$$

This vector is known as the Poynting vector, and it can be seen from eqns (8.21) that the direction of energy flow is reversed for a wave travelling in the opposite direction because the phase of **H** relative to **E** is reversed.

The value of $\mathbf{E}\wedge\mathbf{H}$ gives the instantaneous rate of energy flow. In a periodic wave the values of **E** and **H** at any point are oscillating functions of the time, and the mean rate of energy flow is found by averaging $\mathbf{E}\wedge\mathbf{H}$ over a complete period. In general the mean rate of energy flow is (cf. Problem 7.13) per unit area

$$\overline{\mathrm{Re}(\mathbf{E})\wedge\mathrm{Re}(\mathbf{H})} = \tfrac{1}{2}\,\mathrm{Re}(\mathbf{E}\wedge\mathbf{H}^*) \tag{8.23b}$$

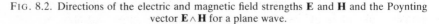

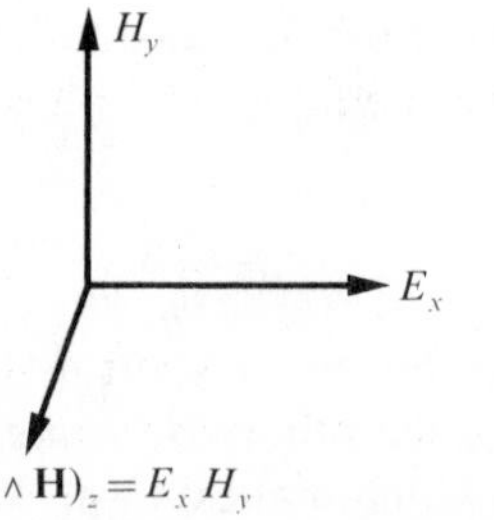

FIG. 8.2. Directions of the electric and magnetic field strengths **E** and **H** and the Poynting vector $\mathbf{E}\wedge\mathbf{H}$ for a plane wave.

where $\mathbf{H}^*$ is the complex conjugate of $\mathbf{H}$; this expression must be used when the phase difference between $\mathbf{E}$ and $\mathbf{H}$ is not 0 or π.

The Poynting vector is of greater significance than might be thought from the way in which it has been introduced above, and we shall now derive it for the general case. Suppose there is a region of space where an electric field of strength $\mathbf{E}$ causes a current flow of density $\mathbf{J}$. Then the power dissipated per unit volume is $\mathbf{E.J}$, and the total power dissipated will be

$$\int \mathbf{E.J}\,\mathrm{d}\tau = \int (\mathbf{E}.\mathrm{curl}\,\mathbf{H})\,\mathrm{d}\tau - \int \mathbf{E}.(\partial\mathbf{D}/\partial t)\,\mathrm{d}\tau,$$

where we have substituted for $\mathbf{J}$ the expression given by eqn (8.7). Now, using the vector identity

$$\mathrm{div}(\mathbf{E}\wedge\mathbf{H}) = \mathbf{H}.\mathrm{curl}\,\mathbf{E} - \mathbf{E}.\mathrm{curl}\,\mathbf{H}$$

and the expression for curl $\mathbf{E}$ given by eqn (8.3), we find that the power dissipated is

$$-\int \mathbf{E}.(\partial\mathbf{D}/\partial t)\,\mathrm{d}\tau - \int \mathbf{H}.(\partial\mathbf{B}/\partial t)\,\mathrm{d}\tau - \int \mathrm{div}(\mathbf{E}\wedge\mathbf{H})\,\mathrm{d}\tau.$$

The first two terms may be written as

$$-\int \left(\mathbf{E}\frac{\partial\mathbf{D}}{\partial t} + \mathbf{H}\frac{\partial\mathbf{B}}{\partial t}\right)\mathrm{d}\tau,$$

and this is the rate at which the stored energy increases, since the work done in infinitesimal changes $\mathrm{d}\mathbf{D}$, $\mathrm{d}\mathbf{B}$ at constant $\mathbf{E}$, $\mathbf{H}$ is

$$\int (\mathbf{E}\,\mathrm{d}\mathbf{D} + \mathbf{H}\,\mathrm{d}\mathbf{B})\,\mathrm{d}\tau.$$

If ϵ, μ are independent of $\mathbf{E}$, $\mathbf{H}$ and of time, the first two terms may be written as

$$-\frac{1}{2}\frac{\partial}{\partial t}\int (\mathbf{D.E} + \mathbf{B.H})\,\mathrm{d}\tau,$$

which we may interpret as the rate at which the energy stored in the electromagnetic field diminishes. The last term may be transformed using eqn (A.8) of Appendix A into the surface integral

$$-\int (\mathbf{E}\wedge\mathbf{H})\,.\,\mathrm{d}\mathbf{S}$$

taken over the surface bounding the volume under consideration. It represents the rate at which energy flows into the volume under consideration, and we may interpret the vector $\mathbf{E}\wedge\mathbf{H}$ as the rate at which energy flows across unit area of the boundary. Strictly speaking, only the integral of this quantity over a closed surface has been shown to represent the energy flow, but in most cases the vector $\mathbf{E} \wedge \mathbf{H}$ does represent the flow of energy per unit area at each point. An obvious exception to this would be a region with an electrostatic and a steady magnetic field arising from different sources, such as a charged capacitor between the poles of a permanent magnet.

8.4. Plane waves in conducting media

In a medium with a finite conductivity σ the field equations are the same as (8.8)–(8.11) except for an additional term $+\sigma\mathbf{E}$ on the right-hand side of eqn (8.11). These equations may be solved quite generally in the same way as in § 8.1, and elimination of $\mathbf{H}$ gives the differential equation

$$\nabla^2\mathbf{E} = \mu\epsilon\frac{\partial^2\mathbf{E}}{\partial t^2} + \mu\sigma\frac{\partial\mathbf{E}}{\partial t}. \tag{8.24}$$

An exactly similar equation holds for $\mathbf{H}$, and it may be shown that each of these equations represents a damped wave motion where the amplitude decays as the wave progresses owing to the extra term in $\partial\mathbf{E}/\partial t$ or $\partial\mathbf{H}/\partial t$. We shall not pursue this general solution, but specialize to the case of a plane wave travelling parallel to the z-axis. Then all partial derivatives $(\partial/\partial x)$ and $(\partial/\partial y)$ vanish, and the same methods as used in § 8.2 show that both E_z and H_z are zero, so that we again have a purely transverse wave. For a plane-polarized wave with a single component of $\mathbf{E}$ parallel to the x-axis, the field equations reduce to

$$\left.\begin{aligned} \partial E_x/\partial z &= -\mu(\partial H_y/\partial t), \\ -\partial H_y/\partial z &= \epsilon(\partial E_x/\partial t) + \sigma E_x. \end{aligned}\right\} \tag{8.25a}$$

These equations show that, as before, E_x is associated with H_y. (There is an independent solution, polarized in the perpendicular direction, where E_y is associated with H_x.) Elimination of H_y would give a wave equation for E_x of the same type as the general equation above, but a simpler method of solution is available on assuming that we are dealing with waves of a single frequency. It will appear that the velocity of propagation is now dependent on the frequency, so that it is necessary to consider each frequency separately in any case; we therefore introduce this restriction at the beginning, and take E_x and H_y to vary in time as $\exp \mathrm{j}\omega t$. We will also try a solution where the variation with z takes the form of a complex exponential, so that both E_x and H_y vary as

$$\exp\{\mathrm{j}\omega(t - \mathbf{n}z/c)\},$$

where $\mathbf{n}$ is a complex refractive index. Then our equations become

$$\left.\begin{aligned} j\omega(\mathbf{n}/c)E_x &= j\omega\mu H_y, \\ j\omega(\mathbf{n}/c)H_y &= (j\omega\epsilon+\sigma)E_x. \end{aligned}\right\} \tag{8.25b}$$

Elimination of E_x or H_y shows that these equations are satisfied provided that

$$\mathbf{n}^2 = \mu_r\epsilon_r - j(\sigma\mu_r/\omega\epsilon_0), \tag{8.26}$$

where we have substituted $(\mu_0\epsilon_0)$ for $1/c^2$.

To separate the real and imaginary parts of $\mathbf{n}$ we write it as

$$\mathbf{n} = n - jk, \tag{8.27}$$

giving

$$\left.\begin{aligned} n^2 - k^2 &= \mu_r\epsilon_r \\ nk &= (\sigma\mu_r/2\omega\epsilon_0). \end{aligned}\right\} \tag{8.28}$$

To understand the significance of n and k we note that the wave is propagated as

$$\exp\{j\omega(t-\mathbf{n}z/c) = \exp(-\omega kz/c)\exp j\omega(t-nz/c),$$

showing that the value of k determines the rate at which the amplitude of the wave decays (k appears in the argument of the real exponential) while n determines the phase velocity in the medium. Thus a complex refractive index $\mathbf{n}$ means that the wave is being absorbed as it proceeds, because the finite conductivity of the medium gives rise to a power loss through the Joule heating.

The intrinsic impedance of our medium is defined as E_x/H_y, so that

$$\mathbf{Z}_0 = \frac{c\mu}{\mathbf{n}} = \left(\frac{\mu}{\epsilon - j\sigma/\omega}\right)^{\frac{1}{2}}. \tag{8.29}$$

Thus the impedance is a complex quantity, but is independent of z or t. This means that the ratio of the amplitudes of E_x and H_y is everywhere the same, but there is a phase difference between them. Since also

$$\mathbf{Z}_0 = \frac{c\mu(n+jk)}{(n^2+k^2)}$$

we see that the phase difference ϕ is given by the relation $\tan\phi = k/n$. $(\mathbf{E}\wedge\mathbf{H})$ is also complex, and the mean power flow must be calculated using eqn (8.23b).

The full solution of eqns (8.28) to find n and k gives rather complicated expressions, but these simplify greatly for the case of a good conductor. For a metal the conduction current is enormously greater than the displacement current at all frequencies up to those of ultraviolet light, as will be seen by comparing the values of σ and $\omega\epsilon$, which occur in the second of eqns (8.25b). For most metals σ is $10^7\ \Omega^{-1}\ \mathrm{m}^{-1}$ or greater, while for light of wavelength 500 nm, $\omega\epsilon \times 10^{15} \times 8{\cdot}85 \times \epsilon_r \approx 3{\cdot}5 \times 10^4\ \epsilon_r$. We do not know what ϵ_r is for a metal, but provided it is not very much greater than the ordinary values found in dielectric substances, the displacement current even for visible light will be much smaller than the conduction current. For lower frequencies the inequality increases, and a good approximation is obtained by omitting the displacement current. This is equivalent to setting $\epsilon_r = 0$, and eqns (8.28) then give

$$n = k = (\sigma\mu_r/2\omega\epsilon_0)^{\frac{1}{2}}.$$

If we introduce a quantity δ such that $n = k = c/\omega\delta$, the field components in the metal are propagated as

$$\exp(-z/\delta)\exp \mathrm{j}(\omega t - z/\delta), \tag{8.30}$$

from which it can be seen that δ has the dimensions of a length. δ is known as the 'skin depth', and the amplitude of the wave falls to $1/\mathrm{e}$ of its initial value in a distance δ, while the apparent wavelength in the metal is $2\pi\delta$. The equation for δ is (writing $\omega = 2\pi f$)

$$\delta = (\tfrac{1}{2}\sigma\omega\mu)^{-\frac{1}{2}} = (\pi\sigma f\mu)^{-\frac{1}{2}} \tag{8.31}$$

and the intrinsic impedance of our metal can be written as

$$\mathbf{Z}_0 = (1+\mathrm{j})/(\sigma\delta) = (1+\mathrm{j})(\rho/\delta), \tag{8.32}$$

where ρ is the resistivity. The magnitude of the skin depth decreases with the inverse half-power of the frequency, and of the permeability and conductivity of the metal. An idea of its order of magnitude is obtained from the fact that for copper it has the approximate values: $6{\cdot}6 \times 10^{-5}$ m at a frequency of 1 MHz, $6{\cdot}6 \times 10^{-7}$ m at 10^4 MHz (a wavelength of 3×10^{-2} m), and $2{\cdot}7 \times 10^{-9}$ m at 6×10^8 MHz (green light). Thus the wavelength of the radiation in the metal is very small compared with the wavelength in free space; and Z_0 is also much smaller than the value for free space; for copper it is about $0{\cdot}026(1+\mathrm{j})\Omega$ at 10^4 MHz.

8.5. The skin effect

Since an electromagnetic wave varies in phase very rapidly inside a metal, one must ask whether this affects the distribution of alternating current inside a conductor. An electromagnetic field is associated with such a current, and Problem 8.1 shows that there is a flow of energy into the surface of a conductor which is just equal to the energy dissipated as Joule heat. This implies that an electromagnetic wave is passing through the surface, and from what has been found about the behaviour of such a wave we should expect it to be very rapidly attenuated inside. The current associated with it would therefore flow only in a thin skin of thickness of the order of δ, a phenomenon known as the 'skin effect'. For steady currents $\omega=0$ and $\delta=\infty$, so that the current distribution would be uniform, but as the frequency rises δ decreases and the current is increasingly confined to the surface. For this reason thin tubes are just as good conductors of high-frequency currents as solid rods, but the resistance is of course higher than for a steady current in a solid rod owing to the smaller effective cross-section. We will now compute the high-frequency resistance of a wire. Neglecting displacement currents, and writing **J** for $\sigma\mathbf{E}$ in eqn (8.24), we have

$$\nabla^2\mathbf{J}=\mu\sigma(\partial\mathbf{J}/\partial t)=\mathrm{j}\omega\mu\sigma\mathbf{J} \tag{8.33}$$

for a current oscillating with frequency $\omega/2\pi$. For a cylindrical wire we should express the operator ∇^2 in cylindrical coordinates and the general solution involves Bessel functions with a complex argument. The values we have found for δ in copper show that at high frequencies it will ordinarily be very small compared with the radius a of the wire. In this case we may obtain an approximate solution by neglecting the curvature of the surface of the wire and regarding it as a thin tube which can be split and unrolled to form an infinite flat strip of width $b=2\pi a$, in which a current flows parallel to the surface of the strip in the long direction. If (as in Fig. 8.3) we take this direction as the x-axis and the normal to the surface of the strip as the z-axis, then the current density has a component σE_x in the x direction which varies with z and t as given by eqn (8.30), so that

$$J_x=J_0\exp(-z/\delta)\exp\mathrm{j}(\omega t-z/\delta)\},$$

where J_0 is the current density at the surface of the wire ($z=0$). This

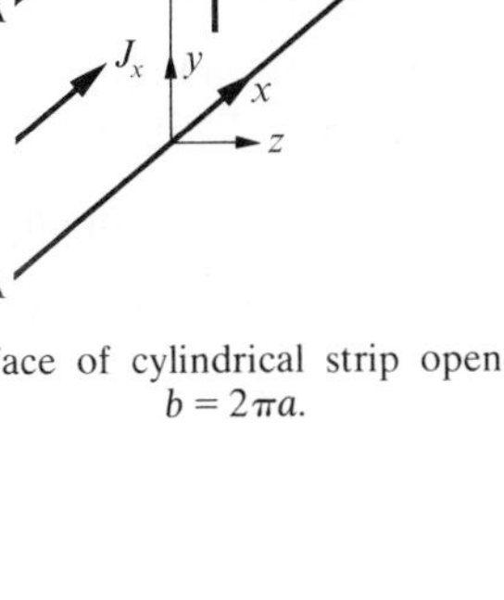

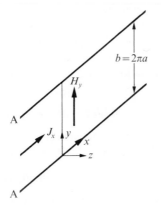

FIG. 8.3. Current flow at surface of cylindrical strip opened out to flat strip of width $b=2\pi a$.

equation shows that the current changes phase as we move into the wire as well as decreasing rapidly in amplitude.

To find the effective resistance R per unit length of the wire we must calculate the total current I flowing in the strip at any instant and the power dissipation per unit length P, and set $P=\frac{1}{2}|I|^2 R$. Now

$$I = b\int_0^\infty J_x \,\mathrm{d}z = bJ_0 \exp \mathrm{j}\omega t\int_0^\infty \exp\{-(1+\mathrm{j})z/\delta\}\,\mathrm{d}z$$

$$= \left(\frac{b\delta J_0}{1+\mathrm{j}}\right)\exp \mathrm{j}\omega t = \frac{1}{\sqrt{2}} b\delta J_0 \exp \mathrm{j}(\omega t - \tfrac{1}{4}\pi).$$

This expression shows that the total current has a different phase from J_0; we require only the amplitude of the total current whose square is

$$|I|^2 = \tfrac{1}{2}(b\delta J_0)^2.$$

The instantaneous power dissipation per unit volume of the metal is ρJ_x^2, and hence the total average dissipation per unit length of the wire is

$$P = b\int_0^\infty \tfrac{1}{2}\rho\,|J_x|^2\,\mathrm{d}z = \tfrac{1}{2}b\rho J_0^2\int_0^\infty \exp(-2z/\delta)\,\mathrm{d}z = \tfrac{1}{4}b\rho\delta J_0^2 = \tfrac{1}{2}\rho\,|I|^2/b\delta,$$

where ρ is the resistivity of the wire. Hence the effective resistance per unit length of the wire is

$$R = 2P/|I|^2 = \rho/b\delta = \rho/2\pi a\delta. \tag{8.34a}$$

This equation shows that the resistance is the same as if there were a current I of uniform density flowing in a thin tube of radius a and thickness equal to δ, and the reason for the name 'skin depth' is thus apparent. The high-frequency resistance of such a wire is greater than the d.c. resistance by a factor $a/2\delta$, when $\delta \ll a$.

This type of calculation may be applied to the more general problem of the power dissipated per unit area of a plane metallic surface when there is a tangential magnetic field whose strength has an amplitude H_0 at the surface. At any point in the metal the current density is

$$J_x = \sigma E_x = -\partial H_y/\partial z \quad \text{(from eqn (8.25a))},$$

assuming that the displacement current can be neglected. It follows that the total current in the metal (per unit width) is given by $\int_0^\infty (-\partial H_y/\partial z)\,\mathrm{d}z = H_0$, the value of the magnetic field strength at the surface. The mean power dissipation per unit area is

$$P = \tfrac{1}{4}\rho\delta J_0^2 = \tfrac{1}{2}\rho H_0^2/\delta. \tag{8.34b}$$

This is just equal to the average value of Poynting's vector at the surface (see Problem 8.1).

8.6. Reflection and refraction of plane waves at the boundary of two dielectrics

The reflection and refraction of light waves at the surface separating two media of different refractive indices is a familiar phenomenon and we now inquire whether electromagnetic theory offers a simple explanation of it. We assume two non-conducting dielectric media, separated by a plane boundary which we take to be the xy plane (the plane $z = 0$). We also choose the direction of the x-axis to be in the plane of incidence, that is, the incident ray lies in the plane $y = 0$, as in Fig. 8.4, making an angle θ with the normal to the boundary. θ is known as the 'angle of incidence'. Then all field components of the incident wave vary with the space and time coordinates as

$$\exp[j\omega\{t - n_1(x \sin\theta + z\cos\theta)/c\},$$

where c/n_1 is the velocity of the wave in the first medium, and $\omega/2\pi$ is its frequency of oscillation.

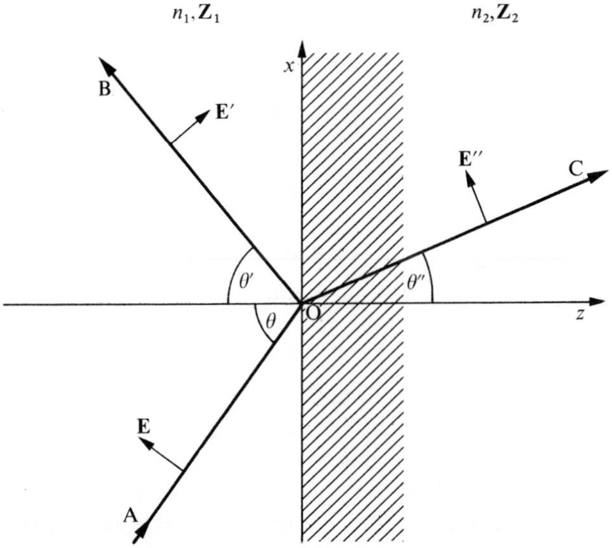

FIG. 8.4. Reflected wave OB and refracted wave OC at the boundary ($z = 0$) between two media. AO is the incident wave, and the electric vectors are in the plane of incidence as shown. The magnetic vector is parallel to the y-axis, and hence normal to the plane of the paper. If $n_2 > n_1$, and the permeability of both media is unity, then the direction of $\mathbf{E}'$ is as shown when the angle of incidence is greater than the Brewster angle $\tan^{-1}(n_2/n_1)$, but in the opposite direction when it is less than the Brewster angle.

When the incident wave falls on the boundary there will in general be both a reflected wave and a transmitted wave. We know nothing about these waves, either as to their frequency or their direction. We have, however, boundary conditions for the components of **E** and **H** at the surface and we assume that the behaviour of **E** and **H** must be the same as derived earlier in Chapters 1 and 4. Hence the tangential components of **E** and **H** must be continuous on the two sides of the boundary at all times and for all values of x and y. The first of these conditions shows that the reflected and transmitted waves must have the same frequency as the incident wave. Secondly, at all points in the boundary plane along a line for which x is constant the field components of the incident wave are constant in amplitude and phase. Our boundary conditions can be satisfied everywhere only if this is true of the reflected and transmitted waves as well, and they must therefore also travel in the plane of incidence. The corresponding rays are shown in Fig. 8.4, making angles θ' and θ'' with the normal, respectively; they are known as the 'angle of reflection' and the 'angle of refraction'. The field components of the two waves must be propagated as

$$\text{(reflected wave)} \qquad \exp j\omega\{t-n_1(x\sin\theta'-z\cos\theta')/c\},$$

$$\text{(transmitted wave)} \qquad \exp j\omega\{t-n_2(x\sin\theta''+z\cos\theta'')/c\},$$

where c/n_2 is the velocity in the second medium. At the boundary plane $(z=0)$, the boundary conditions can only be satisfied everywhere if the arguments of the exponentials for the incident, reflected, and transmitted waves are all identical. Hence

$$n_1 x\sin\theta = n_1 x\sin\theta' = n_2 x\sin\theta'',$$

which gives

$$\theta=\theta',$$

so that the angles of incidence and reflection are equal, and

$$n_1\sin\theta = n_2\sin\theta'', \tag{8.35}$$

which is Snell's law for refraction.

These laws of reflection and refraction are true not only for electromagnetic waves but for all kinds of wave motion, since they depend only on the assumption that the characteristic quantities involved in the wave motion (in this case the electric and magnetic fields) shall be continuous at the boundary.

Intensity relations

In order to obtain the amplitudes of the reflected and refracted waves we must examine the problem in more detail, and match the amplitudes of the tangential components of the electric and magnetic fields on the two sides of the boundary. To do this we must consider the two principal directions of polarization of the incident wave separately. These directions are with the electric field in the plane of incidence and normal to the plane of incidence respectively. For the former of these there are components of **E** parallel to the x- and z-axes, but $E_y = 0$; the only component of **H** is parallel to the y-axis. These statements hold for all the three waves, so that we have (see Fig. 8.4) the following:

incident wave

$$\left.\begin{aligned} E_x &= A\cos\theta \exp j\omega\{t-n_1(x\sin\theta+z\cos\theta)/c\} \\ E_z &= -A\sin\theta \exp j\omega\{t-n_1(x\sin\theta+z\cos\theta)/c\} \\ H_y &= (A/Z_1)\exp j\omega\{t-n_1(x\sin\theta+z\cos\theta)/c\}, \end{aligned}\right\} \tag{8.36}$$

reflected wave

$$\left.\begin{aligned} E'_x &= A'\cos\theta \exp j\omega\{t-n_1(x\sin\theta-z\cos\theta)/c\} \\ E'_z &= A'\sin\theta \exp j\omega\{t-n_1(x\sin\theta-z\cos\theta)/c\} \\ H'_y &= -(A'/Z_1)\exp j\omega\{t-n_1(x\sin\theta-z\cos\theta)/c\} \end{aligned}\right\} \tag{8.37}$$

refracted wave

$$\left.\begin{aligned} E''_x &= A''\cos\theta'' \exp j\omega\{t-n_2(x\sin\theta''+z\cos\theta'')/c\} \\ E''_z &= -A''\sin\theta'' \exp j\omega\{t-n_2(x\sin\theta''+z\cos\theta'')/c\} \\ H''_y &= (A''/Z_2)\exp j\omega\{t-n_2(x\sin\theta''+z\cos\theta'')/c\}, \end{aligned}\right\} \tag{8.38}$$

where Z_1, Z_2 are the intrinsic impedances of the two media respectively.

On making the tangential components of **E** and **H** (that is, E_x and H_y) continuous at the plane $z = 0$, we obtain the relations

$$A\cos\theta + A'\cos\theta = A''\cos\theta'' \tag{8.39}$$

and

$$(A-A')/Z_1 = A''/Z_2, \tag{8.40}$$

which give

$$\frac{A'}{A} = \frac{Z_2\cos\theta''-Z_1\cos\theta}{Z_2\cos\theta''+Z_1\cos\theta} \tag{8.41}$$

and

$$\frac{A''}{A} = \frac{2Z_2\cos\theta}{Z_2\cos\theta''+Z_1\cos\theta}. \tag{8.42}$$

If the media are such that $\mu_1 = \mu_2 = 1$, as is the case for light of visible wavelengths, $Z_1/Z_2 = n_2/n_1 = \sin\theta/\sin\theta''$, and these formulae can be written in the form

$$\frac{A'}{A} = \frac{\sin 2\theta''-\sin 2\theta}{\sin 2\theta''+\sin 2\theta}, \tag{8.43}$$

$$\frac{A''}{A} = \frac{4\sin\theta''\cos\theta}{\sin 2\theta''+\cos 2\theta}. \tag{8.44}$$

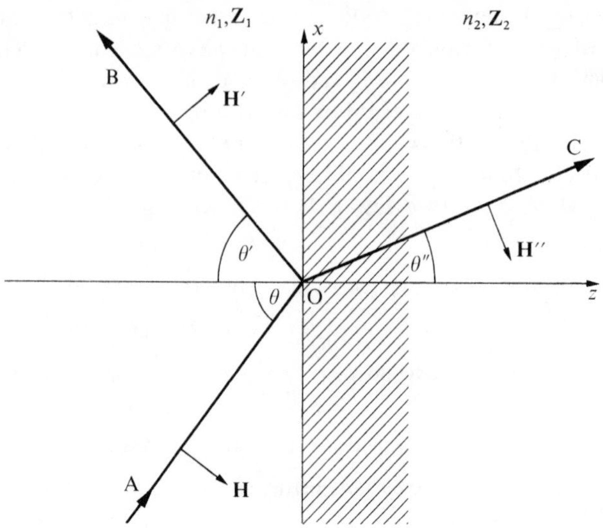

FIG. 8.5. Incident wave AO, reflected wave OB, and refracted wave OC at the boundary between two media. The electric vector is normal to the plane of incidence and hence normal to the plane of the paper; the magnetic vector is in the plane of incidence. The theory shows that if $n_2 > n_1$, and the permeability of both media is unity, the actual direction of $\mathbf{H}'$ is opposite to that given in the figure for all values of θ.

The condition that there shall be no reflection ($A' = 0$) is that

$$\sin 2\theta = \sin 2\theta'',$$

which is satisfied if 2θ and $2\theta''$ are supplementary angles; that is

$$\theta + \theta'' = \tfrac{1}{2}\pi,$$

so that the reflected and refracted rays are normal to one another. Using Snell's law, we find that this occurs when

$$\tan \theta = n_2/n_1. \tag{8.45}$$

The angle which satisfies this relation is known as the Brewster angle.

The field components of the wave which is plane polarized with its electric vector normal to the plane of incidence are as follows (see Fig. 8.5):

incident wave

$$\left.\begin{aligned} E_y &= BF_1 \\ H_x &= -(B/Z_1)\cos\theta\, F_1 \\ H_z &= (B/Z_1)\sin\theta\, F_1, \end{aligned}\right\} \tag{8.46}$$

reflected wave

$$\left.\begin{aligned} E'_y &= B'F_2 \\ H'_x &= (B'/Z_1)\cos\theta\, F_2 \\ H'_z &= (B'/Z_1)\sin\theta\, F_2, \end{aligned}\right\} \tag{8.47}$$

refracted wave

$$\left.\begin{aligned} E''_y &= B''F_3 \\ H''_x &= -(B''/Z_2)\cos\theta'' F_3 \\ H''_z &= (B''/Z_2)\sin\theta'' F_3, \end{aligned}\right\} \tag{8.48}$$

where for simplicity we have written F_1, F_2, F_3 for the three complex exponentials representing the propagation of the waves, which are the same as in eqns (8.36), (8.37), and (8.38) respectively.

The boundary conditions for E_y and H_x now give

$$B+B' = B'',$$
$$(B-B')\cos\theta/Z_1 = B''\cos\theta''/Z_2,$$

the solutions of which are

$$\frac{B'}{B} = \frac{Z_2\cos\theta - Z_1\cos\theta''}{Z_2\cos\theta + Z_1\cos\theta''} \tag{8.49}$$

and

$$\frac{B''}{B} = \frac{2Z_2\cos\theta}{Z_2\cos\theta + Z_1\cos\theta''}. \tag{8.50}$$

These equations are similar to those obtained for the other direction of polarization but not identical, so that the magnitudes of the reflected and refracted waves are different. This is still true when we specialize to the case of $\mu_1 = \mu_2 = \mu_0$, when the formulae become

$$\frac{B'}{B} = \frac{\sin(\theta''-\theta)}{\sin(\theta''+\theta)}, \tag{8.51}$$

$$\frac{B''}{B} = \frac{2\sin\theta''\cos\theta}{\sin(\theta''+\theta)}. \tag{8.52}$$

Since in general θ'' is not equal to θ, eqn (8.51) shows that there is no angle at which the reflected wave for this direction of polarization has zero amplitude. Thus if we start with unpolarized light, which consists of light containing a superposition of many components with their electric vectors in random orientations, and reflect it from a dielectric such as glass at the Brewster angle, the reflected wave will only contain one direction of polarization, that with the electric vector normal to the plane of incidence. This phenomenon can therefore be used to produce plane polarized light, and originally it was used to define the plane of polarization of the reflected light at the Brewster angle as being the plane of the reflected ray. Our treatment shows that this is in fact the plane of the magnetic vector in the reflected light, not of the electric vector.

The fact that the expressions for A'/A and A''/A are real shows that the phase changes at the boundary between two perfect dielectrics are always 0 or π. Inspection of the equations shows that for a wave with the electric vector in the plane of incidence, passing from a medium of lower refractive index to one of higher refractive index, there is a change of sign in the tangential component E_x (but not in E_z, H_y) of the reflected wave when the angle of incidence is less than the Brewster angle, whereas the reverse is the case when it is greater than this angle. For the wave with the electric vector normal to the plane of incidence, there is a change of phase in E_y and H_z in the reflected wave (assuming $n_2 > n_1$), but not in H_x, whatever the angle of incidence.

Eqns (8.43), (8.44), (8.51), and (8.52) are known as Fresnel's formulae, after their discoverer. For normal incidence they give indeterminate results, but it is readily seen from eqns (8.41) and (8.42) (or (8.49) and (8.50)) that we have then

$$\frac{A'}{A} = \frac{Z_2 - Z_1}{Z_2 + Z_1} = \frac{n_1 - n_2}{n_1 + n_2} \tag{8.53}$$

and

$$\frac{A''}{A} = \frac{2Z_2}{Z_2 + Z_1} = \frac{2n_1}{n_1 + n_2}, \tag{8.54}$$

where the expressions on the extreme right apply only when

$$\mu_2 = \mu_1 = \mu_0.$$

The 'reflecting power' of the surface is equal to $(A'/A)^2$, since the power in the incident and reflected waves is proportional to the square of the amplitude. If $Z_2 > Z_1$, $A'' > A$ so that the amplitude of the electric vector is greater in the transmitted wave than the incident wave. Since the power varies as (amplitude of electric vector)^2/(impedance), the transmitted power is less than the incident power, and it is readily verified that the difference is equal to the reflected power.

Total internal reflection

When the refractive index n_2 of the second medium is less than n_1, application of Snell's law (eqn (8.35))

$$\sin\theta'' = (n_1/n_2)\sin\theta$$

to find the angle of refraction leads to values of $\sin\theta''$ greater than unity when $\sin\theta$ is greater than n_2/n_1. Since there is no real angle for which the sine is greater than unity, we conclude that there is no refracted wave and that all the energy is reflected. This is confirmed by inspection of eqns (8.41) and (8.49), for when $\sin\theta''>1$, $\cos\theta''=(1-\sin^2\theta'')^{\frac{1}{2}}$ is a purely imaginary quantity, and the expressions for A'/A and B'/B are each of the form $(a-\mathrm{j}b)/(a+\mathrm{j}b)$ whose modulus is unity. Thus the reflection is total, but there is a change of phase on reflection which is different for the two directions of polarization. If the incident wave is plane-polarized in a direction which is not in or normal to the plane of incidence, the two components in these planes of the reflected wave will not be in phase and the wave will therefore be elliptically polarized.

The fact that there is no 'refracted wave' does not mean that there is no disturbance in the second medium, for eqns (8.42) and (8.50) show that A''/A and B''/B are finite. In order to find what kind of wave is propagated in the second medium, we write $\sin\theta''=\cosh\gamma$ (since $\cosh\gamma$ is always greater than one), and then

$$\cos\theta''=(1-\sin^2\theta'')^{\frac{1}{2}}=\mathrm{j}(\cosh^2\gamma-1)^{\frac{1}{2}}=\pm\mathrm{j}\sinh\gamma.$$

Hence the field components in the second medium are propagated as

$$\begin{aligned}\exp \mathrm{j}\omega\{t-n_2(x\cosh\gamma-\mathrm{j}z\sinh\gamma)/c\}&\\ &=\exp\{(-\omega n_2 z\sinh\gamma)/c\}\exp \mathrm{j}\omega\{t-(n_2 x\cosh\gamma)/c\}\\ &=\exp\{(-2\pi z\sinh\gamma)/\lambda_2\}\exp \mathrm{j}\omega\{t-(n_1 x\sin\theta)/c\} \quad (8.55)\end{aligned}$$

where λ_2 is the wavelength of the radiation in the second medium. Eqn (8.55) shows that the wave is rapidly attenuated on the far side of the boundary, since at a distance $z=\lambda_2$ its amplitude falls by

$$\exp(-2\pi\sinh\gamma).$$

For a given value of x, there is no phase change as we proceed in the z direction, but there is a phase change as we move along the boundary in the x direction. This is because a wave front obliquely incident on the boundary arrives at points of greater x at a later time.

If all the incident energy is reflected, no energy can be transmitted in the second medium. If we compute the mean value of Poynting's vector for a direction normal to the boundary (that is, the value of $\frac{1}{2}\operatorname{Re}(E_xH_y^*)$ or $\frac{1}{2}\operatorname{Re}(-E_yH_x^*)$ according to the direction of polarization) in the second medium, we find that it is zero, because H_y is $\frac{1}{2}\pi$ out of phase with E_x. This means that no energy is transported away from the boundary on the average. Energy does flow into the second medium, since the stored energy must be finite if the field components are finite, but the flow is in the opposite direction during a later part of the cycle and the stored energy is returned to the first medium.

8.7. Reflection from the surface of a metal

When an electromagnetic wave is incident on the plane surface of a conducting medium, the amplitudes of the reflected and transmitted waves can be calculated in a manner essentially similar to that used for dielectrics. The refractive index and intrinsic impedance of a conducting medium are now complex numbers, and the formulae deduced for dielectrics may be taken over by use of complex values for n_2 and Z_2. From eqns (8.41) and (8.49) it is then apparent that the phase change on reflection is not in general 0 or π, so that a plane-polarized wave will become elliptically polarized after reflection. Here we shall only calculate the reflecting power of a metal for a wave falling on it from free space at normal incidence, using eqn (8.53). For a good conductor

$$\mathbf{Z}_2 = (1+\mathrm{j})/\sigma\delta \quad \text{from eqn (8.32)},$$

while for free space

$$\mathbf{Z}_1 = \mu_0 c,$$

where δ is the skin depth in the metal and c is the velocity of light in free space. Thus the reflection coefficient is

$$\frac{\mathrm{A}'}{\mathrm{A}} = \frac{\mathbf{Z}_2 - \mathbf{Z}_1}{\mathbf{Z}_2 + \mathbf{Z}_1} = \frac{(1+\mathrm{j}) - \mu_0\sigma c\delta}{(1+\mathrm{j}) + \mu_0\sigma c\delta},$$

where the complex value shows that there is a phase change even at normal incidence, and the reflecting power is

$$\left|\frac{\mathrm{A}'}{\mathrm{A}}\right|^2 = \left|\frac{(1-t)+\mathrm{j}}{(1+t)+\mathrm{j}}\right|^2 = \frac{2+t^2-2t}{2+t^2+2t},$$

where we have written t for the quantity $\mu_0\sigma c\delta$. For a metal, where $\sigma \approx 10^7\,\Omega^{-1}\,\mathrm{m}^{-1}$, $t \approx 4\times10^9\delta$, and is thus much greater than unity at all frequencies up to that of visible light ($f \approx 10^{15}$ Hz). Hence approximately

$$|(\mathrm{A}'/\mathrm{A})|^2 = 1 - 4/t = 1 - 2{\cdot}1\times10^{-5}(\mu_r f/\sigma)^{\frac{1}{2}}. \tag{8.56}$$

This formula shows that metals should be almost perfect reflectors of electromagnetic radiation for all frequencies up to those of visible light. For copper at a frequency of 10^{10} Hz (a wavelength of 3×10^{-2} m), the reflecting power differs from unity by about $2{\cdot}7\times10^{-4}$, and this formula would lead us to expect that it should only fall below unity by about 0·09 at a frequency of 10^{15} Hz (a wavelength of 300 nm). The fact that copper appears strongly coloured shows that this formula fails for optical frequencies, when electrons in the atom other than those responsible for the conductivity begin to play a role. In addition, the effective conductivity of a metal at high frequencies is less than the low frequency value (see § 12.10).

The high reflection coefficient of a metal arises from the fact that its intrinsic impedance is very much smaller than that of free space. For copper at 10^{10} Hz, Z_2 is $0{\cdot}026(1+j)\Omega$, while for free space $Z_1=376{\cdot}7\ \Omega$. Thus to satisfy the boundary conditions the wave must be almost totally reflected with a phase change in the electric vector but not in the magnetic vector, making **E** almost zero at the metal surface, and **H** almost twice the amplitude due to the incident wave alone.

8.8. The pressure due to radiation

When a plane electromagnetic wave travels through a conducting medium in the z direction, a conduction current flows of density $J_x=\sigma E_x$ (assuming the wave to be linearly polarized parallel to the x-axis). Associated with the wave is a magnetic flux density B_y, which will exert a force on the current whose magnitude on a volume element $d\tau$ is

$$dF_z=(\mathbf{J}\wedge\mathbf{B})_z\ d\tau=J_xB_y\ d\tau$$

in the positive z direction. This force is in the direction in which the energy is travelling, and gives rise to a 'radiation pressure'. From eqn (8.7), $J_x=-\partial H_y/\partial z-\partial D_x/\partial t$, since all other components vanish in a plane wave, and hence

$$\begin{aligned}dF_z/d\tau&=(-\partial H_y/\partial z-\partial D_x/\partial t)B_y\\&=-(\partial H_y/\partial z)B_y-\partial(D_xB_y)/\partial t+(\partial B_y/\partial t)D_x.\end{aligned}$$

But from eqn (8.3), $\partial B_y/\partial t=-\partial E_x/\partial z$, so that

$$\begin{aligned}dF_z/d\tau&=-\left(D_x\frac{\partial E_x}{\partial z}+B_y\frac{\partial H_y}{\partial z}\right)-\frac{\partial}{\partial t}(D_xB_y)\\&=-\frac{\partial U}{\partial z}-\epsilon\mu\frac{\partial(E_xH_y)}{\partial t}\end{aligned}$$

in a medium where D, B are linearly proportional to E, H. In the steady state the second term $\partial(E_xH_y)/\partial t$ vanishes when averaged over a cycle, and the significance of the minus sign before $(\partial U/\partial z)$ is that the force is in the forward direction provided that the stored energy U diminishes as the wave travels onwards, as it necessarily will in a conducting medium. We can interpret the volume force as due to the gradient of a pressure P, and, from the geometry shown in Fig. 8.6,

$$dF_z=-(\partial P/\partial z)\ dx\ dy\ dz=-(\partial P/\partial z)\ d\tau.$$

Hence

$$P=U\quad\text{(normal incidence).}\qquad(8.57)$$

The pressure is exerted in the direction of Poynting's vector, and can be attributed to a momentum G per unit volume which flows across unit

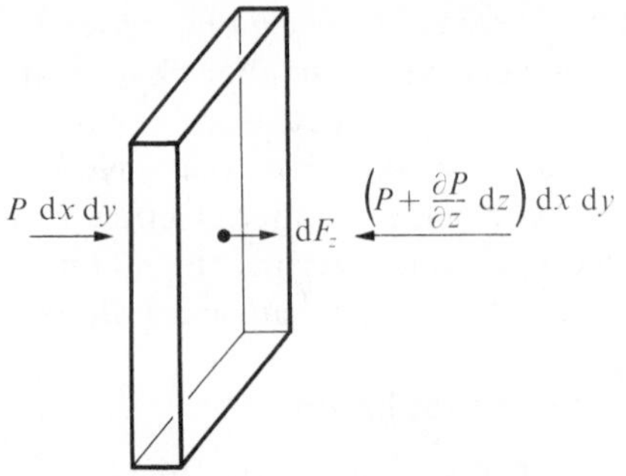

FIG. 8.6. Interpretation of volume force as gradient of a pressure.

area at the rate (see § 8.3)

$$U = vG = v^{-1}\,|\mathbf{E}\wedge\mathbf{H}| \tag{8.58}$$

This result is consistent with quantum and relativity theory by which radiation consists of photons of energy $h\nu$, whose momentum *in vacuo* is $h\nu/c$. For diffuse radiation, consisting of waves travelling in all directions, only one-third of the total energy density will on average be associated with waves travelling normal to the surface, so that we have

$$P = \tfrac{1}{3}U \quad \text{(diffuse radiation).} \tag{8.59}$$

When a wave is totally reflected, its momentum is reversed so that the pressure is doubled; however the energy density is also doubled so that the equation $P = U$ for normal incidence and eqn (8.59) for diffuse radiation still hold. These results are easily obtained from the concept of momentum flow in the electromagnetic wave; other methods are much more complex, as it is necessary to include both the stress at the boundary and the volume force on the conduction current. In the example above there is no change of medium, and we obtain the correct answer from the volume force alone. On the other hand, when an electromagnetic wave is partially reflected at the boundary between two non-conducting dielectrics there is no force on any conduction current and the pressure at the boundary arises entirely from the difference in the Maxwell stress tensor on either side of the boundary. It can again be calculated much more easily from the momentum balance by considering the energy densities in the incident, reflected, and transmitted waves; that the two approaches give the same answer in the case of normal incidence is verified in Problem 8.13.

8.9. Radiation from an oscillating dipole

In Chapter 4 the concept of a vector potential $\mathbf{A}$, such that

$$\mathbf{B} = \text{curl}\ \mathbf{A}, \tag{8.60}$$

was introduced. The vector potential is not of great use in elementary problems, but is of considerable assistance in calculating the radiation from an aerial. The fundamental equations of the electromagnetic field are

$$\operatorname{div} \mathbf{D} = \rho_e, \tag{8.1}$$

$$\operatorname{div} \mathbf{B} = 0, \tag{8.2}$$

$$\operatorname{curl} \mathbf{E} = -\frac{\partial \mathbf{B}}{\partial t}, \tag{8.3}$$

$$\operatorname{curl} \mathbf{H} = \mathbf{J} + \frac{\partial \mathbf{D}}{\partial t}. \tag{8.7}$$

On substituting from eqn (8.60) into (8.3) we have, as in § 5.1,

$$\operatorname{curl}\left(\mathbf{E} + \frac{\partial \mathbf{A}}{\partial t}\right) = 0,$$

the solution of which is

$$\mathbf{E} = -\frac{\partial \mathbf{A}}{\partial t} - \operatorname{grad} V, \tag{8.61}$$

where grad V is the 'constant of integration', V being some scalar function. Since curl grad of a scalar function is zero, any such function satisfies eqn (8.3). In a static problem, where $\mathbf{A}$ is constant, eqn (8.61) reduces to $\mathbf{E} = -\operatorname{grad} V$, where V is the ordinary scalar electrostatic potential, as defined in eqn (1.6).

Eqn (8.60) does not define the vector potential $\mathbf{A}$ completely, since $\mathbf{A}$ is the solution of a differential equation and we can add to this solution any vector whose curl is zero. In eqn (4.49) we added the condition $\operatorname{div} \mathbf{A} = 0$, but for our present problem it is more convenient to generalize this condition in the form

$$\operatorname{div} \mathbf{A} = -\mu\epsilon \frac{\partial V}{\partial t} = -\frac{1}{v^2}\frac{\partial V}{\partial t}, \tag{8.62}$$

where $v = (\mu\epsilon)^{-\frac{1}{2}}$ is the velocity of electromagnetic waves in the medium. This definition reduces to that used previously when V is independent of time, and has the advantage that it enables us to separate the variables V and $\mathbf{A}$. From eqn (8.61) we have (using eqn (8.1))

$$-\nabla^2 V = -\operatorname{div} \operatorname{grad} V = \operatorname{div}\left(\mathbf{E} + \frac{\partial \mathbf{A}}{\partial t}\right)$$

$$= \operatorname{div} \mathbf{E} + \frac{\partial}{\partial t} \operatorname{div} \mathbf{A} = \frac{\rho_e}{\epsilon} - \frac{1}{v^2}\frac{\partial^2 V}{\partial t^2},$$

so that

$$-\nabla^2 V + \frac{1}{v^2}\frac{\partial^2 V}{\partial t^2} = \frac{\rho_e}{\epsilon} \tag{8.63}$$

Again,

$$\text{curl } \mathbf{B} = \text{curl(curl } \mathbf{A}) = \text{grad div } \mathbf{A} - \nabla^2 \mathbf{A} = -\frac{1}{v^2}\text{ grad}\frac{\partial V}{\partial t} - \nabla^2 \mathbf{A},$$

while from eqns (8.7) and (8.61)

$$\begin{aligned}\text{curl } \mathbf{B} = \mu \text{ curl } \mathbf{H} = \mu\left(\mathbf{J} + \frac{\partial \mathbf{D}}{\partial t}\right) &= \mu \mathbf{J} + \mu\epsilon\frac{\partial \mathbf{E}}{\partial t} \\ &= \mu \mathbf{J} - \frac{1}{v^2}\left(\frac{\partial^2 \mathbf{A}}{\partial t^2} + \text{grad}\frac{\partial V}{\partial t}\right)\end{aligned}$$

Identifying these two equations for curl $\mathbf{B}$ gives

$$-\nabla^2 \mathbf{A} + \frac{1}{v^2}\frac{\partial^2 \mathbf{A}}{\partial t^2} = \mu \mathbf{J}. \tag{8.64}$$

For static systems, or systems which vary with time only slowly, eqns (8.63) and (8.64) reduce to the eqns (2.1) and (4.54) obtained earlier. The general solutions of our new equations are also similar to those of the earlier equations, and may be written as

$$V = \frac{1}{4\pi\epsilon}\int\frac{[\rho_e]\,\mathrm{d}\tau}{r}, \tag{8.65}$$

$$\mathbf{A} = \frac{\mu}{4\pi}\int\frac{[\mathbf{J}]\,\mathrm{d}\tau}{r}, \tag{8.66}$$

where the square brackets round ρ_e and $\mathbf{J}$ have the following significance. The values of V and $\mathbf{A}$ at a time t and at a point distance r from the element of volume containing ρ_e and $\mathbf{J}$ are related not to the values of ρ_e and $\mathbf{J}$ at the origin at the same time but to those values which obtained at a time $(t - r/v)$. In other words, the disturbance set up by the values of ρ_e and $\mathbf{J}$ at the origin is propagated with the velocity v and reaches a point distance r away at a time later by r/v. Thus the disturbance at this point is related to what happened at the origin a time r/v earlier, just as the light reaching the earth from a star tells us what was happening on the star, not at the same instant, but at the time when the light left the star. The values of V and $\mathbf{A}$ given by eqns (8.65) and (8.66) are known as 'retarded potentials'.

These equations will now be applied to the case of a short length of wire $\mathbf{s}$ at the origin of coordinates, carrying a current

$$I = I_0 \cos \omega t.$$

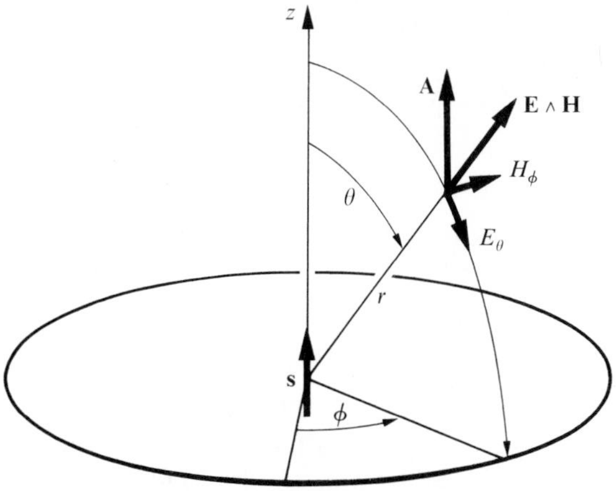

FIG. 8.7. Radiation from a short dipole **s**; at the point (r, θ, ϕ) the magnetic vector potential **A** is parallel to **s**; Poynting's vector $\mathbf{E}\wedge\mathbf{H}$ is along the radius vector **r**, and the field components are E_θ (along a line of longitude) and H_ϕ (along a line of latitude).

Then, since $\mathbf{J}\,\mathrm{d}\tau = I\,\mathrm{d}\mathbf{s}$, the retarded value $[\mathbf{J}]\,\mathrm{d}\tau = [I]\,\mathrm{d}\mathbf{s}$, and if **s** is very short compared with r, so that r does not change significantly during the integration, we may write

$$\mathbf{A} = \frac{\mu}{4\pi r}\int [I]\,\mathrm{d}\mathbf{s} = \frac{\mu}{4\pi r}[I]\mathbf{s}. \tag{8.67}$$

Eqn (8.67) shows that the vector potential **A** is everywhere parallel to **s**, as shown in Fig. 8.7.

The magnetic field strength **H** can now be found at any point, for

$$\begin{aligned}\mathbf{H} = \mathbf{B}/\mu &= (1/\mu)\mathrm{curl}\,\mathbf{A}\\ &= \frac{1}{4\pi}\mathrm{curl}\left([I]\frac{\mathbf{s}}{r}\right)\\ &= \frac{1}{4\pi}\left(\frac{[I]}{r}\mathrm{curl}\,\mathbf{s} - \mathbf{s}\wedge\mathrm{grad}\frac{[I]}{r}\right)\\ &= -\frac{1}{4\pi}\left(\mathbf{s}\wedge\mathrm{grad}\frac{[I]}{r}\right),\end{aligned}$$

since $\mathrm{curl}\,\mathbf{s} = 0$. Now $\mathrm{grad}([I]/r) = \mathbf{r}_1\,\partial([I]/r)/\partial r$, where $\mathbf{r}_1$ is a unit vector in the direction of r, since $[I]$ varies in space only with r. Hence

$$\mathbf{H} = -\frac{1}{4\pi}\{\mathbf{s}\wedge\mathbf{r}_1\}\frac{\partial}{\partial r}\left\{\frac{[I]}{r}\right\},$$

showing that **H** is normal to **s** and to $\mathbf{r}_1$. It follows that it has only one component, H_ϕ, in spherical polar coordinates.

The retarded value $[I] = I_0 \cos \omega(t - r/v)$, and hence

$$\frac{\partial}{\partial r}\left(\frac{[I]}{r}\right) = -\frac{I_0}{r^2}\cos \omega\left(t - \frac{r}{v}\right) + \frac{\omega I_0}{vr}\sin \omega\left(t - \frac{r}{v}\right)$$
$$= -\frac{I_0}{r^2}\cos \omega\left(t - \frac{r}{v}\right) + \frac{2\pi I_0}{r\lambda}\sin \omega\left(t - \frac{r}{v}\right),$$

where λ is the wavelength of the radiation. The magnetic field strength is then given by the sole component

$$H_\phi = \frac{sI_0 \sin\theta}{4\pi r^2}\cos \omega\left(t - \frac{r}{v}\right) - \frac{sI_0 \sin\theta}{2r\lambda}\sin \omega\left(t - \frac{r}{v}\right). \tag{8.68}$$

The first term predominates at short distances ($r \ll \lambda$) from the origin, and will be recognized as just the field strength given by the law of Biot and Savart (eqn (4.41)). It is known as the 'induction field'. At large distances of many wavelengths from the origin only the second term is significant; it falls off as $(\lambda r)^{-1}$ instead of r^{-2} and is known as the 'radiation field'.

The electric field may be found by using eqn (8.3), or it may be found from eqn (8.61) if the scalar potential V is first computed by means of eqn (8.62). We are interested only in its value at a large distance from the origin, where our spherical wave front has very small curvature and over a small region may be taken as a plane wave. As we should expect, **E** is then normal both to **H** and to the direction of propagation, and its only component is

$$E_\theta = Z_0 H_\phi = (\mu/\epsilon)^{\frac{1}{2}} H_\phi.$$

The energy crossing unit area per second is given by Poynting's vector,

$$E_\theta H_\phi = Z_0 H_\phi^2,$$

showing that the energy flow is radially outwards. The mean value averaged over a cycle of oscillation is

$$\overline{E_\theta H_\phi} = \tfrac{1}{8} Z_0 (I_0 s/r\lambda)^2 \sin^2\theta, \tag{8.69}$$

and the total energy radiated in all directions per second is

$$P = -\frac{dW}{dt} = \int_0^\pi 2\pi r^2 (\overline{E_\theta H_\phi}) \sin\theta \, d\theta = \frac{\pi Z_0 I_0^2 s^2}{3\lambda^2}. \tag{8.70}$$

The power radiated may be expressed in terms of a resistance R_r, called the 'radiation resistance', obtained by writing $P = \frac{1}{2} R_r I_0^2$, since this is just the mean power which would be dissipated in a real resistance R_r. For our current element eqn (8.70) gives

$$R_r = \frac{2\pi Z_0}{3}\left(\frac{s}{\lambda}\right)^2, \tag{8.71}$$

showing that the radiation resistance depends only on the square of the ratio of the length of the wire to the wavelength. In free space the numerical value of the radiation resistance R_r is $789(s/\lambda)^2\ \Omega$.

The oscillating current element may consist of two oscillating charges $\pm Q_0 \sin \omega t$, separated by a short distance $\mathbf{s}$, which constitute an oscillating electric dipole of moment $\mathbf{s}Q_0 \sin \omega t = \mathbf{p}_0 \sin \omega t$. At any instant the current is $I = \mathrm{d}Q/\mathrm{d}t = \omega Q_0 \cos \omega t$, showing that $\mathbf{s}I = \omega \mathbf{p}_0 \cos \omega t$. A more realistic example is provided by a charge Q_0 performing a simple harmonic motion with displacement $\mathbf{s} \sin \omega t$. We may write

$$I\mathbf{s} = \int I\, \mathrm{d}\mathbf{s} = \int (\mathrm{d}Q/\mathrm{d}t)(\mathbf{v}\, \mathrm{d}t) = \int \mathbf{v}\, \mathrm{d}Q = \mathbf{v}\, Q_0 = \omega \mathbf{s} Q_0 \cos \omega t = \omega \mathbf{p}_0 \cos \omega t.$$

In each case $\mathbf{s}I_0 = \omega \mathbf{p}_0$, and from (8.70) the total power radiated into free space is

$$P = \frac{\pi Z_0 \omega^2 p_0^2}{3\lambda^2} = \frac{4\pi^3 Z_0 p_0^2}{3\lambda^4} = \frac{Z_0 \omega^4 p_0^2}{12\pi c^2} = \frac{\omega^4}{3c^3}\left(\frac{p_0^2}{4\pi\epsilon_0}\right). \tag{8.72}$$

This shows that the power radiated is proportional to the square of the oscillating dipole moment, and the fourth power of the frequency (or the inverse fourth power of the wavelength).

Since the oscillating current element can be regarded as an oscillating electric dipole, it is interesting to calculate the scalar potential and compare it with our earlier results for a static dipole. Since $\mathbf{A}$ has a component only parallel to the polar axis (which we will call the z-axis), $\operatorname{div} \mathbf{A} = \partial A_z/\partial z$. In the differentiation x, y are constant and

$$(\partial/\partial z)_{x,y} = (z/r)(\partial/\partial r) = \cos\theta(\partial/\partial r) \quad \text{(cf. § 2.2)}$$

Hence, using eqns (8.62) and (8.67),

$$\frac{\partial V}{\partial t} = -v^2 \operatorname{div} \mathbf{A} = -\frac{v^2 \mu I_0 s \cos\theta}{4\pi} \frac{\partial}{\partial r}\left\{r^{-1} \cos \omega\left(t - \frac{r}{v}\right)\right\}.$$

Since r and t are independent variables we may integrate with respect to t inside the differential, giving

$$\begin{aligned} V &= -\frac{v^2 \mu I_0 s \cos\theta}{4\pi\omega} \frac{\partial}{\partial r}\left\{r^{-1} \sin \omega\left(t - \frac{r}{v}\right)\right\} \\ &= -\frac{p_0 \cos\theta}{4\pi\epsilon} \frac{\partial}{\partial r}\left\{r^{-1} \sin \omega\left(t - \frac{r}{v}\right)\right\} \\ &= -\frac{1}{4\pi\epsilon}\left[\mathbf{p}_0 \cdot \operatorname{grad}\left\{r^{-1} \sin \omega\left(t - \frac{r}{v}\right)\right\}\right]. \end{aligned} \tag{8.73}$$

Comparison with eqn (1.11a) shows that this is of the same form except that allowance must be made for the retardation in taking the gradient.

Magnetic dipole radiation

An oscillating magnetic dipole also radiates electromagnetic waves, the radiation field being similar to that of an electric dipole except that the roles of **E** and **H** are interchanged. The lines of **E** (and of **A**) are concentric circles about the axis of the dipole; **H** lies in the meridian plane, and when $r \gg \lambda$ it has just a single component H_θ along a 'line of longitude' (cf. Fig. 8.7). The power radiated into free space from a magnetic dipole of moment $m = m_0 \cos \omega t$ is

$$P = \frac{\omega^4}{3c^2}\left\{\left(\frac{\mu_0}{4\pi}\right) m_0^2\right\}. \tag{8.74}$$

This may be compared with the power from an electric dipole. If this is formed by a current $I_0 \cos \omega t$ in a length s, while the magnetic dipole carries the same current in a loop of area A, the ratio is

$$\frac{\text{power radiated by electric dipole}}{\text{power radiated by magnetic dipole}} = \frac{(I^2 s^2/4\pi\epsilon_0\omega^2)}{(\mu_0 I^2 A^2/4\pi)}$$

$$= \left(\frac{cs}{\omega A}\right)^2 = \left(\frac{\lambda s}{2\pi A}\right)^2. \tag{8.75}$$

This ratio is large if the linear dimensions of s and A are small compared to λ. The same ratio applies in the reception of electromagnetic waves (see Problem 8.8).

Radio antennae

The equations derived above apply only to systems where the linear dimensions are small compared with the wavelength. This does not in general hold for antennae used in radio or television transmission, but the radiative properties may be calculated by methods similar to those used above, the value of **A** being obtained by the use of eqn (8.66). For a single straight wire this gives

$$\mathbf{A} = \frac{\mu}{4\pi}\int \frac{[I]}{r}\, \mathrm{d}\mathbf{s}, \tag{8.76}$$

where the integration is along the wire and I is the current in the element d**s**. Two points must be noted in performing the integration: (1) the current is not in general constant along the wire, since it must fall to zero at the ends; (2) in evaluating the retarded potential, allowance must be made for the phase difference in signals coming from different parts of the wire owing to the change in the distance. The problem is similar to that of the diffraction pattern of a single slit where the amplitude varies over the aperture. The interference concepts used for light waves may be applied in finding the radiation pattern produced by an aerial array consisting of many elements; the problem is similar to that of a diffraction grating.

Problems

8.1. A plane electromagnetic wave propagates parallel to the z-axis and falls at normal incidence onto a metal bounded by the plane $z = 0$. At this surface there exists a tangential oscillating magnetic field of strength $H_y = H_0 \cos \omega t$. Show that the mean value of Poynting's vector just inside the surface is equal to the power per unit area dissipated in heating the metal, as given by eqn (8.34b).

8.2. Show that the superposition of two waves of equal amplitude, one with angular frequency $\omega + \mathrm{d}\omega$ and phase constant $\beta + \mathrm{d}\beta$, the other with $\omega - \mathrm{d}\omega$ and $\beta - \mathrm{d}\beta$, gives a wave of frequency ω and phase constant β, whose amplitude varies as

$$\binom{\sin}{\cos}(t\,\mathrm{d}\omega - x\,\mathrm{d}\beta).$$

Hence a point of maximum amplitude moves as

$$(t\,\mathrm{d}\omega - x\,\mathrm{d}\beta) = \text{constant},$$

or with velocity $u = (\mathrm{d}\omega/\mathrm{d}\beta)$. Since the energy is proportional to the square of the amplitude, u gives the rate at which energy is propagated, known as the 'group velocity'.

Show that in a good conductor, such as a metal, where the displacement current can be neglected, the phase velocity v $(=\omega/\beta)$ is $\omega\delta$, and the group velocity u $(=\mathrm{d}\omega/\mathrm{d}\beta)$ is $2\omega\delta$, where δ is the skin depth.

8.3. Show from eqn (8.41) that the condition for there to be no reflected wave at the boundary between two insulators (ϵ_1, μ_1; ϵ_2, μ_2) when the electric vector is in the plane of incidence requires that

$$\epsilon_1 \tan\theta = \epsilon_2 \tan\theta''.$$

This equation shows that at this particular angle the lines of electric field are refracted as in the electrostatic case (cf. Fig. 1.13) so that the boundary conditions are satisfied without any reflected wave. Similarly, when the magnetic vector is in the plane of incidence, the condition for no reflection is

$$\mu_1 \tan\theta = \mu_2 \tan\theta''$$

so that the lines of magnetic field are refracted as in the magnetostatic case.

Verify that in general the boundary conditions for the normal components of **D** and **B** are automatically satisfied in the theory of § 8.6 when the conditions for the tangential components of **E** and **H** are satisfied.

8.4. The rate at which solar energy falls on the earth's surface is approximately $1.4\times10^3\ \mathrm{J\,m^{-2}\,s^{-1}}$. Calculate the r.m.s. values of the electric and magnetic field strengths at the earth's surface, and the pressure exerted on it, assuming it to behave as a perfect absorber.

(*Answer*: $E = 730\ \mathrm{V\,m^{-1}}$; $H = 1{\cdot}9\ \mathrm{A\,m^{-1}}$; pressure, $4{\cdot}7\times10^{-6}\ \mathrm{N\,m^{-2}}$.

8.5. Show that by the introduction of a complex relative permittivity $\epsilon_r = \epsilon' - \mathrm{j}\epsilon''$, where $\epsilon'' = \sigma/\omega\epsilon_0$, the eqns (8.25a) may be written in the same form as for a non-conducting dielectric. Show that this leads to the relation $\mathbf{n}^2 = \mu_r\epsilon_r$ and that $\epsilon''/\epsilon' = \tan\delta$, where $\tan\delta$ is the loss tangent of the dielectric (see Problem 7.12).

8.6. A slightly imperfect dielectric has a small loss tangent. Show that in the first approximation the velocity of electromagnetic waves is the same as if tan δ were zero, but the power falls as $\exp(-\sigma Z_0 z)$ in travelling a distance z, where Z_0 is the intrinsic impedance of the medium neglecting the conductivity σ. This result has the simple interpretation: in a thickness dz the power dissipated per unit cross-section is $\sigma E^2\,\mathrm{d}z$, while the incident power $P = E^2/Z_0$. Hence

$$-\mathrm{d}P/\mathrm{d}z = \sigma Z_0 P.$$

Show also that

$$\sigma Z_0 = 2\pi(\text{loss tangent of the dielectric})/\lambda,$$

where λ is the wavelength of the radiation in the dielectric.

8.7. When a plane wave is incident on a conducting wire the electric field set up in the wire is equal to the tangential component of the electric field strength in the wave. Show that in a short straight wire the power picked up is proportional to $\sin^2\theta$, where θ is the angle between the wire and the direction of travel of the wave. This shows that the directional properties of the wire are the same for receiving as for transmitting (cf. eqn (8.69)).

8.8. When a plane wave falls on a small plane loop of wire, the e.m.f. induced in the loop is determined by the rate of change of magnetic flux through it. Show that the directional properties of the loop are the same as those of the short wire, if θ is measured from the normal to the plane of the loop, but the planes of the electric and magnetic vectors in the wave must be interchanged. This is consistent with the fact that a loop carrying an alternating current behaves as an oscillating magnetic dipole.

Show that the ratio of the e.m.f. set up in a short wire of length s to that in a small loop of area A is $\lambda s/2\pi A$, where λ is the wavelength of the radiation. If the linear dimensions of the loop are roughly the same as those of the wire ($A \approx s^2$), this shows that the loop is a much poorer aerial when $s \ll \lambda$.

8.9. A transmitter radiates a power P from a short dipole aerial. Show that the r.m.s. electric field strength at a distance D in the equatorial plane of the dipole aerial is

$$E = (3Z_0P/8\pi D^2)^{\frac{1}{2}},$$

where Z_0 is the intrinsic impedance of free space.

If $P = 1$ kW, show that the field strength at a distance of 10 km is about 0·021 V m^{-1}.

8.10. A transmitter radiates a power P from a short horizontal dipole aerial located at a height H above the sea. Show that the signal received at a target whose distance is D and height above the sea is h ($D \gg H, h$) is a maximum when $h = D\lambda/4H$, where λ is the wavelength of the transmitter. The sea may be assumed to act as a flat, perfectly reflecting (conducting) surface.

Show that if the height of the target is very much smaller than this value, the power incident on unit area of it (assuming it to lie in the equatorial plane of the dipole) is $6\pi PH^2h^2/D^4\lambda^2$. This equation shows that the effect of the sea is to make the power fall off with the inverse fourth power of the distance instead of the inverse square; it shows also the improvement gained by using short wavelengths.

If $P = 1$ kW, $D = 10$ km, $H = h = \lambda = 10$ m, show that the r.m.s. electric field strength at the target is about $2{\cdot}7\times10^{-4}$ V m^{-1}.

8.11. Obtain eqn (8.56) by the use of eqn (8.34b) and the law of conservation of energy.

8.12. At high frequencies the magnetic field falls inside each lamination of a transformer owing to the skin effect; the effective permeability at high frequencies is found by calculating the total flux in a lamination in phase with that at each surface of the lamination. Show that the effective radio-frequency permeability is less than the static value by a factor δ/a when the skin depth δ is very much smaller than the thickness a of the lamination.

What thickness of lamination is required to make $\delta = a$ for a material of relative permeability 10^5 and resistivity $5\times10^{-7}\,\Omega$ m at a frequency of 50 Hz?

(*Answer*: 0·16 mm)

8.13. A plane wave falls at normal incidence on the boundary ($z = 0$) between two media with permittivity and permeability (ϵ_1, μ_1) and (ϵ_1, μ_2) respectively. If the field strength components at the boundary are E_x, H_y, the stress is (by Maxwell's theory)

$$\tfrac{1}{2}\{(\epsilon_1-\epsilon_2)E_x^2+(\mu_1-\mu_2)H_y^2\}.$$

Show that this is equal to the sum of the momentum flows per unit area in the incident and reflected waves less that in the transmitted wave, that is, to

$$\frac{E^2+E'^2}{Z_1v_1}-\frac{E''^2}{Z_2v_2},$$

where E, E', and E'' are the electric intensities in the incident, reflected, and transmitted waves respectively, and Z_1, v_1; Z_2, v_2 are the impedance of and phase velocity in the two media. *Hint.* Note that $E_x = E''$, $H_y = E''/Z_2$ and show that both expressions can be reduced to

$$\tfrac{1}{2}E''^2\left(\frac{1}{Z_1v_1}+\frac{Z_1}{Z_2^2v_1}-\frac{2}{Z_2v_2}\right).$$

9. Filters, transmission lines, and waveguides

9.1. Elements of filter theory

A FREQUENT requirement in radio and telephony is the separation of two signals of different frequencies. Any circuit whose impedance varies with frequency can be used for this purpose, a simple example already considered being the tuned circuit, which can be employed to accept or reject a narrow band of frequencies centred on its natural resonant frequency. This acts as a 'band-pass' or 'band-stop' filter. Another type of filter may be required to pass all frequencies up to a certain value, and stop all higher frequencies; this is a 'low-pass' filter. The reverse case is a 'high-pass' filter, which rejects all low frequencies up to a certain value, and passes all higher frequencies.

The action of a filter can be understood by considering a simple example, the low-pass filter shown in Fig. 9.1. A common use of such a filter is to remove the ripple voltage from the output of a rectifier unit which is converting an a.c. voltage to a steady voltage. The output is applied to the terminals AB of the filter, which is required to pass the steady voltage component on to the terminals CD, but not the alternating component. If the latter has a frequency of, say, 100 Hz, and the capacitance C is chosen to have a low impedance at 100 Hz while the inductance L has a high impedance, only a small fraction of the input voltage at this frequency will appear at the terminals CD, because the inductance and second capacitance act as a voltage divider. The fraction is approximately $X_C/X_L = (1/\omega C)/\omega L = 1/\omega^2 LC$, and if $C = 10\ \mu F$, $L = 25$ H, the fraction is about 1/100. On the other hand, if there is no leakage in the capacitor, the full steady voltage component will be passed on to CD. Thus the filter accepts the signal of zero frequency, and partially rejects the signal at 100 Hz frequency. Better rejection is obtained by

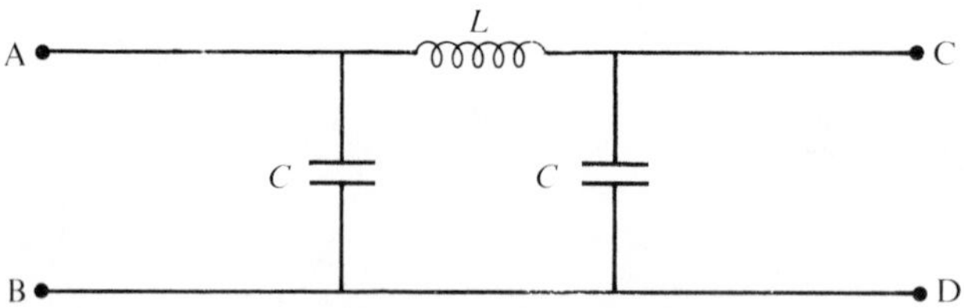

FIG. 9.1. Simple low-pass filter section.

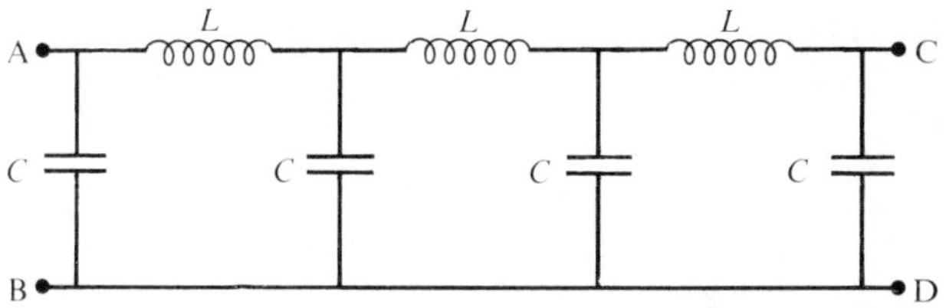

FIG. 9.2. Chain of low-pass filter sections.

adding more sections of this kind, as in Fig. 9.2. This arrangement is called a step- or ladder-type filter. It is clear that evaluation of the currents and voltages in the different elements by application of Kirchhoff's laws would be very laborious, and it is preferable to proceed in a different way, making use of the recurrent nature of the elements.

We shall begin by considering a uniform filter, consisting of a chain of similar sections, as in Fig. 9.3, which are repeated indefinitely, forming an infinite chain. If a generator is applied at some point earlier in the chain, currents will flow in the various sections; let the currents in successive sections be I_{n-1}, I_n, I_{n+1}. Application of Kirchhoff's law to the central section gives

$$\mathbf{Z}_2(I_n - I_{n-1}) + \mathbf{Z}_1 I_n + \mathbf{Z}_2(I_n - I_{n+1}) = 0$$

or

$$-\mathbf{Z}_2 I_{n-1} + (\mathbf{Z}_1 + 2\mathbf{Z}_2) I_n - \mathbf{Z}_2 I_{n+1} = 0. \tag{9.1}$$

We may write $I_n = aI_{n-1}$, where a is a real or complex number; then in an infinite chain where we cannot distinguish between sections we must also have $I_{n+1} = aI_n$. Eqn (9.1) then gives

$$\tfrac{1}{2}(a + 1/a) = 1 + \mathbf{Z}_1/2\mathbf{Z}_2. \tag{9.2}$$

This equation determines the attentuation constant a. We shall confine our attention to the case where $\mathbf{Z}_1/\mathbf{Z}_2$ is real, which corresponds to $\mathbf{Z}_1$ and $\mathbf{Z}_2$ being both pure resistances or pure reactances. Then a can be either real or complex, but not a purely imaginary quantity. The real roots arise when the right-hand side of eqn (9.2) lies outside the range $+1$ to -1, that

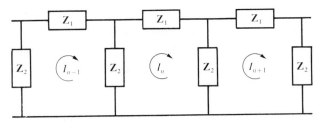

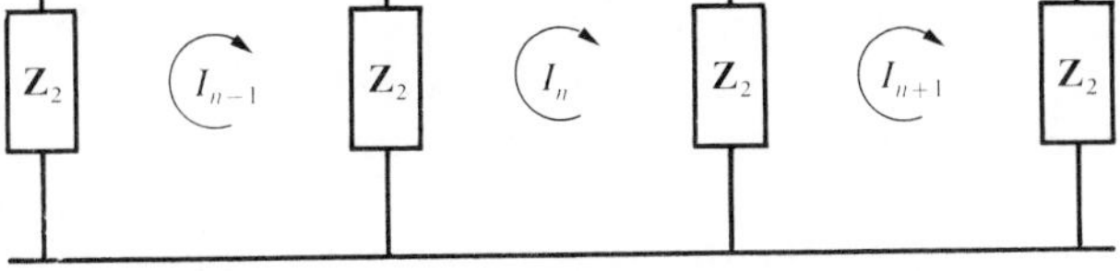

FIG. 9.3. Generalized type of uniform ladder filter.

is, $\mathbf{Z}_1/4\mathbf{Z}_2$ lies outside the range 0 to -1. We consider separately the three cases where it is greater than 0, between 0 and -1, and less than -1.

(*i*) $(\mathbf{Z}_1/4\mathbf{Z}_2)>0$

a is real and positive. Since the network is 'passive' (that is, it contains no power-generating elements), the currents must decrease as we move away from the generator attached to one end of the filter. We take $a<1$ for a wave travelling towards the right (that is a generator attached to the left-hand end of the filter) and $a>1$ for a wave travelling towards the left. This interpretation is consistent with the fact that the roots of eqn (9.2) are reciprocal; thus the wave is attenuated by the same amount per section in the direction in which it progresses, whichever way it is going. The significance of the positive sign of a is that the wave is attenuated without change of phase. If we write $a=e^{-\alpha}$, where α is the attenuation constant per section, eqn (9.2) becomes

$$\cosh\alpha = 1+\mathbf{Z}_1/2\mathbf{Z}_2. \tag{9.3}$$

(*ii*) $0>(\mathbf{Z}_1/4\mathbf{Z}_2)>-1$

a is complex, with modulus unity, so that we may write $a=\exp(-j\beta)$. The wave is not attenuated at all, but suffers a phase change by an angle β in each section, where

$$\cos\beta = 1+\mathbf{Z}_1/2\mathbf{Z}_2. \tag{9.4}$$

(*iii*) $(\mathbf{Z}_1/4\mathbf{Z}_2)<-1$

a is then real and negative, so that the wave is attenuated with a phase change of π in successive sections. If we write $a=-e^{-\alpha}$, eqn (9.2) becomes

$$-\cosh\alpha = 1+\mathbf{Z}_1/2\mathbf{Z}_2. \tag{9.5}$$

Here we have already spoken of the current as part of a 'wave' and this terminology needs some justification. If the current in one section alternates at a given frequency, so that we can write $I_n = I_n^0 \exp j\omega t$, then the current in a later section $(n+m)$ will be

$$I_{n+m} = a^m I_n^0 \exp j\omega t.$$

If we write $a=\exp(-\gamma)$, the current in the section $n+m$ will be

$$I_{n+m} = I_n^0 \exp(-\gamma m)\exp(j\omega t) = I_n^0 \exp(j\omega t-\gamma m),$$

which is similar to the expression for a wave motion in a continuous medium except that m, the number of sections, which can only be an integer, replaces x, the distance travelled in the medium. If γ is real, we write it as α, and we have an attenuated wave, but if γ is a pure imaginary quantity $j\beta$ we have an unattenuated wave with a phase change β per section. The values of α and β are given by eqns (9.3)–(9.5). The

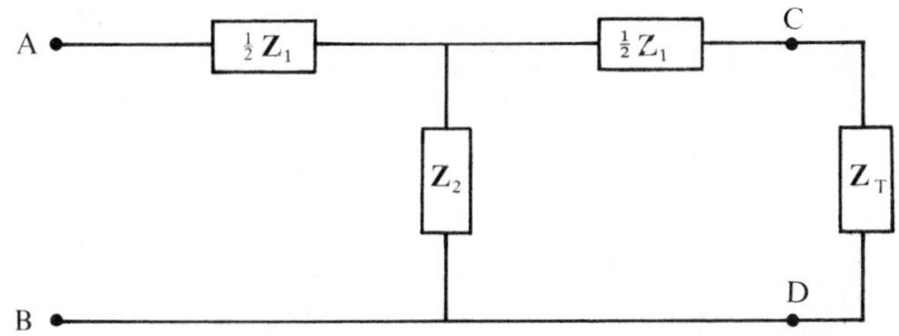

FIG. 9.4. T-section ABCD terminated by its iterative impedance.

reciprocal roots are obtained by changing the sign of γ, corresponding to a wave travelling in the opposite direction. The third case, (*iii*) above, where the attenuated wave changes sign in successive sections, could not arise in a continuous medium. The behaviour of the filter is determined immediately from the value of $(\mathbf{Z}_1/4\mathbf{Z}_2)$. If it lies within the range 0 to -1, we have a 'pass band' with no attenuation; outside this range we have a 'stop band'.

The word 'section' has been used so far without any precise definition of its meaning. The uniform chain can be regarded as made up of similar sections joined together, but two types of section can be obtained by cutting the chain in different ways. These two types are shown in Figs 9.4 and 9.5 and for obvious reasons are known as T-sections and π-sections respectively. It is clear that a succession of either type of section joined together gives the same chain filter (see Fig. 9.6), so that the transmission characteristics derived earlier apply to either type of section.

If a generator of voltage V is connected to a semi-infinite chain and the current I drawn from it is measured, a definite value of the ratio V/I is obtained, known as the 'characteristic' or 'iterative' impedance of the chain. It is obvious that the same value would be obtained if a number of sections were removed and the impedance of the semi-infinite chain measured at a later point. Again, the chain may be severed at some point and the infinite tail replaced by an impedance equal to the iterative impedance without altering the impedance measured at the input terminals. This gives a method of calculating the iterative impedance. Fig. 9.4

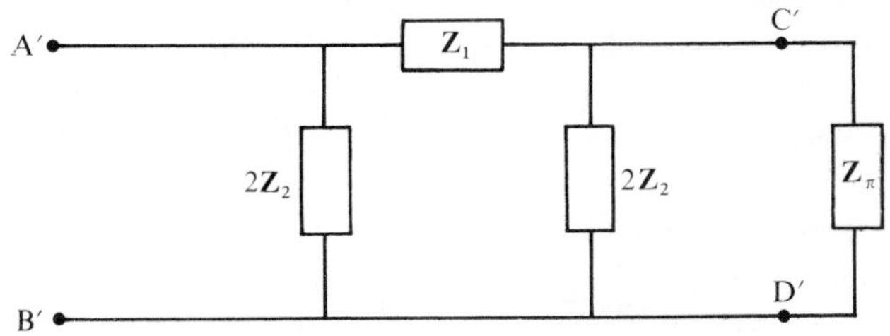

FIG. 9.5. π-section A′B′C′D′ terminated by its iterative impedance.

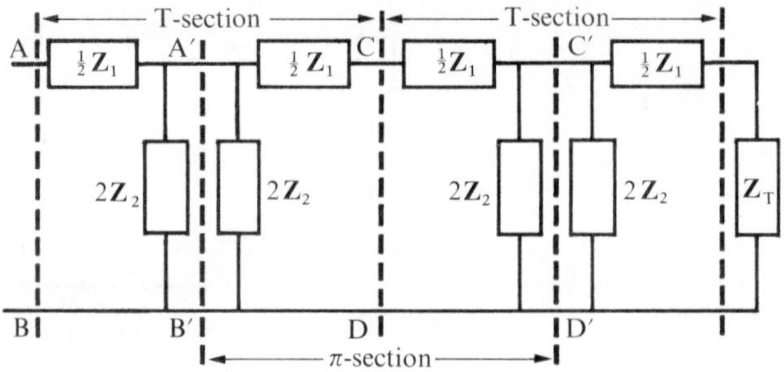

FIG. 9.6. Filter of T-sections such as ABCD terminated by the iterative impedance $\mathbf{Z}_T$. It may also be divided into π-sections terminated by the network to the right of C′D′.

shows a T-section terminated by an impedance $\mathbf{Z}_T$; the impedance measured at the input terminals AB can be calculated by standard methods, and on equating this to $\mathbf{Z}_T$ we have

$$\mathbf{Z}_T = \tfrac{1}{2}\mathbf{Z}_1 + \left(\frac{1}{\mathbf{Z}_2} + \frac{1}{\frac{1}{2}\mathbf{Z}_1 + \mathbf{Z}_T}\right)^{-1}$$

Solution of this equation gives

$$\mathbf{Z}_T = (\mathbf{Z}_1\mathbf{Z}_2 + \tfrac{1}{4}\mathbf{Z}_1^2)^{\frac{1}{2}} = (\mathbf{Z}_1\mathbf{Z}_2)^{\frac{1}{2}}(1+\mathbf{Z}_1/4\mathbf{Z}_2)^{\frac{1}{2}}. \tag{9.6}$$

This is the value of the iterative impedance $\mathbf{Z}_T$ for a T-section. By applying the same method to a π-section, as shown in Fig. 9.5, we find the iterative impedance of a π-section to be

$$\mathbf{Z}_\pi = (\mathbf{Z}_1\mathbf{Z}_2)^{\frac{1}{2}}(1+\mathbf{Z}_1/4\mathbf{Z}_2)^{-\frac{1}{2}}. \tag{9.7}$$

Thus

$$\mathbf{Z}_T\mathbf{Z}_\pi = \mathbf{Z}_1\mathbf{Z}_2. \tag{9.8}$$

The importance of the iterative impedance lies in the fact that sections of different kinds may be joined together to form a non-uniform filter, without disturbing a wave travelling down the filter, provided that they have the same iterative impedance. If sections with different iterative impedances are used a 'reflection' of the wave occurs at the junction and the simple filter theory is no longer applicable. Similarly, the filter must be terminated by a load equal to its iterative impedance in order to avoid a reflection at the output terminals. Any such reflection will diminish the power dissipated in the load, since part of the incident energy will be reflected.

A filter correctly terminated by the iterative impedance $\mathbf{Z}_T$ is shown in Fig. 9.6, where it is regarded as made up of T-sections (ABCD). Alternatively, we may regard it as made up of π-sections (A′B′C′D′), which must still be correctly terminated. Thus the impedance of the portion to the

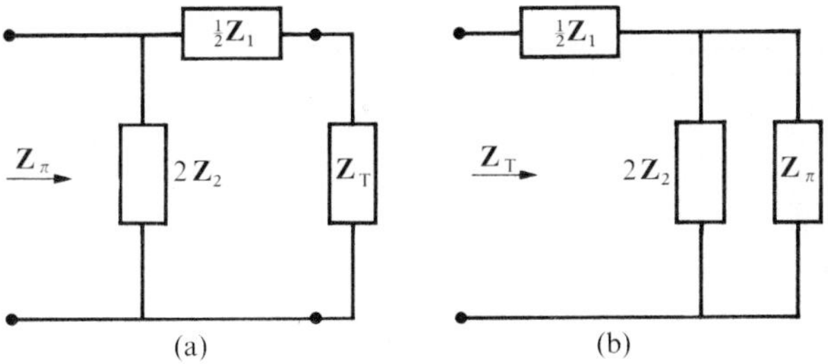

FIG. 9.7. Half-sections used as transformer (a) converting $\mathbf{Z}_T$ to $\mathbf{Z}_\pi$; (b) $\mathbf{Z}_\pi$ to $\mathbf{Z}_T$.

right of C′D′, shown separately in Fig. 9.7(a), must be $\mathbf{Z}_\pi$. Hence the half-section acts as a transformer from the impedance $\mathbf{Z}_T$ to $\mathbf{Z}_\pi$, as may also be verified by direct calculation. Similarly, the other half-section shown in Fig. 9.7(b) transforms the impedance $\mathbf{Z}_\pi$ to $\mathbf{Z}_T$. To a limited extent such half-sections can be used to match a load to a filter, but their especial importance comes in the design of composite filters (see Problem 9.4).

9.2. Some simple types of filter

Low-pass filter

The simplest type of low-pass filter is that shown in Fig. 9.2 with capacitances in the shunt arms and inductances in the series arms. Then $\mathbf{Z}_1 = \mathrm{j}\omega L$ and $\mathbf{Z}_2 = 1/(\mathrm{j}\omega C)$, giving

$$\mathbf{Z}_1/4\mathbf{Z}_2 = -\omega^2 LC/4.$$

The cut-off frequency ω_0 is obtained by setting this equal to -1, giving $\omega_0 = 2(LC)^{-\frac{1}{2}}$. Frequencies below this are passed without attenuation while higher frequencies are attenuated. In the pass band the phase change is given by

$$\cos\beta = 1-\omega^2 LC/2 = 1-2(\omega/\omega_0)^2,$$

showing that β changes from 0 to π as the frequency increases from zero to the cut-off value. The characteristic impedance of a T-section is

$$\mathbf{Z}_T = \left(\frac{L}{C}-\frac{\omega^2 L^2}{4}\right)^{\frac{1}{2}} = \left(\frac{L}{C}\right)^{\frac{1}{2}}\left(1-\frac{\omega^2}{\omega_0^2}\right)^{\frac{1}{2}},$$

which is resistive in the pass band, but varies from $(L/C)^{\frac{1}{2}}$ to zero at the cut-off frequency. In the stop band $\mathbf{Z}_T$ is a pure reactance and the attenuation is given by

$$-\cosh\alpha = 1-\omega^2 LC/2,$$

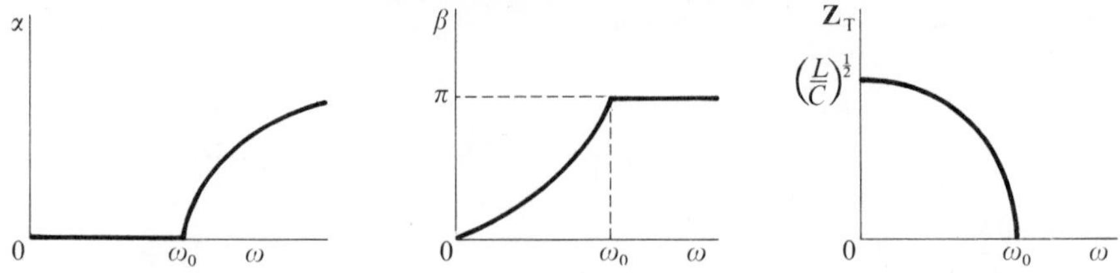

FIG. 9.8. Variation of α, β, and the real part of $\mathbf{Z}_T$ for a simple low-pass filter.

which rises with frequency from zero at $\omega = \omega_0$. The negative sign shows that currents in successive sections are reversed in direction.

The general behaviour of α, β, and $\mathbf{Z}_T$ is illustrated in Fig. 9.8. The variation of $\mathbf{Z}_\pi$ is easily found from the relation

$$\mathbf{Z}_\pi = \mathbf{Z}_1\mathbf{Z}_2/\mathbf{Z}_T = (L/C)\mathbf{Z}_T^{-1}$$

As an example, we may take the problem of smoothing the output of a rectifier considered at the beginning of this chapter. Then $C = 10\ \mu$F, $L = 25$ H, and the ripple frequency is 100 Hz. The attenuation per section at this frequency is given by

$$-\cosh\alpha = -\tfrac{1}{2}(e^{\alpha}+e^{-\alpha}) = 1-\tfrac{1}{2}\omega^2 LC = (1-49{\cdot}3).$$

Hence $e^{\alpha} \approx 100$, or the attenuation in power is approximately 10^4 per section. If we have n such inductances and capacitances, it is very much better to join them as a ladder filter of n sections than to lump them all into one section. The latter would reduce the ripple voltage by a factor $e^{\alpha} \approx \omega^2 n^2 LC = 100n^2$ approximately, while the ladder type filter will reduce it by $e^{n\alpha} \approx (100)^n$. Thus three sections would reduce the ripple voltage by 10^6, while the single section with three inductances in series and three capacitances in parallel gives only a factor 900.

The high-pass filter

As would be expected, the simplest high-pass filter is formed by putting capacitances in the series arms and inductances in the shunt arms, and a typical T-section is shown in Fig. 9.9.

Then $\mathbf{Z}_1 = 1/\mathrm{j}\omega C$, $\mathbf{Z}_2 = \mathrm{j}\omega L$, and

$$\mathbf{Z}_1/4\mathbf{Z}_2 = -1/4\omega^2 LC.$$

This lies between zero and -1 for $\omega \geqslant \tfrac{1}{2}(LC)^{-\frac{1}{2}}$, the latter being the critical frequency ω_0. The phase change per section in the pass band is given by

$$\cos\beta = 1-1/2\omega^2 LC = 1-2(\omega_0/\omega)^2,$$

and the modulus of this expression gives also the value of $\cosh\alpha$ in the stop band.

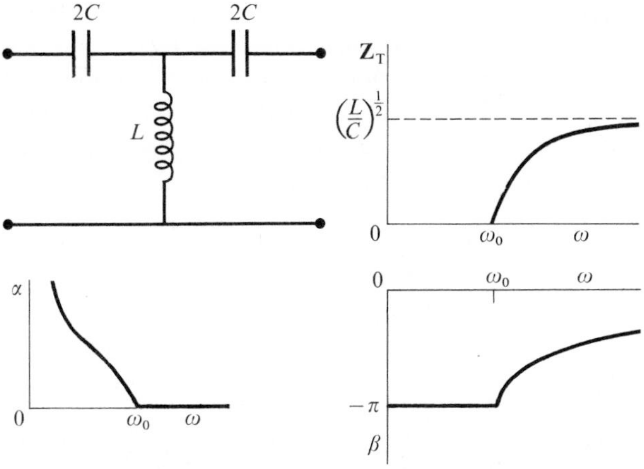

FIG. 9.9. T-section of simple high-pass filter, with variation of real part of $\mathbf{Z}_T$, α, and β.

The iterative impedance is

$$\mathbf{Z}_T = \left(\frac{L}{C} - \frac{1}{4\omega^2 C^2}\right)^{\frac{1}{2}} = \left(\frac{L}{C}\right)^{\frac{1}{2}}\left(1 - \frac{\omega_0^2}{\omega^2}\right)^{\frac{1}{2}},$$

which is imaginary in the stop band, and rises from zero at the critical frequency to a limiting value of $(L/C)^{\frac{1}{2}}$ in the pass band. The behaviour of α, β, and $\mathbf{Z}_T$ is illustrated in Fig. 9.9.

Band-pass filters

A simple type of band-pass filter is shown in Fig. 9.10. Qualitatively its behaviour can be seen as follows. At very low frequencies the impedance

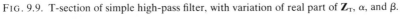

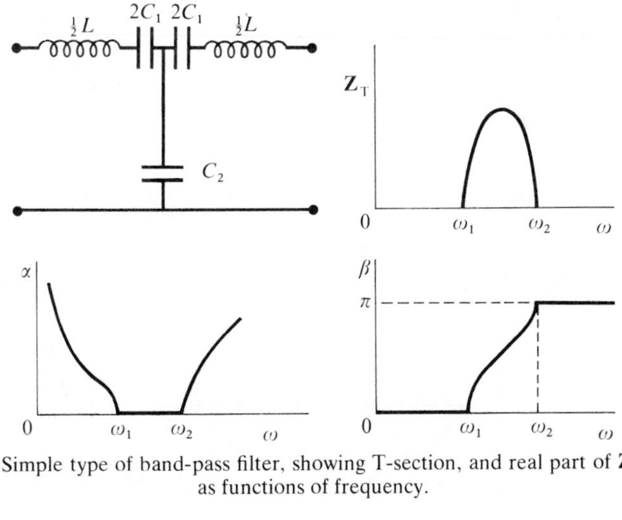

FIG. 9.10. Simple type of band-pass filter, showing T-section, and real part of $\mathbf{Z}_T$, α, and β as functions of frequency.

in the series arm will be dominated by the capacitance, so that the section will act as a simple capacitance-type attenuator (see Problem 9.1). At frequencies higher than the resonant frequency of the combination L and C_1, $\omega L > (\omega C_1)^{-1}$ and the impedance of the series arm is inductive; then the section behaves as a low-pass filter, the highest frequencies again being stopped. Quantitatively, the analysis is

$$\mathbf{Z}_1 = \mathrm{j}\omega L + 1/\mathrm{j}\omega C_1,$$

$$\mathbf{Z}_2 = 1/\mathrm{j}\omega C_2,$$

$$\mathbf{Z}_1/4\mathbf{Z}_2 = \frac{1}{4}\left(\frac{C_2}{C_1} - \omega^2 LC_2\right),$$

which is positive at zero frequency, but tends to $-\infty$ as $\omega \to \infty$. When it lies between 0 and -1, we have a pass band whose lowest frequency is

$$\omega_1 = (LC_1)^{-\frac{1}{2}}, \quad \text{where } \mathbf{Z}_1/4\mathbf{Z}_2 = 0,$$

and highest frequency is

$$\omega_2 = \left(\frac{C_2 + 4C_1}{LC_1C_2}\right)^{\frac{1}{2}}, \quad \text{where } \mathbf{Z}_1/4\mathbf{Z}_2 = -1.$$

The behaviour of α, β, and $\mathbf{Z}_T$ is shown in Fig. 9.10.

Disadvantages of the simple filter

A simple filter, consisting of a chain of similar sections, suffers from two principal disadvantages:

(1) the iterative impedance varies with frequency in the pass band, making it impossible to terminate the filter correctly throughout the pass band;

(2) the attenuation in the stop band varies with frequency, being low near the cut-off frequencies.

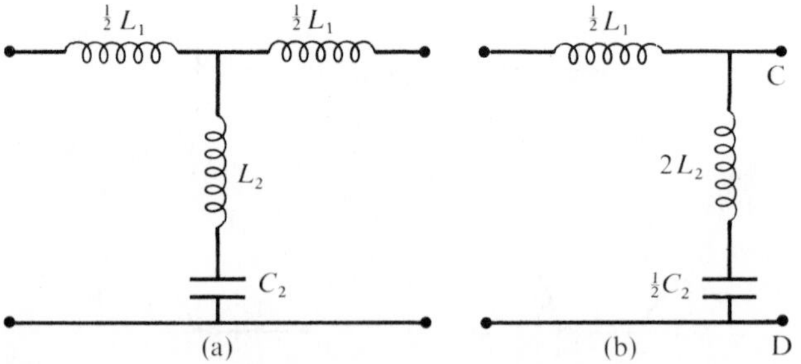

FIG. 9.11. T-section and terminal half-section of an m-derived filter. The values must obey the equations $L_1 = mL$, $C_2 = mC$, $L_2 = L(1-m^2)/4m$.

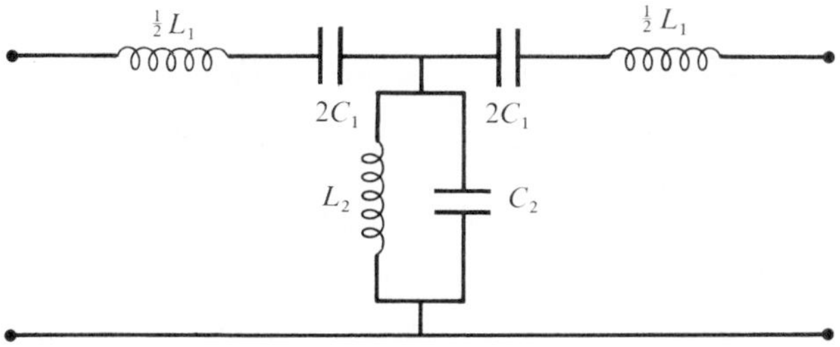

FIG. 9.12. T-section of constant-k band-pass filter ($L_2/C_1 = L_1/C_2 = k^2$).

These drawbacks can be reduced by using a composite filter, containing sections of different types, instead of a uniform filter. For example, the attenuation just above the cut-off frequency of a low-pass filter can be made high by using a section with a parallel tuned circuit in the series arm, or one with a series tuned circuit in the shunt arm (as in Fig. 9.11), adjusted to resonate at a frequency just above the cut-off frequency. If there were no resistive loss in the components, this would give infinite attenuation at the resonant frequency. Such a section will have low attenuation at the high frequencies, but it can be combined with a simple low-pass section so that the composite filter has a sufficiently high attenuation throughout the stop band. In such a composite filter each section must have the same iterative impedance at all frequencies, or reflections will occur at the junctions between sections. This can be achieved by the use of 'm-derived' filters, an example of which is given in Problem 9.3. Half-sections may also be used to make the impedance more constant in the pass band (see Problem 9.4). The simple low-pass and high-pass filters considered earlier belong to the class of 'k-derived' or 'constant-k' filters, since their impedances obey the relation

$$\mathbf{Z}_1\mathbf{Z}_2 = k^2,$$

where k is a constant independent of frequency. The band-pass filter considered earlier is not of this type, but the T-section shown in Fig. 9.12 obeys this relation provided that $L_2/C_1 = L_1/C_2 = k^2$ (see Problem 9.2).

9.3. Travelling waves on loss-free transmission lines

In the electrical circuits considered hitherto we have been able to identify the circuit elements as inductances, resistances, capacitances, or combinations thereof, known as 'lumped impedances'. These elements are connected together and to generators and detectors by lengths of wire whose effect is assumed to be negligible. This is true only when the lengths of wire involved are very small compared with the wavelength of the radiation flowing along them. When this condition is not fullfilled, the

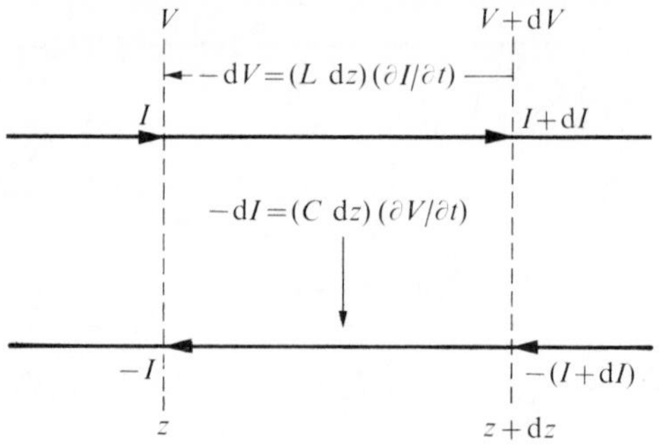

FIG. 9.13. Infinitesimal section of loss-less transmission line, showing current flow and voltage between conductors.

signal changes phase as it flows along the wires, in much the same way as in the pass band of a filter. In fact we may regard the wires as a limiting case of a low-pass filter where the elements are made infinitesimally small, but there are an infinite number of sections per unit length so that the 'distributed impedance' per unit length remains finite. Normally two wires are required to complete the circuit between any two pieces of apparatus, and we shall assume that these take the form either of two parallel wires, or two coaxial cylinders. In an element dz of two such conductors, as shown in Fig. 9.13, the voltage and current will be linearly related to one another, so that the voltage drop in a length dx will be

$$V-(V+\mathrm{d}V)=-\mathrm{d}V=(L\ \mathrm{d}z)(\partial I/\partial t),$$

where L is the series inductance associated with unit length of the two conductors. Since charge is conserved, $+\mathrm{d}I$ is the change in the current flowing along the conductors in a distance dz, and the current flow between the conductors in this element is

$$-\mathrm{d}I=(C\,\mathrm{d}z)(\partial V/\partial t),$$

where C is the shunt capacitance per unit length between the conductors. If there are no losses in the dielectric medium or in the conductors, the basic equations are simply

$$-(\partial V/\partial z)=L(\partial I/\partial t), \qquad -(\partial I/\partial z)=C(\partial V/\partial t). \tag{9.9}$$

These form a pair of simultaneous differential equations which may be solved for either I or V; for the latter we have

$$\frac{\partial^2 V}{\partial z^2}=-\frac{\partial}{\partial z}\left(L\frac{\partial I}{\partial t}\right)=-\frac{\partial}{\partial t}\left(L\frac{\partial I}{\partial z}\right)=LC=\frac{\partial^2 V}{\partial t^2}. \tag{9.10}$$

This is similar to eqn (8.17), and the solution is similar to (8.18):

$$V = F_1(z-vt)+F_2(z+vt). \tag{9.11a}$$

The corresponding solution for the current is

$$Z_0 I = F_1(z-vt)-F_2(z+vt) \tag{9.11b}$$

in which

$$v = (LC)^{-\frac{1}{2}} \tag{9.12}$$

and

$$Z_0 = (L/C)^{\frac{1}{2}}. \tag{9.13}$$

Here Z_0 is the 'characteristic impedance' of the line. It plays a role similar to the intrinsic impedance of a medium for unbounded electromagnetic waves, or the iterative impedance of a filter; it is equal to the limiting values of Z_T and Z_π for a low-pass filter in which the inductance and capacitance per section are allowed to go to zero, keeping their ratio constant.

For a pair of coaxial cylinders, radii a and b $(b > a)$, we have

$$C = \frac{2\pi\epsilon}{\ln(b/a)} \qquad \text{(eqn (1.30))},$$

$$L = \frac{\mu \ln(b/a)}{2\pi} \qquad \text{(eqn (5.18))},$$

while for a pair of parallel cylinders radius a, separation $2d$ $(2d \gg a)$

$$C = \frac{\pi\epsilon}{\ln(2d/a)} \qquad \text{(eqn 2.56)),}$$

$$L = \frac{\mu \ln(2d/a)}{\pi} \qquad \text{(Problem 5.2).}$$

In each case the wave velocity is

$$v = (LC)^{-\frac{1}{2}} = (\mu\epsilon)^{-\frac{1}{2}} = c/(\mu_r\epsilon_r)^{\frac{1}{2}},$$

which is the same as for an electromagnetic wave in the medium when no conductors are present. This is a general result for a pair of parallel lossless conductors, whatever their shape (see below).

From the values of L and C for our two special types of line, we have:

$$\text{coaxial line:} \quad Z_0 = \frac{1}{2\pi}\left(\frac{\mu}{\epsilon}\right)^{\frac{1}{2}} \ln(b/a),$$

$$\text{parallel-wire line:} \quad Z_0 = \frac{1}{\pi}\left(\frac{\mu}{\epsilon}\right)^{\frac{1}{2}} \ln(2d/a).$$

Typical values of Z_0 are: for an air-spaced coaxial line with $b \approx 3a$, $Z_0 \approx 70\ \Omega$; for a parallel-wire line, $2d/a \approx 20$, and $Z_0 \approx 400\ \Omega$. In a coaxial cable a continuous dielectric is used to support the inner conductor, and the ratio b/a is adjusted to make Z_0 some standard value, usually either 70 Ω or 50 Ω.

An important quantity is the power flowing along the line. If we confine ourselves to a wave travelling along towards positive z, the energy flowing past any plane normal to the z axis is, at any instant,

$$P = v(\tfrac{1}{2}LI^2+\tfrac{1}{2}CV^2) = (\tfrac{1}{2}LI^2+\tfrac{1}{2}CV^2)/(LC)^{\frac{1}{2}}$$
$$= \tfrac{1}{2}I^2Z_0+\tfrac{1}{2}V^2/Z_0 = I^2Z_0 = V^2/Z_0 = IV. \tag{9.14}$$

Here we have given all the equivalent expressions for P, and we see that Z_0 behaves like a pure resistance, except that P represents the energy flowing along the line per second rather than the energy dissipated as Joule heating of a real resistance. In a sense the energy stored in a section of the line is being dissipated, since it flows away from that section, and the stored energy would therefore diminish unless it were continually replaced by the energy flowing into it from the previous section. P represents the energy crossing a given point in the line at any instant; the average flow of energy is found by using the r.m.s. values $\tilde{I}$ and $\tilde{V}$ in eqn (9.14). The fact that Z_0 is real shows that the phase difference between V and I is 0 or π, according to whether the wave is travelling towards positive or negative x. From eqn (9.14), P has the same sign as Z_0, showing that the energy flows in opposite directions in the two waves, as we should expect. This result follows also from the direction of Poynting's vector (see Problem 9.5), since V and I are linearly related to the electric and magnetic field strengths **E** and **H**.

This treatment of parallel-wire transmission lines in terms of current and voltage provides a useful, if somewhat artificial, simplification of the mathematics. An alternative and more rigorous approach starts from Maxwell's equations or the wave equation. For two parallel conductors in which current flow is confined to one direction (the z-axis), the magnetic vector potential **A** has just the single component A_z. In the medium between the conductors this must satisfy eqn (8.64) with $\mathbf{J}=0$; we write this in the form

$$\left\{\left(\frac{\partial^2}{\partial x^2}+\frac{\partial^2}{\partial y^2}\right)+\left(\frac{\partial^2}{\partial z^2}-\frac{1}{v^2}\frac{\partial^2}{\partial t^2}\right)\right\}A_z = 0.$$

In the same region the electric potential V must, from eqn (8.63), obey an identical equation. Each equation is satisfied if, separately,

$$\left(\frac{\partial^2}{\partial z^2}-\frac{1}{v^2}\frac{\partial^2}{\partial t^2}\right)(V, A_z) = 0, \tag{9.15a}$$

$$\left(\frac{\partial^2}{\partial x^2}+\frac{\partial^2}{\partial y^2}\right)(V, A_z) = 0. \tag{9.15b}$$

These have solutions of the form

$$\begin{aligned} V &= \phi_1(x, y)F_1(z - vt) + \phi_2(x, y)F_2(z + vt), \\ vA_z &= \phi_1(x, y)F_1(z - vt) - \phi_2(x, y)F_2(z + vt), \end{aligned} \tag{9.16}$$

where the relation between V and A_z satisfies eqn (8.62) with $v = (\mu\epsilon)^{-\frac{1}{2}}$.

These relations show that V and A_z have the same variation in the x, y plane. For a wave travelling, say, in the positive z-direction, each varies as $\phi_1(x, y)$, which is a solution of the time-independent eqn (9.15b) and the same as in the static case. Whatever the cross-sectional shape of the conductors, it can be shown that this leads to the relation $LC = \mu\epsilon$ for the inductance and capacitance per unit length (cf. § 5.2 and Problem 5.2). Such solutions are valid provided that (1) the current flow is unidirectional, and in the direction of propagation; (2) the conductivity of the cylinders is infinite. This second condition follows from the fact that the solutions for V and A_z lead, through eqn (8.61), to

$$E_z = -\partial A_z/\partial t - \partial V/\partial z = 0,$$

a condition that cannot be fullfilled if the conductors have a finite resistivity and current flows in the z direction.

9.4. Terminated loss-free lines

Hitherto we have regarded the two waves travelling in opposite directions along a line as two quite independent solutions of the wave equation. Usually, however, there is only one generator attached to the line, producing a wave travelling, say, towards positive x. If the line is infinite in this direction, there will be no return wave. If the line is terminated in some way, a reflection may occur at the termination and this will generate a return wave, which will not be independent of the incident wave. If the latter has a given frequency, the return wave must have the same frequency in order that the ratio of current to voltage at the termination (assumed to consist of some constant impedance) shall be independent of the time. If $\mathbf{Z}_L$ is the load impedance (as in Fig. 9.14), the fact that V/I must equal $\mathbf{Z}_L$ at this point for all values of the

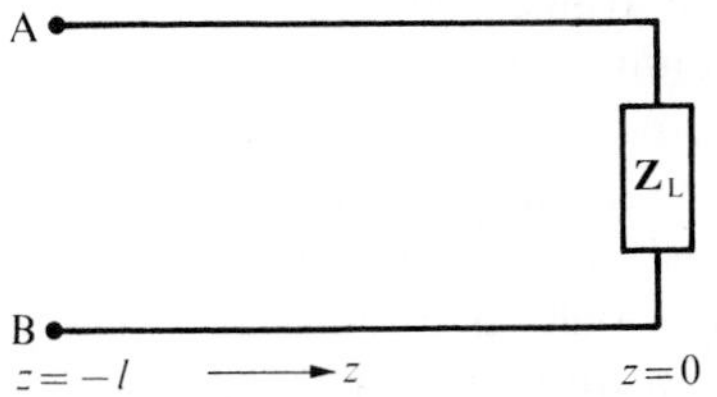

FIG. 9.14. Line terminated by impedance $\mathbf{Z}$ at $z = 0$.

time constitutes the boundary condition, from which one can calculate the magnitude and phase of the reflected wave relative to the incident wave.

If the terminating impedance is not a pure resistance, its value depends on the frequency, and so also will the reflection coefficient. We must therefore assume a wave of a given angular frequency ω. (If the wave is not purely sinusoidal, we must perform a Fourier analysis and treat each harmonic separately.) The equations for current and voltage may then be written as complex exponentials, where, as always, the real or imaginary part must be extracted at the end, according to whether the input voltage is a cosine or sine function. Eqns (9.11) become

$$\begin{aligned} V &= A \exp j\omega(t-z/v)+A' \exp j\omega(t+z/v) \\ Z_0 I &= A \exp j\omega(t-z/v)-A' \exp j\omega(t+z/v) \end{aligned} \tag{9.17}$$

For simplicity we shall take the termination to be at the origin of coordinates $z=0$, noting that all points on the line will then have negative values of z. Inserting the boundary condition, we have

$$\left(\frac{V}{Z_0 I}\right)_{z=0} = \frac{\mathbf{Z}_L}{Z_0} = \frac{A \exp j\omega t + A' \exp j\omega t}{A \exp j\omega t - A' \exp j\omega t} = \frac{(A+A')}{(A-A')}.$$

Hence

$$A'/A = (\mathbf{Z}_L - Z_0)/(\mathbf{Z}_L + Z_0), \tag{9.18}$$

which defines the reflection coefficient A'/A. If $\mathbf{Z}_L$ is complex, A' will be complex (assuming A real, which can always be made true by choosing the zero of the time-scale correctly), showing that there is a phase change in the reflected wave. If $\mathbf{Z}_L$ is a pure resistance, A' is real and there is no phase change (we may exclude a phase change of π by allowing negative values of A').

If the line is open-circuited, the voltage at the end is $2A$, since $A'=A$. If $\mathbf{Z}_L$ is finite, the voltage across it is $A+A'=2A\mathbf{Z}_L/(\mathbf{Z}_L+Z_0)$, showing that the line behaves as a generator of voltage $2A$, with an internal resistance Z_0. The power transferred to the load, if $\mathbf{Z}_L$ is a resistance R, is $P=\tilde{V}^2/R=2A^2R/(R+Z_0)^2$. If $\mathbf{Z}_L=R=Z_0$, the power transferred to the load is a maximum; that is, the load is matched to the generator. Reference to eqn (9.18) above shows that under this condition $A'=0$. Thus maximum power transfer to the load corresponds to no reflected wave; hence the matching condition has a simple physical meaning, since any reflected wave would carry energy away from the load and reduce the power dissipated in it. The power in the incident wave is $\frac{1}{2}A^2/Z_0$, and that in the reflected wave is $\frac{1}{2}A'^2/Z_0$. The difference will be found to equal the power dissipated in the load, as calculated above.

Examination of the formula for A'/A, when $\mathbf{Z}_L$ is a pure resistance R, shows that as R goes from zero to infinity, A'/A changes continuously from -1 to $+1$. When $R < Z_0$, the reflected voltage wave has opposite sign from the incident wave, while the reflected current wave has the same sign. The voltage across R is then less than A, and the current through it greater than A/Z_0. The reverse is true if $R > Z_0$.

When $\mathbf{Z}_L$ is a pure reactance jX, A'/A is complex and its modulus is unity. This is to be expected, since no energy is dissipated in a pure reactance, and the amplitude of the reflected wave must therefore equal that of the incident wave. We may write the reflection coefficient A'/A in this case as $\exp j\delta$, and an algebraic reduction shows that

$$\exp j\delta = (X^2 - Z_0^2 + 2jXZ_0)/(X^2 + Z_0^2).$$

Hence $\tan\delta = 2XZ_0/(X^2 - Z_0^2)$, which may be written in the more convenient form $\tan\frac{1}{2}\delta = Z_0/X$.

In the general case, when $\mathbf{Z}_L$ is complex, the formulae are rather cumbersome, but may be reduced somewhat by writing $\mathbf{Z}_L = Z_1 \exp j\phi$, $A' = A_1 \exp j\delta$. Then, on clearing imaginary terms from the denominator, we find

$$\frac{A_1}{A}\exp j\delta = \frac{Z_1^2 - Z_0^2 + 2jZ_1Z_0\sin\phi}{Z_1^2 + Z_0^2 + 2Z_1Z_0\cos\phi},$$

giving

$$\left.\begin{aligned} \frac{A_1}{A} &= \left(\frac{Z_1^2 + Z_0^2 - 2Z_1Z_0\cos\phi}{Z_1^2 + Z_0^2 + 2Z_1Z_0\cos\phi}\right)^{\frac{1}{2}} \\ \text{and} \qquad \tan\delta &= 2Z_1Z_0\sin\phi/(Z_1^2 - Z_0^2). \end{aligned}\right\} \tag{9.19}$$

The maximum voltage on the line is $A + A_1$, and occurs at a voltage anti-node where the incident and reflected voltages are in phase; the minimum is $A - A_1$ at a node where they are in anti-phase. The ratio $(A + A_1)/(A - A_1)$ is called the voltage standing wave ratio (v.s.w.r.), and measurement of it together with the position of the node (which is related to δ) form the basis of a method of measuring an unknown impedance $\mathbf{Z}_L$ at short wavelengths (see § 22.4).

Transmission line terminated by another line of different impedance

A special case of a terminated line is one with impedance Z_1 joined to another line of different characteristic impedance Z_2. In this case there will be an incident and a reflected wave on the first line, and a transmitted wave on the second line. Our equations are then:

first line:

$$V_1 = A\exp j\omega(t - z/v_1) + A'\exp j\omega(t + z/v_1),$$

$$Z_1I_1 = A\exp j\omega(t - z/v_1) - A'\exp j\omega(t + z/v_1);$$

second line:

$$V_2 = A'' \exp j\omega(t - z/v_2),$$
$$Z_2 I_2 = A'' \exp j\omega(t - z/v_2),$$

where v_1, v_2 are the wave velocities on the first and second lines respectively. If the junction is at $z = 0$, the voltage and current at this point must be the same on the two lines, so that

$$A + A' = A'',$$
$$(A - A')/Z_1 = A''/Z_2,$$

the solutions of which are

$$A'/A = (Z_2 - Z_1)/(Z_2 + Z_1), \qquad A''/A = 2Z_2/(Z_2 + Z_1). \qquad (9.20)$$

The former of these is the same as for a line terminated by a resistance Z_2, as we would expect. The reflected power is the same as in the case of a real resistance Z_2, and the transmitted power the same as would be dissipated in a real resistance. Note that if $Z_2 > Z_1$, the voltage on the second line is greater than that in the incident wave. The power is less, however, since the characteristic impedance is higher.

Eqns (9.20) are identical with the formulae for reflection and transmission of a plane electromagnetic wave at normal incidence at the boundary of two media (eqns (8.41), (8.42), (8.49), and (8.50)). This shows that there is a close analogy between the characteristic impedance of a transmission line and the intrinsic impedance of a medium transmitting an electromagnetic wave. The similarity appears also in the expressions for the power transmitted: in a plane wave we have (power transmitted across unit area) $= E_x^2/Z_0 = Z_0 H_y^2$ (see § 8.3), while for a transmission line $P = V^2/Z_0 = Z_0 I^2$ (eqn (9.14)). The analogous behaviour makes it possible to adapt many of the formulae derived below to the case of plane waves.

Input impedance of terminated lines

When the reflection coefficient due to the load $\mathbf{Z}_L$ used to terminate a line is known, it is a simple matter to calculate the current and voltage, and hence the effective impedance, at any point in the line. This can be done for an arbitrary load $\mathbf{Z}_L$, but we shall limit ourselves to a few of the simpler and more interesting cases.

For a short-circuited line, $\mathbf{Z}_L = 0$ and $A' = -A$ in eqns (9.18), so that at a point $z = -l$ on the line (that is, at the terminals A, B in Fig. 9.14)

$$V = A\{\exp j\omega(t + l/v) - \exp j\omega(t - l/v)\}$$
$$= j2A \exp j\omega t \sin \omega l/v$$
$$= j2A \exp j\omega t \sin 2\pi l/\lambda$$

and

$$Z_0 I = 2A \exp \mathrm{j}\omega t \cos 2\pi l/\lambda.$$

The impedance at this point is then

$$V/I = \mathrm{j}Z_0 \tan 2\pi l/\lambda. \tag{9.21a}$$

This formula shows that a section of short-circuited line behaves as a pure reactance. If l is less than a quarter of a wavelength, then $\tan 2\pi l/\lambda$ is positive and the line behaves like an inductance. If l lies between a quarter- and a half-wavelength, the tangent is negative and the line behaves like a capacitance. These statements hold also if we increase l by an integral number of half-wavelengths.

If the line is open-circuited, $A' = +A$, and the equations for V and $Z_0 I$ are just interchanged. The impedance at $z = -l$ is therefore

$$V/I = -\mathrm{j}Z_0 \cot 2\pi l/\lambda. \tag{9.21b}$$

An open-circuited line, less than a quarter-wavelength long, therefore behaves like a capacitance; if its length lies between a quarter- and a half-wavelength, it behaves like an inductance. If its length is exactly a quarter-wavelength its impedance is zero. Thus if we have an open-circuited line we can cut off a quarter-wavelength and replace it by a short circuit without affecting the conditions earlier on the line. For we then have a short-circuited line of length $(l - \frac{1}{4}\lambda)$, so the impedance at the input terminals becomes $+\mathrm{j}Z_0 \tan(2\pi l/\lambda - \frac{1}{2}\pi) = -\mathrm{j}Z_0 \cot 2\pi l/\lambda$, in agreement with the value found directly from eqn (9.21b).

These results show that a lumped reactance X at the end of a line can be replaced by a suitable additional length of line, either open- or short-circuited. We found earlier that the wave reflected by a reactance has the same amplitude as the incident wave, but a phase change δ, where $\tan \frac{1}{2}\delta = Z_0/X$. If the reactance is replaced by an open- or short-circuited length of line, then the wave reflected from the far end will have the same amplitude as the incident wave, but the phase change arises from the time taken by the wave to travel the extra distance to the end of the line and back. At metre wavelengths, suitable lengths of either coaxial or parallel-wire lines are commonly used as inductances because they have a lower resistance than a coil of wire of the same 'nominal' radio-frequency resistance. By 'nominal' radio-frequency resistance is meant the value which would be calculated from the skin depth for a straight wire. In a closely wound coil there is an additional energy loss because of eddy currents induced by the oscillatory currents in neighbouring turns; this is known as the 'proximity effect', and increases the effective radio-frequency resistance. Its effect is minimized by using straight wires, as in a section of a transmission line.

The transmission line as a transformer

Since the voltage and current are in different ratio at the input terminals of a line from the ratio they bear at the output terminals, it follows that a line can be used as an impedance transformer. The case of greatest interest is that of a line a quarter-wavelength long. If the terminating impedance $\mathbf{Z}_L$ is at $z=0$, then at the point $z=-\frac{1}{4}\lambda$, the voltage and current are

$$V = A \exp j(\omega t+\tfrac{1}{2}\pi)+A' \exp j(\omega t-\tfrac{1}{2}\pi) = j(A-A')\exp j\omega t,$$
$$Z_0 I = j(A+A')\exp j\omega t,$$

and

$$V/I = Z_0(A-A')/(A+A') = Z_0^2/\mathbf{Z}_L, \tag{9.22}$$

showing that the terminating impedance has been transformed to $Z_0^2/\mathbf{Z}_L$. In this respect a quarter-wave behaves like a tuned circuit, which transforms a series resistance R into a parallel resistance L/CR. Since $Z_0^2 = L/C$, the formulae are similar in the two cases.

The quarter-wave transformer may be used to match a resistive load Z_L to a line of impedance Z_0 by inserting immediately before the load a $\frac{1}{4}\lambda$ section of line whose impedance Z_1 is such that $Z_1^2/Z_L = Z_0$. If Z_L and Z_1 are lines of the same dimensions but in different dielectric media, the situation is exactly analogous to the 'blooming' of optical lenses. The fraction of the incident intensity reflected from an air–glass surface is about 4 per cent, and the loss of light in an optical system with 10 or 20 surfaces is serious. The reflection may be reduced by depositing on the surface a quarter-wave thick layer of material of low refractive index, ideally equal to the square-root of the refractive index of the glass. The thickness is adjusted to be correct for the middle of the optical region, and is thus not quite correct for the ends of this region. 'Bloomed' surfaces appear slightly purple, therefore, owing to reflection of the red and blue rays. Quarter-wave films are also used to produce highly reflecting layers (see Problem 9.8). Further details of optical applications are given by Lissberger (1970).

The half-wave transformer is also of interest. It may be regarded as two consecutive quarter-wave transformers, giving an impedance

$$Z_0^2/(Z_0^2/\mathbf{Z}_L) = \mathbf{Z}_L.$$

Alternatively, this result may be obtained directly, since if we move one half-wave along a line, all voltages and currents are the same except for their reversed sign. The half-wave line is therefore a 1:1 transformer. A typical use is that of a connecting link between two pieces of apparatus, which makes the impedance of either appear unchanged. This is often useful at very short wavelengths where connecting wires sufficiently short to give no impedance transformation are not practicable.

9.5. Lossy lines and resonant lines

When loss is present in the conductors or in the medium the approach used above in terms of current and voltage is not rigorous, but it provides a method of analysis which is simple and sufficiently accurate for most purposes when the losses are small. In Fig. 9.15, the distributed impedance **Z** and admittance **Y** per unit length are assumed to be complex, giving the relations

$$-\partial V/\partial z = I\mathbf{Z}, \qquad -\partial I/\partial z = V\mathbf{Y}. \tag{9.23a}$$

Suppose that waves are propagated as

$$\exp(\mathrm{j}\omega t - \gamma z),$$

where γ is complex; as in Chapter 8, we must here consider just one frequency because we shall find that the propagation constants are not independent of frequency. Eqns (9.23a) become

$$\left.\begin{aligned}\gamma V &= I\mathbf{Z} = I(R+\mathrm{j}\omega L),\\ \gamma I &= V\mathbf{Y} = V(G+\mathrm{j}\omega C).\end{aligned}\right\} \tag{9.23b}$$

Elimination of I or V gives

$$-\gamma^2 = -(R+\mathrm{j}\omega L)(G+\mathrm{j}\omega C) = (LC\omega^2 - RG) - \mathrm{j}\omega(RC+GL),$$

and on writing $\gamma = \alpha + \mathrm{j}\beta$ we obtain

$$\beta^2 - \alpha^2 = LC\omega^2 - RG, \qquad 2\beta\alpha = \omega(RC+GL).$$

These equations show that the phase constant β is altered by the presence of loss, and that the wave is attenuated because $\alpha \neq 0$. Both the phase velocity and the attenuation are frequency-dependent.

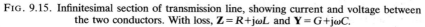

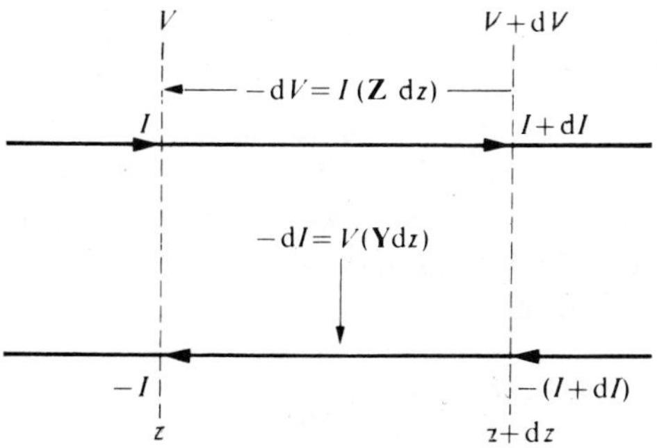

FIG. 9.15. Infinitesimal section of transmission line, showing current and voltage between the two conductors. With loss, $\mathbf{Z} = R + \mathrm{j}\omega L$ and $\mathbf{Y} = G + \mathrm{j}\omega C$.

These equations may be solved exactly for α and β, but it is more instructive to solve them approximately, assuming R, G to be small. Then

$$\beta = \omega(LC)^{\frac{1}{2}}\left\{1+\frac{1}{8\omega^2}\left(\frac{G}{C}-\frac{R}{L}\right)^2\right\}, \qquad \alpha = \frac{1}{2}\left\{R\left(\frac{C}{L}\right)^{\frac{1}{2}}+G\left(\frac{L}{C}\right)^{\frac{1}{2}}\right\}. \tag{9.24}$$

The velocity is approximately

$$\frac{\omega}{\beta} = \frac{1}{(LC)^{\frac{1}{2}}}\left\{1-\frac{1}{8\omega^2}\left(\frac{G}{C}-\frac{R}{L}\right)^2\right\}, \tag{9.25}$$

showing that it is altered only in the second order. The power flowing along the line decays as it travels along as (using $Z_0 = (L/C)^{\frac{1}{2}}$)

$$\exp(-2\alpha z) = \exp\{-(R/Z_0 + GZ_0)z\}.$$

The two terms in the exponential represent just the fraction of the stored energy which is dissipated per unit length in the resistance and conductance respectively. We see that if R/Z_0 and GZ_0 are both small, the line is distortionless in the first approximation, since neither the velocity of the wave nor its attenuation depend on the frequency to this order. In the next approximation, distortion arises from the change in velocity with frequency and this is most serious at low frequencies. At high frequencies (owing to skin effect) the resistance rises, increasing α. If G is negligible, distortion may be reduced by increasing L. On telephone land lines this is accomplished by introducing inductances in series with the line at regular intervals. This also reduces the attenuation, since it increases Z_0, but gives the line a periodic structure so that it behaves like a low-pass filter. The cut-off frequency must be kept above the audio-frequency range, and to do this the spacing of the inductances must be small compared with the shortest wavelength which it is required to transmit. The distortion can also be greatly reduced by transmitting a voice-modulated signal of, say, 100 kHz frequency, instead of the actual voice frequency range of 100 Hz to 10 kHz. Although the bandwidth required is the same, the fractional change in frequency involved is very much smaller.

In the laboratory, the lengths of line used are so short that attenuation is negligible except at the highest frequencies, where R rises owing to the skin effect. To obtain a numerical value, let us take an air-spaced coaxial line at a frequency of 3×10^9 Hz ($\lambda = 0{\cdot}1$ m). Then for copper the skin depth δ is about $1{\cdot}2\times10^{-6}$ m. For a conductor whose dimensions are $a = 2{\cdot}5$ mm, $b = 8$ mm (giving a 70 Ω line),

$$R = \rho\left(\frac{1}{2\pi a\delta}+\frac{1}{2\pi b\delta}\right) = 1{\cdot}2\ \Omega\ \text{m}^{-1}.$$

Then $\alpha = 8\times10^{-3}$ m^{-1}, and the power transmitted along the line will fall by a factor of $1/e$ in a distance of 60 m.

If the space between the conductors is filled with a dielectric, the attenuation due to dielectric loss will be important unless the dielectric is of the highest quality. If its loss tangent is tan δ, then $G = \omega C \tan \delta$, and

$$2\alpha = GZ_0 = \omega(LC)^{\frac{1}{2}} \tan \delta = (\omega/v)\tan \delta = (2\pi/\lambda_1)\tan \delta,$$

where λ_1 is the wavelength in the dielectric. If its relative permittivity is 2·2, and $\tan \delta = 2\times10^{-4}$, $\alpha = 9\times10^{-3}\ \mathrm{m}^{-1}$ at $\lambda = 0{\cdot}1$ m, which is as large as that due to the resistance. For this reason high-frequency cables are often made with some device such as an open spiral of polythene string supporting the centre conductor in order to reduce the amount of dielectric in the position of maximum electric field.

In calculating the loss on the line we have taken no account of energy lost by radiation, though we might expect each element of the conductors to radiate since it carries an alternating current. A transmission line has two conductors carrying equal and opposite currents, however, and in computing the radiation we must allow for the destructive interference between their two radiation patterns. For a parallel-wire line this does not give an exact null, but the maximum phase difference between the signals from the two wires in any direction will differ from π at most by $2\pi(2d)/\lambda$, where $2d$ is the separation between the two conductors. Hence the radiated energy is less than that from a single wire by a factor of the order $(d/\lambda)^2$, and is small if $d \ll \lambda$. The coaxial line gives an exact null because the one current entirely encloses the other, and the magnetic field at any external point is zero. For this reason coaxial lines are to be preferred at wavelengths less than about a metre.

Transmission lines as tuned circuits

If we have a quarter-wave section of loss-free line, short-circuited at one end, then the impedance measured at the other end is infinite. Similarly, if it is open-circuited at one end, then the input impedance at the other end is zero. We have already seen that if a resistance R is connected across one end, the impedance at the other end is

$$Z_0^2/R = L/CR.$$

Thus in all respects the section behaves like a tuned circuit. If R is zero or infinity, the input impedance will only be infinity or zero so long as the line is completely loss-free. This, of course, will never occur in practice, and to assess the properties of the section as a tuned circuit we must include the effect of distributed losses. We can do this by bringing in the attenuation constant α.

We assume that the section is open-circuited at the far end, so that $A' = A$. Then the voltage and current at a point $z = -l$ are (since A is the incident voltage amplitude at $z=0$)

$$V = A\left[\exp\left\{j\left(\omega t+\frac{2\pi l}{\lambda}\right)+\alpha l\right\}+\exp\left\{j\left(\omega t-\frac{2\pi l}{\lambda}\right)-\alpha l\right\}\right],$$

$$Z_0 I = A\left[\exp\left\{j\left(\omega t+\frac{2\pi l}{\lambda}\right)+\alpha l\right\}-\exp\left\{j\left(\omega t-\frac{2\pi l}{\lambda}\right)-\alpha l\right\}\right].$$

The impedance $\mathbf{Z}$ at this point is then given by

$$\frac{\mathbf{Z}}{Z_0}=\frac{1+\exp\{-j4\pi l/\lambda-2\alpha l\}}{1-\exp\{-j4\pi l/\lambda-2\alpha l\}}. \tag{9.26}$$

We now assume that the length of the line l is close to an odd multiple of a quarter-wavelength, and examine how the impedance changes in the neighbourhood of this point. In the complex exponentials of eqn (9.26) the imaginary part of the argument gives a rapid variation and the real part (which is assumed to be small) a slow variation. We therefore treat them separately, and write $l=\{(2n+1)\lambda/4\}+\Delta l$ in the imaginary part only. Then the exponential becomes

$$\begin{aligned}\exp\{-j(2n+1)\pi-j4\pi\Delta l/\lambda-2\alpha l\} &= -\exp(-j4\pi\Delta l/\lambda-2\alpha l)\\ &\approx -(1-j4\pi\Delta l/\lambda-2\alpha l),\end{aligned}$$

where we have assumed that both $\Delta l/\lambda$ and αl are small. Thus the impedance becomes

$$\mathbf{Z}=Z_0\frac{1-(1-j4\pi\Delta l/\lambda-2\alpha l)}{1+(1-j4\pi\Delta l/\lambda-2\alpha l)}\approx Z_0(\alpha l+j2\pi\Delta l/\lambda)=Z_0\alpha l\left(1+j\frac{2\pi}{\alpha\lambda}\frac{\Delta l}{l}\right), \tag{9.27}$$

where only small quantities of the first order have been retained. This equation is of the same form as eqn (7.20) for a series tuned circuit near resonance, which is

$$\mathbf{Z}=R'(1+j2Q_F\,\Delta\omega/\omega_0)$$

since $\Delta l/l=-\Delta\lambda/\lambda=\Delta\omega/\omega_0$ (here the minus sign is introduced in relating $\Delta l/l$ to $\Delta\lambda/\lambda$ because increasing the length of the line has the same effect as shortening the wavelength of the applied radiation). Hence at resonance the quarter-wavelength line (or a line an odd multiple of this length) behaves as a series tuned circuit with a resistance $Z_0\alpha l$ and a quality factor $Q_F=\pi/\alpha\lambda$.

In making measurements at short wavelengths it is always advisable to keep the generator frequency constant, if possible, and vary the element under test, since this avoids errors due to detectors, connecting lines, etc., being frequency sensitive. Eqn (9.27) shows that we may conveniently measure Q_F by finding the fractional change in length of the line $(\Delta l/l)$ required to move between the points at which the impedance rises to $\sqrt{2}$ of its minimum value.

To find a numerical value for Q_F we take the value $\alpha = 8\times10^{-3}\ \text{m}^{-1}$ found earlier for an air-spaced coaxial line at 0·1 m wavelength. This gives $Q_F = 4000$, which is very much higher than can be obtained normally with a lumped circuit at medium radio-frequencies. Q_F is independent of the number of quarter-wavelengths in the section, but the series resistance $R' = Z_0\alpha l = 1{\cdot}4\times10^{-2}(2n+1)\ \Omega$. Thus R' increases when we make n larger, but Q_F does not change. This is because Q_F depends on the ratio of the stored energy to the energy dissipated, and both of these increase as the length of the line increases.

The impedance of a short-circuited quarter-wave line can readily be calculated from the above, since in this case $A' = -A$, and the formulae are the same, if we interchange V and Z_0I. This gives

$$\frac{I}{V} = \mathbf{Y} = \frac{\alpha l}{Z_0}\left(1 + \mathrm{j}\frac{2\pi}{\alpha\lambda}\frac{\Delta l}{l}\right), \qquad (9.28)$$

showing that at resonance the impedance is $Z_0/\alpha l = Z_0^2/R' = L/CR'$, the same expression as for a parallel tuned circuit. For a single quarter-wavelength of the coaxial line considered previously, the parallel impedance is 350 kΩ, showing that the line makes a good load for an oscillator or amplifier. In practice the line may be rather shorter than a quarter-wavelength, so that it behaves as an inductance which can be adjusted to resonate with the stray capacitance of the circuit.

9.6. Guided waves—propagation between two parallel conducting planes

When an electromagnetic wave is launched from an aerial into free space (or a non-conducting medium) its amplitude falls off inversely with the distance owing to the spreading out of the wave in a spherical wave front. There is no dissipation of energy, but the power flowing through unit area normal to the wave front falls off according to the inverse square law. On the other hand, a wave sent along a coaxial line suffers no diminution in amplitude, apart from that due to resistive losses, because it is confined to the space between the conductors and does not spread out. Such a wave is a guided wave, and in the absence of loss it was shown that with two parallel conductors a solution of Maxwell's equations can be found giving a purely transverse wave (no components of **E** or **H** in the

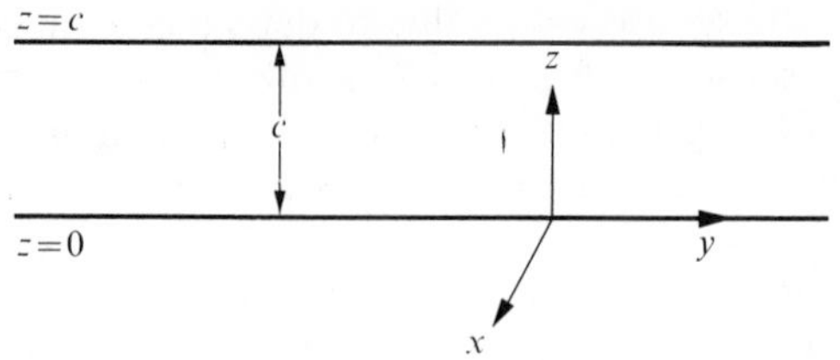

FIG. 9.16. Coordinate system for propagation between two parallel planes.

direction of propagation) which is freely propagated at all frequencies with the same velocity as a wave in the unbounded medium. With a single hollow conductor this is no longer the case, though the fact that it is possible to see down a metal tube shows that some form of electromagnetic wave can be propagated through it. Such a wave is again a guided wave, since it must move in the direction of the tube. As a preliminary to studying propagation through such a tube (known as a 'waveguide'), we shall investigate the problem of propagation between two parallel infinite perfectly conducting planes, separated by a distance c, as shown in Fig. 9.16.

A Cartesian coordinate system may be defined by taking the conductors to be the planes $z=0$ and $z=c$, and assuming that the wave is propagated parallel to the x-axis, which is normal to the plane of the paper. The boundary conditions now demand that any components of the electric field strength (E_x or E_y) tangential to the planes must vanish at the planes $z=0$ and $z=c$. We try to find the simplest possible solution of Maxwell's equations consistent with these demands, and begin by assuming a purely transverse plane wave in which both E_x and H_x are zero, such as we would have in the absence of the conductor. If this wave is polarized with its electric vector normal to the planes (that is, $E_y=0$) it is readily seen that the boundary conditions are satisfied automatically, and a wave of this polarization is possible. The solutions are of the same form as for a wave in the unbounded medium, and the velocity is also the same. On the other hand, a plane wave in which the field components do not vary with z is obviously impossible if the electric vector is parallel to the planes, since E_y must be zero at $z=0$ and $z=c$ and will be zero everywhere unless we allow it to vary with z. We therefore examine whether it is possible to have a wave in which E_y is finite, but E_z and E_x are both zero, so that the electric field is purely transverse. The field components of $\mathbf{H}$ can then be computed from Maxwell's equations. We shall assume that the wave is propagated as $\exp(j\omega t-\gamma x)$, so that we can replace differentiation with respect to t and x by multiplication by $j\omega$ and $-\gamma$ respectively. Then, if the medium between the planes has permittivity ϵ, magnetic permeability μ, and zero conductivity, the curl eqns (8.3) and

(8.7) give the following components:

$$\left.\begin{aligned} -j\omega\mu H_x &= -\partial E_y/\partial z, \\ -j\omega\mu H_y &= 0, \\ -j\omega\mu H_z &= -\gamma E_y, \end{aligned}\right\} \tag{9.29}$$

$$\left.\begin{aligned} 0 &= \partial H_z/\partial y - \partial H_y/\partial z, \\ j\omega\epsilon E_y &= \partial H_x/\partial z + \gamma H_z, \\ 0 &= -\gamma H_y - \partial H_x/\partial y, \end{aligned}\right\} \tag{9.30}$$

where we have already assumed $E_x = E_z = 0$. Eqns (9.29) show immediately that $H_y = 0$, but H_x cannot be zero unless $\partial E_y/\partial z$ is zero, and this is not allowed by the boundary conditions. Hence the wave will not be purely transverse, but will have a component of **H** in the direction of propagation. On putting $H_y = 0$ in eqns (9.30), we see that $\partial H_z/\partial y$ and $\partial H_x/\partial y$ are both zero, so that there is no variation in the y direction, and examination of the components of div $\mathbf{E} = 0$ shows this to be true also of E_y. The remaining components of eqns (9.29) and (9.30) therefore reduce to

$$\left.\begin{aligned} j\omega\mu H_x &= \partial E_y/\partial z, \\ j\omega\mu H_z &= \gamma E_y, \\ j\omega\epsilon E_y &= \partial H_x/\partial z + \gamma H_z. \end{aligned}\right\} \tag{9.31}$$

Elimination of H_x and H_z between these three equations gives

$$-\omega^2\mu\epsilon E_y = \partial^2 E_y/\partial z^2 + \gamma^2 E_y$$

or

$$\partial^2 E_y/\partial z^2 = -(\gamma^2 + \omega^2/v^2)E_y, \tag{9.32}$$

where $v = (\mu\epsilon)^{-\frac{1}{2}}$ is the velocity of an electromagnetic wave in the unbounded medium (in the absence of the conducting planes we could put $\partial/\partial z = 0$ and obtain this result directly from eqn (9.32), since γ must then be an imaginary quantity $j\beta$, and $v = \omega/\beta$). The solution of eqn (9.32) can be written in the form

$$E_y = A\sin(\pi nz/c) + B\cos(\pi nz/c),$$

and the boundary conditions $E_y = 0$ at $z = 0$ and $z = c$ require that $B = 0$ and n must be an integer. To satisfy eqn (9.32) we must have

$$(n\pi/c)^2 = \gamma^2 + \omega^2/v^2$$

and hence

$$\gamma = \left\{\left(\frac{n\pi}{c}\right)^2 - \left(\frac{\omega}{v}\right)^2\right\}^{\frac{1}{2}}. \tag{9.33}$$

Here γ is either real or purely imaginary according to whether the quantity inside the square-root is positive or negative. It is clear that at low frequencies γ will be real, and the wave will then be attenuated. At sufficiently high frequencies γ will be imaginary, and waves will be freely transmitted without attenuation; thus the system acts as a high-pass filter. The condition for free transmission of waves is that

$$\omega/v > n\pi/c.$$

Since $\omega/v = 2\pi/\lambda_0$, where λ_0 is the wavelength of the radiation in the unbounded medium, this condition may be written in the form

$$\lambda_0 < 2c/n.$$

Hence $2c/n$ is the cut-off wavelength λ_c, and only radiation of shorter wavelength is freely transmitted. In the pass band we may write $\gamma = \mathrm{j}\beta = \mathrm{j}(2\pi/\lambda_g)$, where λ_g is the apparent wavelength of the radiation in the guide, that is, the distance between successive points along the x-axis where the phase differs by 2π. Eqn (9.33) then reduces to

$$\frac{f^2}{v^2} = \frac{1}{\lambda_0^2} = \frac{1}{\lambda_g^2} + \frac{1}{\lambda_c^2}. \tag{9.34}$$

This is known as the 'waveguide equation', and it is found to hold for any shape of waveguide, although it has here been deduced only for a simple special case. The cut-off wavelength λ_c depends on the shape and dimensions of the waveguide, and on the mode of propagation (that is, in the present case, on the values of c and n respectively).

Eqn (9.34) shows that the wavelength in the guide is always greater than that in the unbounded medium λ_0. When $\lambda_0 \ll \lambda_c$, λ_g approaches λ_0, while when $\lambda_0 \rightarrow \lambda_c$, λ_g tends to infinity. The phase velocity in the guide behaves in the same way as λ_g, since

$$v_g = \omega/\beta = f\lambda_g, \tag{9.35}$$

showing that v_g is always greater than the velocity in the unbounded medium. If the guide contains no material medium, the phase velocity will be greater than the velocity of light ($f\lambda_0$) in a vacuum, since $\lambda_g > \lambda_0$. This does not mean that energy is transmitted with a velocity greater than that of light, since we have dispersion: the phase velocity depends on the frequency, and does not equal the group velocity $u_g = \mathrm{d}\omega/\mathrm{d}\beta$. From eqn (9.33)

$$-\gamma^2 = \beta^2 = \left(\frac{\omega}{v}\right)^2 - \left(\frac{n\pi}{c}\right)^2,$$

and hence

$$2\beta(\mathrm{d}\beta/\mathrm{d}\omega) = 2\omega/v^2,$$

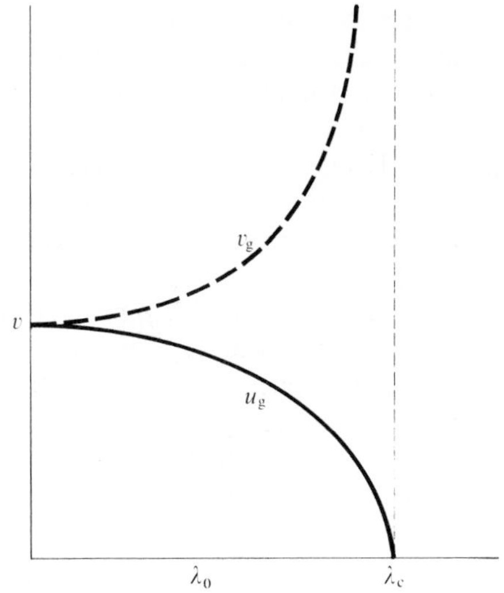

FIG. 9.17. Variation of the phase velocity v_g (broken line) and group velocity u_g (full line) in a waveguide.

giving

$$\left(\frac{\omega}{\beta}\right)\left(\frac{d\omega}{d\beta}\right)=v_g u_g=v^2. \tag{9.36}$$

Since v_g is always greater than v, it follows that u_g is always less than v, and is thus always less than the velocity of light. The behaviour of u_g and v_g is illustrated in Fig. 9.17; these relations hold for all waveguides, since they depend only on the waveguide equation, (9.34).

The propagation of waves between two parallel conducting planes may be considered in another way which is illuminating, particularly in respect of the group and phase velocity. The wave motion may be regarded as consisting of an ordinary plane wave, with the same properties as a wave in the unbounded medium, which is multiply reflected from the two planes. The normal to the wave front is assumed to make an angle θ with the normal to the conducting planes, as in Fig. 9.18. From Figs 8.4, 8.5, if

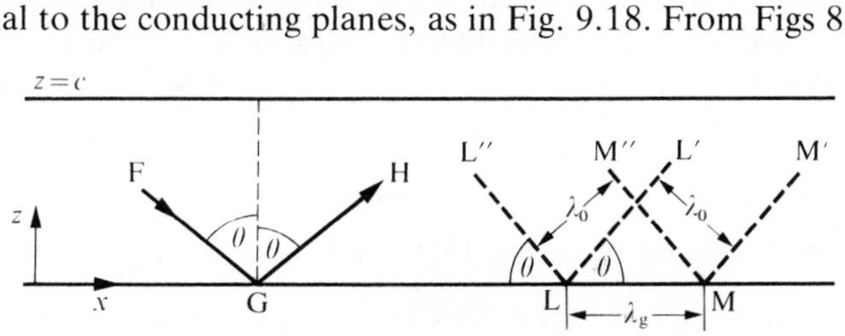

FIG. 9.18. Reflection of ordinary plane waves between two parallel planes. FG, GH normals to wave fronts incident and reflected at plane $z=0$. LL′, MM′ incident wave fronts; LL″, MM″ reflected wave fronts, differing in phase by 2π. LM gives the guide wavelength λ_g.

we replace θ by $\pi-\theta$, the components of such incident and reflected waves are given by eqns (8.36) and (8.37), if the electric vector is in the plane of incidence, or eqns (8.46) and (8.47) if it is normal to the plane of incidence. In either case we must satisfy the boundary conditions, that the tangential components of **E** must be zero all over the planes $z=0$ and $z=c$. For the former case (electric vector in plane of incidence) the first of these conditions gives $A'=-A$ in order to make the x component of **E** zero at $z=0$; the same condition is obtained at $z=c$ if we take $\cos\theta=0$ ($\theta=\frac{1}{2}\pi$). This makes the x component of **E** zero everywhere, but E_z and H_y are finite, so that we have a simple plane wave moving in the x direction; this wave is purely transverse, and moves with the same velocity as in the unbounded medium.

When the electric vector is normal to the plane of incidence, to make the y component of **E** zero at $z=0$, we must take $B'=-B$ in eqns (8.46) and (8.47). At other points the amplitude of the y component of **E** is then (replacing θ by $\pi-\theta$)

$$\begin{aligned} E_y &= B(F_1-F_2) \\ &= B\exp\{j\omega(t-x\sin\theta/v)\}\{\exp(j\omega z\cos\theta/v)-\exp(-j\omega z\cos\theta/v)\} \\ &= 2jB\sin(\omega z\cos\theta/v)\exp j\omega(t-x\sin\theta/v), \end{aligned} \tag{9.37}$$

where we have written v for the velocity in the unbounded medium. At the plane $z=c$ the field component given by eqn (9.37) is zero provided that

$$\omega c\cos\theta/v = 2\pi c\cos\theta/\lambda_0 = n\pi$$

or

$$\cos\theta = \lambda_0(n/2c) = \lambda_0/\lambda_c, \tag{9.38}$$

where λ_c is the cut-off wavelength as previously defined, and λ_0 is the wavelength in the unbounded medium. This equation shows clearly that no wave is possible for $\lambda_0>\lambda_c$, for then there is no real value of θ which satisfies it.

The wavelength λ_0 is defined as the normal distance between two wave fronts such as LL′ and MM′ in the plane waves where the phase differs by 2π; these wave fronts have an intercept LM on the plane $z=0$, and the length of this intercept, which gives the apparent wavelength λ_g of a wave propagated in the x direction, is $\lambda_0/\sin\theta$. Hence

$$\sin\theta = \lambda_0/\lambda_g, \tag{9.39}$$

and on combining this with eqn (9.38) we have

$$1/\lambda_0^2 = (\sin^2\theta+\cos^2\theta)/\lambda_0^2 = 1/\lambda_g^2+1/\lambda_c^2,$$

which is the 'waveguide equation' already derived (eqn (9.34)). We see that it follows from the fact that only one angle θ is possible for the direction of our multiply reflected plane wave in order to satisfy both the boundary conditions. The energy flow travels with velocity v in the plane wave in a direction normal to the plane wave front (that is, along FG or GH); the component of this velocity in the x direction is $v \sin\theta$, and this is the speed u_g at which the energy flows in the guided wave. On the other hand, the phase velocity v_g of the guided wave, from eqn (9.37), is $v/\sin\theta$; hence $u_g v_g = v^2$, as shown earlier (eqn (9.36)). (It should be noted that the possibility of v_g being greater than v is not peculiar to electromagnetic waves; the effect can be observed by watching the movement parallel to a reflecting boundary of a crest in any wave motion, as, for example, in water waves being reflected at an angle from a breakwater.) The behaviour of our guided wave when λ_0 is equal to the cut-off wavelength λ_c can be understood if we remember that this requires $\theta = 0$; that is, the wave motion is an ordinary plane wave being reflected at normal incidence between the two planes. Then no energy is propagated in the x direction, so that $u_g = 0$; but the phase at a given value of z is independent of x, so that the apparent phase velocity v_g in the x direction is infinite.

9.7. Waveguides

The type of wave we have been considering, propagated between two parallel planes, has the following components:

$$\left.\begin{aligned} E_y &= A\sin(2\pi z/\lambda_c)\sin(\omega t - 2\pi x/\lambda_g),\\ H_x &= -A(\lambda_0/Z_1\lambda_c)\cos(2\pi z/\lambda_c)\cos(\omega t - 2\pi x/\lambda_g),\\ H_z &= A(\lambda_0/Z_1\lambda_g)\sin(2\pi z/\lambda_c)\sin(\omega t - 2\pi x/\lambda_g).\end{aligned}\right\} \qquad (9.40)$$

These components may either be obtained from eqns (8.46) and (8.47) (for example, E_y is found by taking the real part of eqn (9.37) and writing $2B = -A$), or by taking E_y as the appropriate solution of eqn (9.32) and using eqns (9.31); Z_1 is the intrinsic impedance of the medium between the conducting planes. These equations show that for a given value of λ_c, the component H_x in the direction of propagation diminishes in amplitude as λ_0 is decreased, so that the wave approaches a purely transverse wave travelling along the x-axis. As $\lambda_0 \to \lambda_c$, $H_z \to 0$ since $\lambda_0/\lambda_g \to 0$, giving in the limit a transverse wave travelling along the z-axis.

Since the only electric field component is in the y direction, it is possible to insert conducting planes normal to the y-axis without introducing any new boundary conditions. We have then a closed rectangular waveguide, as shown in Fig. 9.19(a), bounded by the perfectly conducting planes $z = 0$, $z = c$: $y = 0$, $y = b$. The field components within the guide are given by eqns (9.40), and are zero outside. This type of wave is designated TE_{0n} (or H_{0n}); TE means 'transverse electric', indicating that

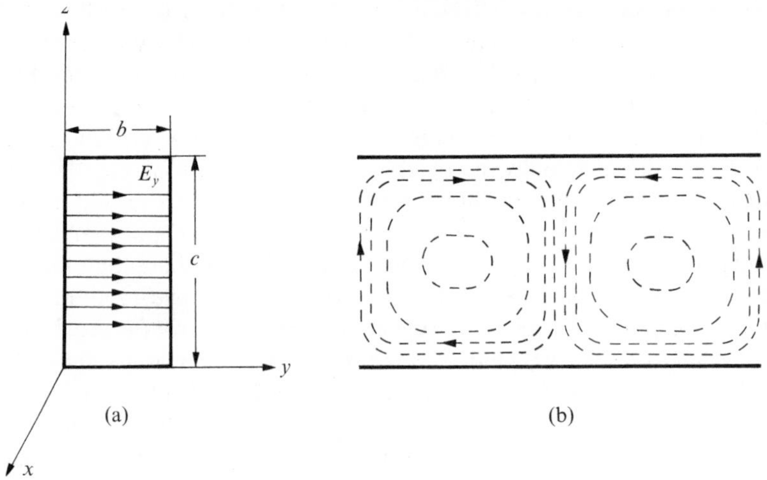

FIG. 9.19. Rectangular waveguide with TE_{01} mode. (a) Lines of electric field ———. (b) Lines of magnetic field - - - - - -.

there is no electric field component in the direction of propagation; the subscripts 0, n mean that there is no variation in the y direction of any field component, while in the z direction they vary as sin or $\cos(\pi nz/c)$. The simplest wave ($n = 1$) is shown in Fig. 9.19, and this is also the mode with the largest cut-off wavelength $\lambda_c = 2c$. It is therefore used as the standard mode for waveguide propagation, and the guide dimensions are chosen so that $2c > \lambda_0 > c$ for the wavelength it is desired to propagate. No higher mode (with $n = 2$ or more) can then be propagated; this has the advantage that waves of higher modes, set up by a local disturbance of the field (due to discontinuities or changes in the guide dimensions), decay exponentially along the guide. By making the dimension b less than $\lambda_0/2$, no TE mode can be propagated with the electric vector polarized in the z direction; and it can be shown that all other modes have still smaller cut-off wavelengths, and so cannot be propagated.

The field components in the TE_{01} mode are shown in Fig. 9.19. E_y is a maximum in the centre of the guide, and zero at the planes $z = 0$ and $z = c$; it varies sinusoidally with z, with just one half-period of variation (modes with higher values of n make n half-periods of variation, and so require a correspondingly larger value of c for a given λ_0). The components of the magnetic field are everywhere tangential to the boundaries, and the lines of magnetic field are shown in Fig. 9.19(b), where the guide is viewed looking down on the broad face. The lines of magnetic field encircle the points at which $\partial E_y/\partial t$ is greatest, that is, the points where the displacement current is greatest. This corresponds to Maxwell's equation curl $\mathbf{H} = \partial\mathbf{D}/\partial t$, which implies that a displacement current (a changing electric displacement) is encircled by lines of magnetic field. The flow of conduction current is always normal to the magnetic field at the metal surface. Thus on the narrow face it is everywhere parallel to the y-axis, and on the broad face it is parallel to the x-axis at the centre. Narrow slits can be cut in the metal without causing energy to leak out, providing that they do not interrupt any current flow. This is used in the construction of a standing-wave detector (cf. § 22.4), where the electric field strength inside the guide is detected by a small probe moving along a narrow slot cut along the centre of the broad face.

Study of the propagation characteristics of waves in guides of other than rectangular shape involves the use of more complex mathematics and we mention only the cylindrical waveguide. This involves the solution of the wave equation or Maxwell's equations in cylindrical coordinates, and requires the use of Bessel functions. Waves may be designated as TM_{mn} or TE_{mn}, corresponding to no magnetic or no electric field in the direction of propagation; the first subscript indicates that the field components vary as cos or sin $m\phi$, where ϕ is the azimuthal angle, while the second gives the number of values of the radius at which the electric field components other than the radial component E_r are zero. The simplest modes are TE_{11}, which is rather similar to the TE_{01} mode in rectangular guide; the electric field is purely transverse, and distributed as shown in Fig. 9.20(a); the cut-off wavelength is $1{\cdot}707d$, where d is the diameter of the guide. The TM_{01} mode has a transverse magnetic field whose lines of force are circular; the cut-off wavelength is $1{\cdot}308d$, and the mode is similar to that in a coaxial line except that the conduction current in the centre conductor is replaced by displacement current, with lines of **E** running down the centre and turning outwards to terminate on the wall as in Fig. 9.20(b). The TE_{01} mode is rather similar, but with the lines of electric field and magnetic field interchanged; the electric field has closed circular lines of force and is purely transverse, while the magnetic field is greatest down the axis (see Fig. 9.20(c)); the cut-off wavelength is $0{\cdot}820d$.

Cavity resonators

If a length of waveguide is closed by conducting walls at each end, it will resonate at wavelengths such that the distance between the end walls is a multiple of half a guide wavelength. In the case of a rectangular guide closed by conducting walls normal to the x-axis a distance a apart, the distance a must be $\frac{1}{2}l\lambda_g$ where l is an integer, in order that the electric field (which is tangential to the end walls) may be zero at the two ends. From eqn (9.34) the wavelength λ_0 in the unbounded medium at which

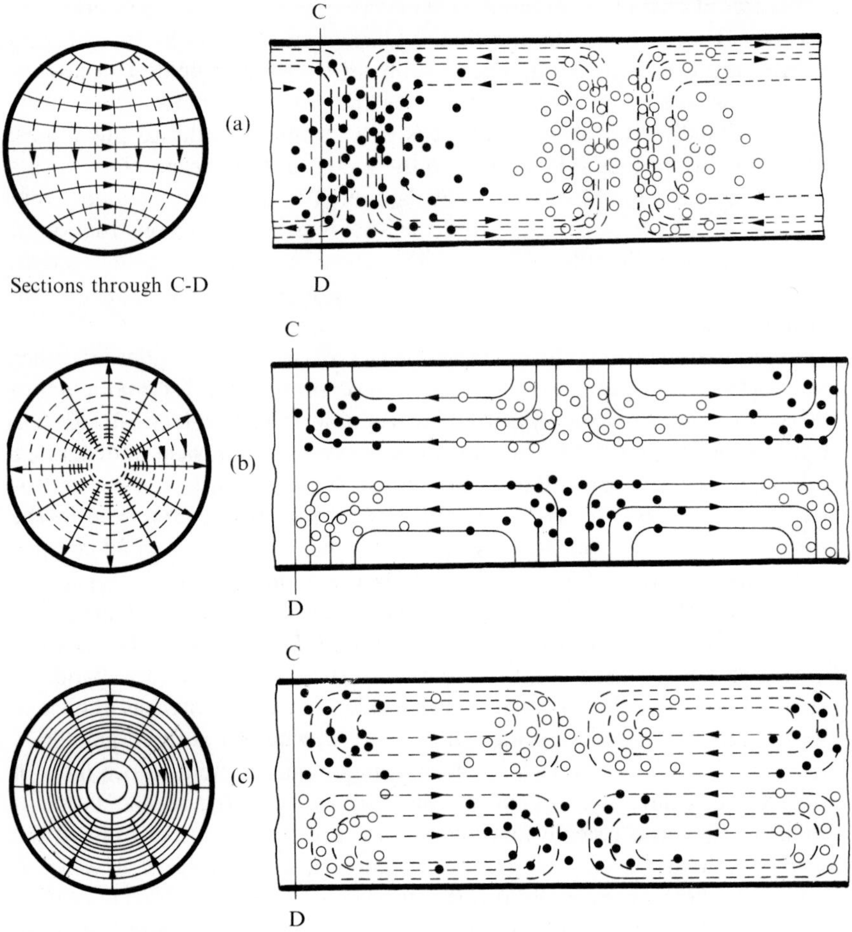

——— Lines of electric field ----- Lines of magnetic field

• towards observer ○ away from observer

FIG. 9.20. Approximate configurations of electric and magnetic fields in a cylindrical waveguide. Propagation is directed away from the observer or to the right. (a) TE_{11} or H_{11} mode; (b) TM_{01} or E_{01} mode; (c) TE_{01} or H_{01} mode. (After Southworth 1936, by courtesy of Bell Telephone Laboratories.)

the rectangular cavity will resonate in a TE_{0n} mode is thus given by

$$\frac{1}{\lambda_0^2} = \left(\frac{l}{2a}\right)^2 + \left(\frac{n}{2c}\right)^2. \tag{9.41}$$

This is a special case of the more general formula for a TE_{mn} mode, for which

$$\frac{1}{\lambda_0^2} = \left(\frac{l}{2a}\right)^2 + \left(\frac{m}{2b}\right)^2 + \left(\frac{n}{2c}\right)^2 \tag{9.42}$$

for a rectangular cavity of dimensions a, b, c; this formula may be recognized as that used in the theory of heat radiation in computing the resonant modes of a hollow rectangular cavity (see Problem 9.12).

The most important quantity for any resonant system is its quality factor Q_F; this may be found for a waveguide cavity by using the relation (see § 7.3)

$$Q_F = \omega(\text{energy stored})/(\text{energy dissipated per second}).$$

The total stored energy may be computed by integrating the energy density in the cavity, while the total energy dissipated in the metallic walls (owing to their finite resistivity) can be found by using eqn (8.34b). The order of magnitude of Q_F can readily be found without carrying through the details of the integration in the following way. If the amplitude of the strength of the oscillating magnetic field in the cavity is H_0, the stored energy $\approx \frac{1}{2}\mu_0 H_0^2 V$, where V is the volume of the cavity (assumed to be evacuated), while the energy lost at the walls $\approx \frac{1}{2}\rho H_0^2 A/\delta$, where A is the total wall area, and ρ and δ the resistivity and skin depth in the wall. Hence

$$Q_F \approx \omega(\tfrac{1}{2}\mu_0 H_0^2 V)/(\tfrac{1}{2}\rho H_0^2 A/\delta) = (V/A)(\omega\delta\mu_0/\rho) \approx V/A\delta, \qquad (9.43)$$

assuming the magnetic permeability of the wall to be unity, and using eqn (8.31). This result shows that at a given wavelength the value of Q_F increases with the linear dimensions of the resonator; while at different wavelengths, if the linear dimensions are scaled in proportion to the wavelength, Q_F varies as $\lambda_0^{\frac{1}{2}}$, since δ varies as $\lambda_0^{\frac{1}{2}}$, and V/A as λ_0. For a given wavelength and size of cavity Q_F does not vary greatly with the mode of resonance, with one exception. The TE_{01} mode in a cylindrical cavity has rather a high Q_F, and there is no radial flow of current on the end walls; for this reason it is used in wavemeters (see § 22.5) where one end is a movable plunger. A good contact between this and the cylindrical wall is not essential to a high Q_F, since there is no current flow across the contact. A value of 10^4 may be obtained for Q_F at wavelengths of 10^{-2} m, and the sharpness of resonance is thus rather higher than for a resonant coaxial line, mainly because there is no centre conductor with its rather high current density to contribute to the dissipation of energy.

References

LISSBERGER, P. H. (1970). *A. Rep. Prog. Phys.* **33,** 197.

SOUTHWORTH, G. C. (1936). *Bell Syst. tech. J.* **15,** 287 or (1937) *Proc. Inst. Radio Engrs,* **25,** 237.

Problems

9.1. Show that if $\mathbf{Z}_1$ and $\mathbf{Z}_2$ in a simple filter are both pure resistances or pure capacitances the filter acts as an attenuator at all frequencies. Calculate the attenuation per section when both $\mathbf{Z}_1$ and $\mathbf{Z}_2$ are pure resistances of 100 Ω and find the iterative impedance of a T-section.

(*Answers*: Power falls by factor 6·8 per section; $\mathbf{Z}_T = 112\ \Omega$.)

9.2. A filter where $\mathbf{Z}_1\mathbf{Z}_2 = k^2$, a constant independent of frequency, is called a constant-k filter. Show that the simple low-pass filters of § 9.2 are of this type, but the band-pass filter of Fig. 9.10 is not.

Show that the filter section of Fig. 9.12 is a band-pass section of the constant-k type provided that $L_1C_1 = L_2C_2$, when $k^2 = L_2/C_1 = L_1/C_2$. Sketch the variation with frequency of the quantity $\mathbf{Z}_1/4\mathbf{Z}_2$.

9.3. The values of the components in the m-derived T-section shown in Fig. 9.11(a) obey the relations $(m<1)$ $L_1 = mL$, $C_2 = mC$, $L_2 = L(1-m^2)/4m$.
Show that the section behaves as a low-pass filter with the following properties: (a) the cut-off frequency f_0 is independent of m; (b) the iterative impedance is $\mathbf{Z}_T = (L/C - \omega^2L^2/4)^{\frac{1}{2}}$, and is thus the same as that of a simple low-pass filter section in § 9.2; (c) the attenuation in the stop band is infinite at a frequency $f = f_0(1-m^2)^{-\frac{1}{2}}$.

9.4. A chain of the T-sections of Fig. 9.11(a) is terminated by the half-section shown in Fig. 9.11(b), where the values of the components obey the same relations as in the preceding problem. Prove that the impedance at the terminals CD is

$$\mathbf{Z} = \left(\frac{L}{C}\right)^{\frac{1}{2}} \frac{\{1-(1-m^2)f^2/f_0^2\}}{(1-f^2/f_0^2)^{\frac{1}{2}}}.$$

If $m = 0{\cdot}6$, show that this does not depart by more than 4 per cent from the value $(L/C)^{\frac{1}{2}}$ for frequencies up to 85 per cent of the cut-off frequency f_0. Thus the use of a half-section as a transformer gives a more uniform impedance in the pass band.

9.5. Find expressions for the electric and magnetic field strengths in a coaxial transmission line carrying a current I and a voltage V, and show by integrating Poynting's vector over the cross-section between the two conductors that the power flowing along the line is IV.

If the conductors have a finite resistivity, compute the power flowing into them per unit length by means of eqn (8.34b), and show that this gives the same attenuation as calculated in § 9.5.

9.6. In an infinite transmission line a leak develops at one point whose resistance is just equal to the characteristic impedance of the line. Show that of the power in the incident wave one-ninth is reflected, four-ninths is transmitted, and four-ninths is dissipated in the leak.

9.7. A length of lossless transmission line is first short-circuited at one end and then open-circuited; the impedance measured at the other end is $\mathbf{Z}_1$ in the first case and $\mathbf{Z}_2$ in the second. Show that $\mathbf{Z}_1\mathbf{Z}_2 = Z_0^2$, where Z_0 is the characteristic impedance of the line. This is a convenient way of measuring Z_0 for a cable of unknown electrical length.

9.8. A film of magnesium fluoride (MgF_2) (refractive index = 1·38) one-quarter wavelength thick is deposited on crown glass (refractive index = 1·52). Show that the reflected intensity is reduced to about 1 per cent.

Alternate layers, each $\lambda/4$ thick, of transparent media with refractive indices n_1, n_2 are deposited onto a substrate of index n_0. With x layers of each material show that for incident light the surface appears to have a refractive index $n_0(n_2/n_1)^{2x}$. If $n_0 = 1{\cdot}52$, $n_1 = 1.38$, and $n_2 = 2.35$ (ZnS), how many layers are needed to produce reflection of at least 99·99 per cent of the incident energy?

(*Answer*: $x = 10$.)

9.9. A quarter-wavelength, air-spaced, parallel-wire transmission line is found to be in resonance with an oscillator when its length is 0·25 m. When a capacitance of 1 pF is connected across the open end, it is found that the length of the line must be reduced to 0·125 m to obtain resonance. Show that the characteristic impedance of the line is approximately 530 Ω.

9.10. Show, either by the use of equations similar to (9.29) and (9.30) but with the assumptions $H_x = H_z = 0$, or by the use of eqns (8.36) and (8.37), that a wave can be propagated between two parallel conducting planes with the following field components:

$$H_y = (A/Z_1)\cos(2\pi z/\lambda_c)\cos(\omega t - 2\pi x/\lambda_g),$$
$$E_x = (A\lambda_0/\lambda_c)\sin(2\pi z/\lambda_c)\sin(\omega t - 2\pi x/\lambda_g),$$
$$E_z = -(A\lambda_0/\lambda_g)\cos(2\pi z/\lambda_c)\cos(\omega t - 2\pi x/\lambda_g),$$

where λ_c and λ_g have the same values as for the transverse electric wave derived in § 9.6. This wave is a transverse magnetic wave, and may be designated as TM_{0n}; note that it cannot exist in a closed rectangular guide because the tangential components of **E** must then vanish at the walls $y = 0$ and $y = b$. The lowest transverse magnetic wave then possible would be TM_{11}, with components H_y, H_z, E_x, E_z each varying sinusoidally in both the y and z directions.

9.11. A hollow cubical box of side a resonates in the TE_{101} mode (that is $l = n = 1$ in eqn (9.41)). Calculate the energy stored and energy dissipated per second, and show that the value of $Q_F = a/2\delta$, where δ is the skin depth in the metal walls at the resonant frequency.

9.12. A hollow rectangular box is bounded by perfectly conducting planes at $x = 0$, $x = a$; $y = 0$, $y = b$; $z = 0$, $z = c$. Show that the standing wave system

$$E_x = A_x \cos \alpha x \sin \beta y \sin \gamma z \exp j\omega t,$$
$$E_y = A_y \sin \alpha x \cos \beta y \sin \gamma z \exp j\omega t,$$
$$E_z = A_z \sin \alpha x \sin \beta y \cos \gamma z \exp j\omega t$$

satisfies the boundary conditions provided that $\alpha a = l\pi$, $\beta b = m\pi$, $\gamma c = n\pi$, and that the wave equation is satisfied if

$$\frac{1}{4\pi^2}(\alpha^2+\beta^2+\gamma^2) = \left(\frac{l}{2a}\right)^2 + \left(\frac{m}{2b}\right)^2 + \left(\frac{n}{2c}\right)^2 = \left(\frac{\omega}{2\pi v}\right)^2 = \frac{1}{\lambda_0^2}$$

how also that, to satisfy div $\mathbf{D} = 0$,

$$\alpha A_x + \beta A_y + \gamma A_z = 0.$$

(In general there can be only two independent amplitudes, corresponding to the two possible polarizations of an electromagnetic wave.)

Appendix A: Vectors

A.1. Definition of scalar and vector quantities

MANY physical quantities are completely defined by magnitude alone. Examples are temperature, time, or length. These are called scalar quantities. They obey the ordinary laws of algebra, and are represented in the text by a symbol printed in italic type. Other physical quantities are not completely defined unless the direction as well as the magnitude is given. Such quantities are called vectors, and are printed in bold-face type in the text.

Quantities such as velocity, force, or acceleration which involve a translation are known as 'polar vectors', to distinguish them from 'axial vectors' which are connected with rotations. They behave differently under reflection, as illustrated in Fig. 6.1.

A.2. Vector addition and subtraction

A vector may be represented graphically by an arrow pointing in the direction of the vector and of length equal to its magnitude. The addition of two vectors **P** and **Q** is shown in Fig. A.1 (a), in which the vectors form two sides of a parallelogram. The vector **R** defined by the equation $\mathbf{P}+\mathbf{Q}=\mathbf{R}$ is the diagonal of this parallelogram, and its magnitude and direction can be found by trigonometry if **P** and **Q** are known. Similarly, the vector $\mathbf{D}=\mathbf{P}-\mathbf{Q}$ is obtained from Fig. A.1 (b). In the special case that the vectors **P**, **Q** are parallel, **R** is equal to the scalar sum of **P** and **Q**, and **D** is equal to the scalar difference, and **R** and **D** are parallel to **P** and **Q**.

The converse process is often useful. That is, a vector **P** (Fig. A.2) can be resolved into two vectors **Q** and **R** such that **P** is the diagonal of a parallelogram, and **Q** and **R** are two adjacent sides. Generally, **Q** and **R** are chosen to be at right-angles, so that the parallelogram is then a rectangle. **Q** and **R** are called the components of **P**. **P** may be resolved into three components parallel to the axes of Cartesian coordinates *x*, *y*, and *z*.

A.3. Multiplication of vectors

1. Multiplication of a vector **P** by a scalar quantity m changes the magnitude of the vector by the factor m, but the direction is unaltered. Multiplication by $-m$ gives a vector of magnitude $m\mathbf{P}$ in the opposite direction, that is, the vector $-m\mathbf{P}$. If **i**, **j**, and **k** are vectors of unit length parallel to the axes x, y, and z, we can write

$$\mathbf{P}=\mathbf{i}P_x+\mathbf{j}P_y+\mathbf{k}P_z,$$

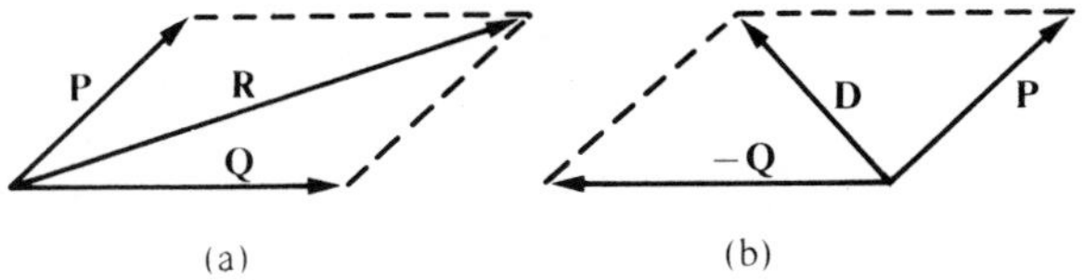

FIG. A.1.

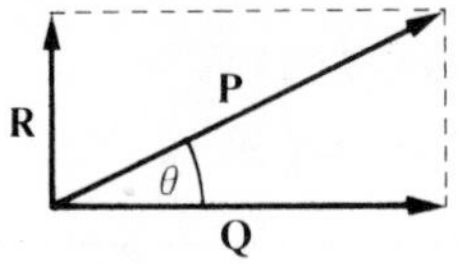

FIG. A.2. If **Q**, **R** are at right-angles, then $Q = P\cos\theta$, $R = P\sin\theta$.

where P_x, P_y, and P_z are scalar quantities giving the magnitude of the components of **P** parallel to the three axes (see Fig. A.3). Since $\mathbf{Q} = \mathbf{i}Q_x + \mathbf{j}Q_y + \mathbf{k}Q_z$ it follows that $\mathbf{P}+\mathbf{Q} = \mathbf{i}(P_x+Q_x)+\mathbf{j}(P_y+Q_y)+\mathbf{k}(P_z+Q_z)$.

2. *The scalar product.* The scalar product of two vectors **P** and **Q** is written **P.Q** and is a scalar quantity numerically equal to the magnitude of one vector multiplied by the component of the other parallel to the direction of the first one. If the angle between **P** and **Q** is θ,

$$\mathbf{P}.\mathbf{Q} = PQ\cos\theta = \mathbf{Q}.\mathbf{P}$$

and

$$\mathbf{P}.(\mathbf{Q}+\mathbf{R}+\mathbf{S}+\ldots) = \mathbf{P}.\mathbf{Q}+\mathbf{P}.\mathbf{R}+\mathbf{P}.\mathbf{S}+\ldots$$

The scalar product of two perpendicular vectors is zero. Therefore, for the unit vectors **i**, **j**, and **k**, we have

$$\mathbf{i}.\mathbf{j} = \mathbf{j}.\mathbf{k} = \mathbf{k}.\mathbf{i} = 0$$

and

$$\mathbf{i}.\mathbf{i} = \mathbf{j}.\mathbf{j} = \mathbf{k}.\mathbf{k} = 1.$$

An example of a scalar product is the work dW done on a charge q in moving a distance d**s** in a region where the electric field strength is **E**, which is

$$\mathrm{d}W = -q\mathbf{E}.\mathrm{d}\mathbf{s}. \tag{A.1}$$

3. *The vector product.* The vector product of two vectors **P** and **Q** is defined as a vector perpendicular to both **P** and **Q** of magnitude $PQ\sin\theta$, where θ is the angle between **P** and **Q**. If **P** is perpendicular to **Q**, the vector product is PQ but if **P** and **Q** are parallel the vector product is zero.

The direction of the vector product $(\mathbf{P}\wedge\mathbf{Q})$ is that in which a right-handed screw would move if turned from the first vector **P** towards the second vector **Q**, as shown in Fig. A.4. Hence we have

$$(\mathbf{P}\wedge\mathbf{Q}) = -(\mathbf{Q}\wedge\mathbf{P}).$$

Also,

$$\mathbf{P}\wedge(\mathbf{Q}+\mathbf{R}+\mathbf{S}+\ldots) = (\mathbf{P}\wedge\mathbf{Q})+(\mathbf{P}\wedge\mathbf{R})+(\mathbf{P}\wedge\mathbf{S})+\ldots$$

The formula for a vector product in terms of the vector components may be conveniently expressed as a determinant. For the unit vectors along a set of

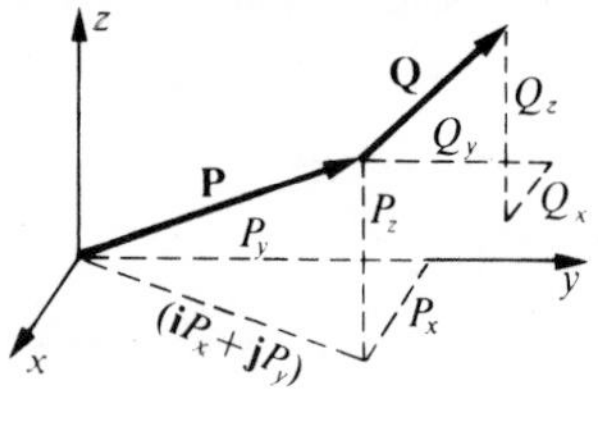

FIG. A.3.

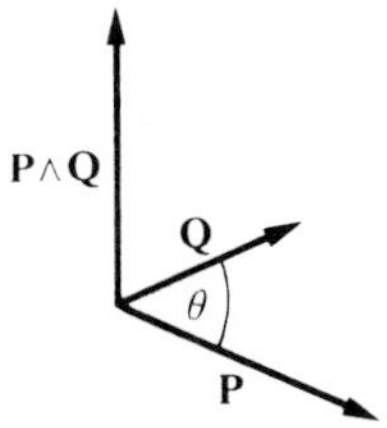

FIG. A.4. The vector **P** ∧ **Q** is normal to the plane containing **P** and **Q**.

right-handed Cartesian coordinates, we have

$$\mathbf{i}\wedge\mathbf{i}=\mathbf{j}\wedge\mathbf{j}=\mathbf{k}\wedge\mathbf{k}=0,$$

$$\mathbf{i}\wedge\mathbf{j}=\mathbf{k}=-\mathbf{j}\wedge\mathbf{i},\ \text{etc.}$$

Hence

$$\begin{aligned}\mathbf{P}\wedge\mathbf{Q}&=(\mathbf{i}P_x+\mathbf{j}P_y+\mathbf{k}P_z)\wedge(\mathbf{i}Q_x+\mathbf{j}Q_y+\mathbf{k}Q_z)\\&=\mathbf{i}(P_yQ_z-P_zQ_y)+\mathbf{j}(P_zQ_x-P_xQ_z)+\mathbf{k}(P_xQ_y-P_yQ_x),\end{aligned}$$

which can be written as

$$\mathbf{P}\wedge\mathbf{Q}=\begin{vmatrix}\mathbf{i} & \mathbf{j} & \mathbf{k}\\ P_x & P_y & P_z\\ Q_x & Q_y & Q_z\end{vmatrix}.$$

If **P** and **Q** are polar vectors, then (**P** ∧ **Q**) is an axial vector.

An example of the use of a vector product is the equation for the force d**F** on an element d**s** of a wire carrying a current I in a magnetic flux density **B**. The force is normal to d**s** and to **B**, and of magnitude $I\ \mathrm{d}s\ B\sin\theta$. It is specified both in magnitude and direction by the vector equation

$$\mathrm{d}\mathbf{F}=I(\mathrm{d}\mathbf{s}\wedge\mathbf{B}).$$

Products of three vectors can be evaluated from the foregoing rules. The scalar triple product

$$\mathbf{P}.(\mathbf{Q}\wedge\mathbf{R})=\text{scalar product of }\mathbf{P}\text{ and }(\mathbf{Q}\wedge\mathbf{R})$$

is a scalar quantity equal in magnitude to the volume of the parallelepiped whose sides are constructed from the three vectors **P**, **Q**, **R** (see Fig. A.5). Clearly,

$$\mathbf{P}.(\mathbf{Q}\wedge\mathbf{R})=(\mathbf{P}\wedge\mathbf{Q}).\mathbf{R},$$

and this is often written simply as **PQR**. We have

$$\mathbf{PQR}=\mathbf{QRP}=\mathbf{RPQ}=-\mathbf{PRQ}=-\mathbf{QPR}=-\mathbf{RQP}.$$

The change of sign on inverting the order of any two of the vectors follows also from the determinantal form

$$\mathbf{PQR}=\begin{vmatrix}P_x & P_y & P_z\\ Q_x & Q_y & Q_z\\ R_x & R_y & R_z\end{vmatrix}.$$

The formula for the vector triple product may be expressed in the form

$$\mathbf{P}\wedge(\mathbf{Q}\wedge\mathbf{R})=\mathbf{Q}(\mathbf{P}.\mathbf{R})-\mathbf{R}(\mathbf{P}.\mathbf{Q}).$$

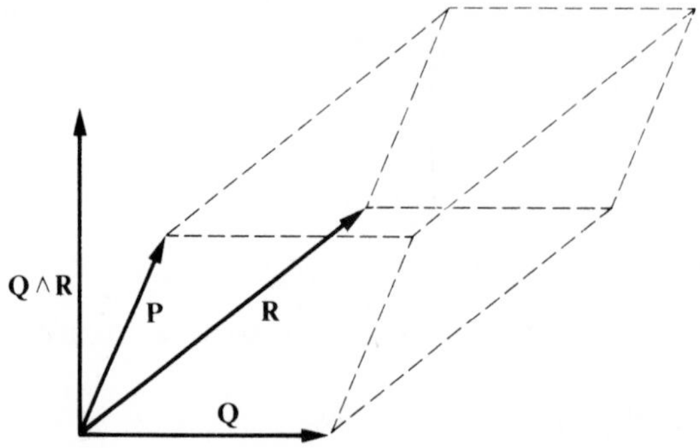

FIG. A.5. The scalar triple product $\mathbf{P}.(\mathbf{Q}\wedge\mathbf{R})$ is numerically equal to the volume of the parallelepiped whose sides are the vectors **P**, **Q**, and **R**.

This may be verified by expressing the vectors in terms of their components along three Cartesian axes.

A.4. Differentiation and integration of vectors

Vector quantities are often expressed as functions of scalar variables. For example, the electric field strength **E** can be expressed as a function of the position coordinates x, y, and z. The vector may be differentiated and integrated with respect to these variables. The differential of **P** with respect to a scalar variable u is defined as

$$\frac{d\mathbf{P}}{du}=\lim_{\Delta u\to 0}\frac{\mathbf{P}(u+\Delta u)-\mathbf{P}(u)}{\Delta u}.$$

When a force **F** acts for a small distance **ds**, the work done is $dW=\mathbf{F}.\mathbf{ds}$ and if the total work done over a finite distance is required, we can write

$$W=\int \mathbf{F}.\mathbf{ds}=\int F\cos\theta\, ds,$$

where ds is the component of **ds** parallel to **F** at any point.

This integral occurs frequently and is called the line integral of **F** along the curve. The line integral along the curve AB is illustrated in Fig. A.6. If the integration is carried out round a closed path, returning to the original point A, it is written $\oint \mathbf{F}.\mathbf{ds}$.

The surface integral $\int \mathbf{F}.\mathbf{dS}$ is also important. $\mathbf{F}.\mathbf{dS}$ is the flux through the element of area **dS** due to **F**, and the integral over a surface gives the total flux through that surface. If the vector $\mathbf{F}=\mathbf{v}$ represents the velocity of flow of a fluid, $\int \mathbf{v}.\mathbf{dS}$ gives the total volume of fluid passing through the area S in unit time. If **F** is the electric displacement **D**, the integral gives the number of lines of displacement crossing the surface S.

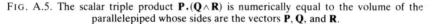

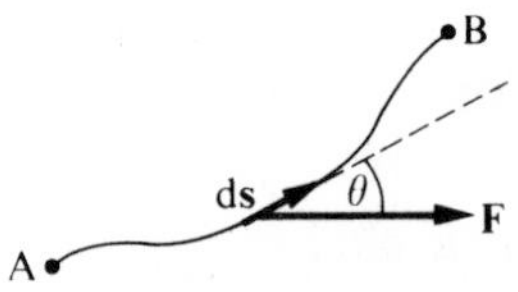

FIG. A.6.

In many problems in physics a scalar quantity is used which is a single-valued function of the position coordinates of the system. For example, in electrostatics the electric potential V is a function of x, y, and z in a Cartesian coordinate system. The change in potential corresponding to an infinitesimal displacement **ds** is given by Taylor's theorem, that is,

$$dV = (\partial V/\partial x)\,dx + (\partial V/\partial y)\,dy + (\partial V/\partial z)\,dz$$

and

$$\mathbf{ds} = \mathbf{i}\,dx + \mathbf{j}\,dy + \mathbf{k}\,dz.$$

The rate of change of V with the displacement **s** is expressed in terms of a new quantity grad V which is defined by the equation

$$dV = (\text{grad } V).\mathbf{ds}, \qquad \text{(A.2)}$$

where

$$\text{grad } V = \mathbf{i}(\partial V/\partial x) + \mathbf{j}(\partial V/\partial y) + \mathbf{k}(\partial V/\partial z).$$

grad V is a vector quantity and is an abbreviation for 'the gradient of V'. When grad V is parallel to **ds**, dV is a maximum, so that grad V is in the direction of the greatest rate of change of V with respect to the coordinates, and is normal to an equipotential surface. From Eqn (A.1) the work done on unit charge in moving a distance **ds** in a field of strength **E** is $-\mathbf{E}.\mathbf{ds}$ and this equal to $-dV$. Therefore we have $\mathbf{E} = -\text{grad } V$ and the electric field strength is equal to the gradient of the potential at any point, and is in the direction of the maximum rate of change of potential with respect to the space coordinates.

The operator $\mathbf{i}(\partial/\partial x) + \mathbf{j}(\partial/\partial y) + \mathbf{k}(\partial/\partial z)$ is often denoted by the symbol ∇ (pronounced 'del'), so that

$$\text{grad } V \equiv \nabla V. \qquad \text{(A.3)}$$

The operator ∇ can be regarded as a vector operator, which operates on both scalar and vector quantities, and forms scalar and vector products. Thus eqn (A.2) can be written

$$dV = (\nabla V).\mathbf{ds}. \qquad \text{(A.4)}$$

In general, any scalar potential function ϕ, which is finite, single-valued, and free from discontinuities (these conditions must apply also to the first and second derivatives of ϕ with respect to the space coordinates), can be related to a field of force **F**, where

$$\mathbf{F} = -\text{grad } \phi,$$

so that once ϕ is everywhere determined, **F** is known at all points. Also, the line integral of **F** between any two points A and B is independent of the path taken between those points since

$$\int_A^B \mathbf{F}.\mathbf{ds} = -\int_A^B (\text{grad } \phi).\mathbf{ds} = -\int_A^B d\phi = \phi_A - \phi_B$$

by analogy with eqn (A.2). Similarly, the line integral round a closed path is zero.

A.5. The divergence of a vector

The divergence of a vector **P** is written div **P**. It is an operator used to describe the excess flux leaving an element of volume in space. The flux may be flow of liquid in hydrodynamics, heat in a thermal field of varying temperature, or electric

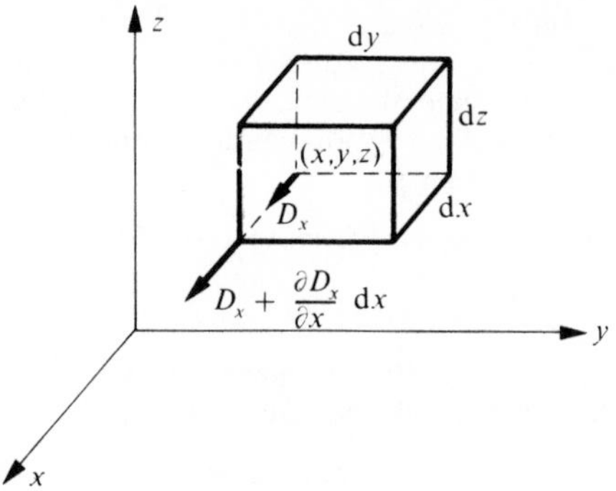

FIG. A.7.

flux. In the latter case, the excess flux leaving the volume element is related to the total charge enclosed by Gauss's theorem. In Fig. A.7 there is a varying electric charge density ρ_e throughout space. Gauss's theorem is applied to a volume element $dx\,dy\,dz$ at the point (x, y, z) in a Cartesian coordinate system. The total charge enclosed is $\rho_e\,dx\,dy\,dz$. The total flux through the faces normal to the x-axis is

$$\left[\left\{D_x + \frac{\partial}{\partial x}(D_x)\,dx\right\} - D_x\right] dy\,dz = \frac{\partial D_x}{\partial x}\,dx\,dy\,dz,$$

where D_x is the component of the electric displacement parallel to the x-axis at the point (x, y, z). Writing similar expressions for the flux through the other two pairs of faces, Gauss's theorem becomes

$$\frac{\partial D_x}{\partial x} + \frac{\partial D_y}{\partial y} + \frac{\partial D_z}{\partial z} = \rho_e,$$

where D_x, D_y, and D_z are the components of the electric displacement along the three axes at (x, y, z).

The expression on the left-hand side of this equation is written div **D**, and is the divergence of the vector **D** at this point.

Now using the operator ∇, we have

$$\nabla.\mathbf{D} = \left(\mathbf{i}\frac{\partial}{\partial x} + \mathbf{j}\frac{\partial}{\partial y} + \mathbf{k}\frac{\partial}{\partial z}\right).(\mathbf{i}D_x + \mathbf{j}D_y + \mathbf{k}D_z)$$

$$= \frac{\partial D_x}{\partial x} + \frac{\partial D_y}{\partial y} + \frac{\partial D_z}{\partial z}.$$

Therefore

$$\nabla.\mathbf{D} \equiv \operatorname{div}\mathbf{D}. \tag{A.5}$$

The divergence of a vector is a scalar quantity, since it represents the net amount of flux, or the number of lines of displacement, coming out of a volume element. If div $\mathbf{D} = 0$, the total flux entering the element $dx\,dy\,dz$ is balanced by that leaving it. A vector satisfying this condition is said to be *solenoidal.*

A.6. The curl of a vector

The curl (or rotation) of a vector **P** is written curl **P**. It arises in problems where a line integral of a vector round a closed path is related to the flux through the

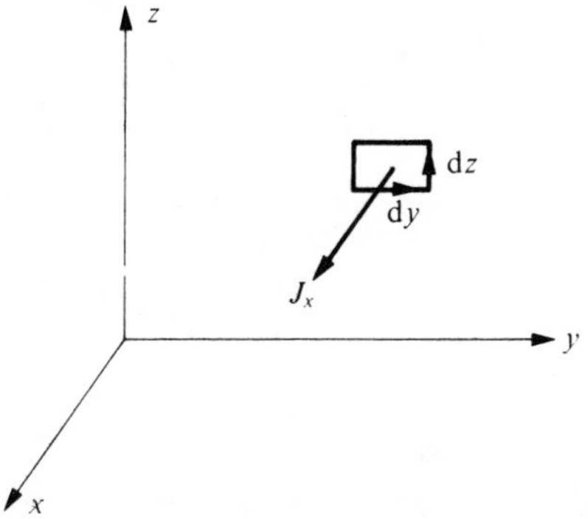

FIG. A.8. Application of Ampère's law in Cartesian coordinates.

surface enclosed by the path of the line integral. For example, Ampère's law for the magnetic field due to a current is

$$\oint \mathbf{H}.\mathrm{d}\mathbf{s} = \int \mathbf{J}.\mathrm{d}\mathbf{S}.$$

Let us apply this equation to an element dy dz at the point (x, y, z) in a Cartesian coordinate system (Fig. A.8). For the x component J_x, of the current, the line integral **H** in the yz plane is positive in an anticlockwise direction, and we have

$$\begin{aligned} J_x \,\mathrm{d}y\,\mathrm{d}z &= \left(H_y - \frac{\partial H_y}{\partial z}\frac{\mathrm{d}z}{2}\right)\mathrm{d}y + \left(H_z + \frac{\partial H_z}{\partial y}\frac{\mathrm{d}y}{2}\right)\mathrm{d}z \\ &\quad - \left(H_y + \frac{\partial H_y}{\partial z}\frac{\mathrm{d}z}{2}\right)\mathrm{d}y - \left(H_z - \frac{\partial H_z}{\partial y}\frac{\mathrm{d}y}{2}\right)\mathrm{d}z \\ &= \left(\frac{\partial H_z}{\partial y} - \frac{\partial H_y}{\partial z}\right)\mathrm{d}y\,\mathrm{d}z, \end{aligned}$$

where H_y and H_z are the components of **H** parallel to the y- and z-axes respectively. Therefore

$$J_x = \frac{\partial H_z}{\partial y} - \frac{\partial H_y}{\partial z},$$

and similarly

$$J_y = \frac{\partial H_x}{\partial z} - \frac{\partial H_z}{\partial x} \quad \text{and} \quad J_z = \frac{\partial H_y}{\partial x} - \frac{\partial H_x}{\partial y}.$$

These equations are written

$$J_x = \mathrm{curl}_x\,\mathbf{H}, \qquad J_y = \mathrm{curl}_y\,\mathbf{H}, \qquad J_z = \mathrm{curl}_z\,\mathbf{H}$$

or simply

$$\mathrm{curl}\,\mathbf{H} = \mathbf{J},$$

where curl **H** is a vector quantity whose components are expressed by means of the determinant

$$\mathrm{curl}\,\mathbf{H} = \begin{vmatrix} \mathbf{i} & \mathbf{j} & \mathbf{k} \\ \dfrac{\partial}{\partial x} & \dfrac{\partial}{\partial y} & \dfrac{\partial}{\partial z} \\ H_x & H_y & H_z \end{vmatrix}, \qquad \text{(A.6)}$$

that is,

$$\text{curl}\,\mathbf{H} = \mathbf{i}\left(\frac{\partial H_z}{\partial y} - \frac{\partial H_y}{\partial z}\right) + \mathbf{j}\left(\frac{\partial H_x}{\partial z} - \frac{\partial H_z}{\partial x}\right) + \mathbf{k}\left(\frac{\partial H_y}{\partial x} - \frac{\partial H_x}{\partial y}\right).$$

Also

$$\begin{aligned}\nabla\wedge\mathbf{H} &= \left(\mathbf{i}\frac{\partial}{\partial x} + \mathbf{j}\frac{\partial}{\partial y} + \mathbf{k}\frac{\partial}{\partial z}\right)\wedge(\mathbf{i}H_x + \mathbf{j}H_y + \mathbf{k}H_z)\\ &= \mathbf{i}\left(\frac{\partial H_z}{\partial y} - \frac{\partial H_y}{\partial z}\right) + \mathbf{j}\left(\frac{\partial H_x}{\partial z} - \frac{\partial H_z}{\partial x}\right) + \mathbf{k}\left(\frac{\partial H_y}{\partial x} - \frac{\partial H_x}{\partial y}\right)\\ &= \text{curl}\,\mathbf{H}.\end{aligned} \tag{A.7}$$

A.7. The divergence theorem

S in Fig. A.9 is a closed surface in a region where there exists a vector field **F**. The flux through an element of area **dS** is **F.dS**, and the total flux through the surface is $\int \mathbf{F}.\mathbf{dS}$. The total flux diverging from an element of volume $d\tau$ inside S is, from § A.5 above, div **F** $d\tau$, where **F** is the value of the force field at this point. The integral $\int \text{div}\,\mathbf{F}\,d\tau$ throughout the whole volume enclosed by S must give the total flux through the surface, since for any two adjacent volume elements the flux through a common face gives equal positive and negative contributions. Hence

$$\int \mathbf{F}.\mathbf{dS} = \int \text{div}\,\mathbf{F}\,d\tau, \tag{A.8}$$

which is the theorem of divergence. Again, the vector field **F** must be a well-behaved function. Conversely if the surface integral of a vector **F** is equal to the volume integral of a scalar function m over the volume enclosed by the surface, whatever the surface, then we may conclude that

$$m = \text{div}\,\mathbf{F}.$$

A.8. Stokes's theorem

In Fig. A.10 the line integral of the vector **H** is taken round a closed path bounding an unclosed surface S. This integral is $\oint \mathbf{H}.\mathbf{ds}$. If the surface is divided up into small elements of area **dS** then, from § A.6,

$$\oint \mathbf{H}.\mathbf{dl} = \text{curl}\,\mathbf{H}.\mathbf{dS},$$

where $\oint \mathbf{H}.\mathbf{dl}$ is the line integral of **H** round one small element of area **dS**. If this equation is now summed over all the elementary areas, all the boundaries within

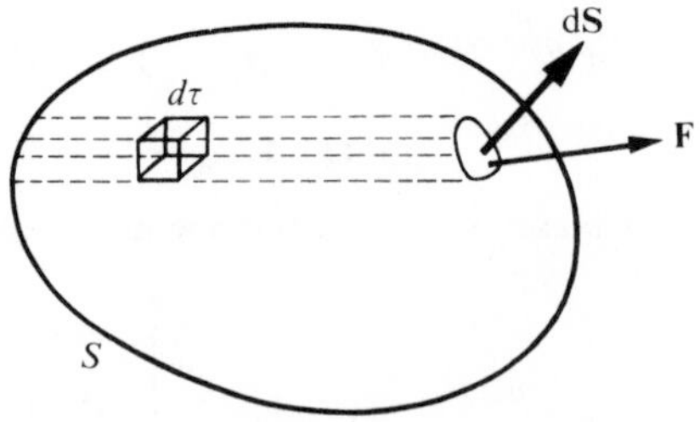

FIG. A.9. Illustrating the divergence theorem.

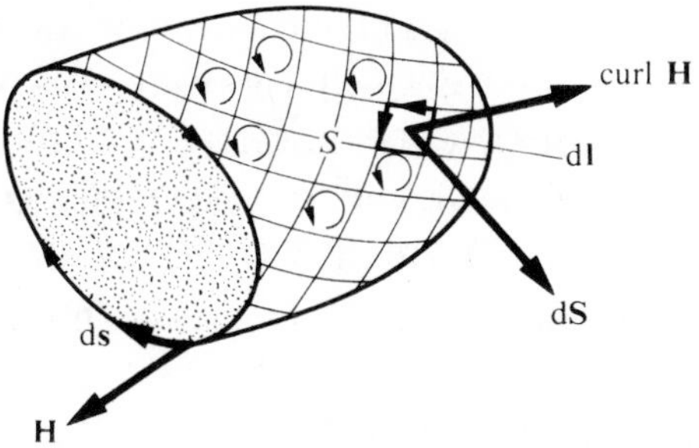

FIG. A.10. Illustrating Stokes's theorem.

the surface will cancel out on the left-hand side, and the result is the line integral round the circuit bounding the surface. Therefore

$$\oint \mathbf{H}\cdot d\mathbf{s} = \int \text{curl}\,\mathbf{H}\cdot d\mathbf{S}. \tag{A.9}$$

This is Stokes's theorem. It is necessary for **H** and its derivatives to be well-behaved continuous functions, but in the cases normally arising in electromagnetism, these conditions are satisfied.

Conversely, if the line integral of **H** round a closed curve is equal to the surface integral of **P** over a surface bounded by the curve, irrespective of what curve or surface are used, then $\mathbf{P} = \text{curl}\,\mathbf{H}$.

A.9. Some useful vector relations

The vector operator ∇ has been used above in expressions for grad $V = \nabla V$, div $\mathbf{D} = \nabla\cdot\mathbf{D}$ and curl $\mathbf{H} = \nabla\wedge\mathbf{H}$. Another useful operator has the form $(\mathbf{P}\cdot\text{grad}) = (\mathbf{P}\cdot\nabla)$; it operates on a vector

$$\begin{aligned}(\mathbf{P}\cdot\nabla)\mathbf{Q} &= \left(P_x\frac{\partial}{\partial x} + P_y\frac{\partial}{\partial y} + P_z\frac{\partial}{\partial z}\right)\mathbf{Q}\\ &= \mathbf{i}\left(P_x\frac{\partial Q_x}{\partial x} + P_y\frac{\partial Q_x}{\partial y} + P_z\frac{\partial Q_x}{\partial z}\right) +\\ &\quad +\mathbf{j}\left(P_x\frac{\partial Q_y}{\partial x} + P_y\frac{\partial Q_y}{\partial y} + P_z\frac{\partial Q_y}{\partial z}\right) +\\ &\quad +\mathbf{k}\left(P_x\frac{\partial Q_z}{\partial x} + P_y\frac{\partial Q_z}{\partial y} + P_z\frac{\partial Q_z}{\partial z}\right).\end{aligned} \tag{A.10}$$

If **Q** satisfies the relation curl $\mathbf{Q} = 0$, then $\partial Q_x/\partial y = \partial Q_y/\partial x$, etc., and (A.10) can be written with components $\mathbf{i}[P_x(\partial Q_x/\partial x) + P_y(\partial Q_y/\partial x) + P_z(\partial Q_z/\partial x)]$, etc. This type of vector can be used to represent the force on a dipole in a non-uniform field (cf. eqns (1.15) and (4.18b)).

The operator ∇ may operate on a product of two quantities; a number of useful relations of this type are:

$$\text{grad}(mV) = \nabla(mV) = m\,\text{grad}\,V + V\,\text{grad}\,m = m\nabla V + V\nabla m. \tag{A.11}$$

$$\text{div}\,(m\mathbf{P}) = \nabla\cdot(m\mathbf{P}) = m\,\text{div}\,\mathbf{P} + \mathbf{P}\cdot\text{grad}\,m = m\,\nabla\cdot\mathbf{P} + \mathbf{P}\cdot\nabla m. \tag{A.12}$$

$$\text{curl}\,(m\mathbf{P}) = \nabla\wedge(m\mathbf{P}) = m\,\text{curl}\,\mathbf{P} - \mathbf{P}\wedge\text{grad}\,m = m\,\nabla\wedge\mathbf{P} - \mathbf{P}\wedge\nabla m. \tag{A.13}$$

$$\text{grad}(\mathbf{P}.\mathbf{Q}) = \nabla(\mathbf{P}.\mathbf{Q}) = (\mathbf{P}.\text{grad})\mathbf{Q}+(\mathbf{Q}.\text{grad})\mathbf{P}+\mathbf{P}\wedge\text{curl }\mathbf{Q}+\mathbf{Q}\wedge\text{curl }\mathbf{P}$$
$$= (\mathbf{P}.\nabla)\mathbf{Q}+(\mathbf{Q}.\nabla)\mathbf{P}+\mathbf{P}\wedge(\nabla\wedge\mathbf{Q})+\mathbf{Q}\wedge(\nabla\wedge\mathbf{P}). \quad (A.14)$$

$$\text{div}(\mathbf{P}\wedge\mathbf{Q}) = \nabla.(\mathbf{P}\wedge\mathbf{Q}) = \mathbf{Q}.\text{curl }\mathbf{P}-\mathbf{P}.\text{curl }\mathbf{Q} = \mathbf{Q}.\nabla\wedge\mathbf{P}-\mathbf{P}.\nabla\wedge\mathbf{Q}. \quad (A.15)$$

$$\text{curl}(\mathbf{P}\wedge\mathbf{Q}) = \nabla\wedge(\mathbf{P}\wedge\mathbf{Q}) = (\mathbf{Q}.\text{grad})\mathbf{P}-(\mathbf{P}.\text{grad})\mathbf{Q}+\mathbf{P}(\text{div }\mathbf{Q})-\mathbf{Q}(\text{div }\mathbf{P})$$
$$= (\mathbf{Q}.\nabla)\mathbf{P}-(\mathbf{P}.\nabla)\mathbf{Q}+\mathbf{P}(\nabla.\mathbf{Q})-\mathbf{Q}(\nabla.\mathbf{P}). \quad (A.16)$$

The operator ∇ occurs twice in a number of relations. If V is a scalar function, div grad $V = \nabla.(\nabla V)$, and from eqns (A.3) and (A.5)

$$\nabla.(\nabla V) = \left(\mathbf{i}\frac{\partial}{\partial x}+\mathbf{j}\frac{\partial}{\partial y}+\mathbf{k}\frac{\partial}{\partial z}\right).\left(\mathbf{i}\frac{\partial V}{\partial x}+\mathbf{j}\frac{\partial V}{\partial y}+\mathbf{k}\frac{\partial V}{\partial z}\right)$$
$$= \frac{\partial^2 V}{\partial x^2}+\frac{\partial^2 V}{\partial y^2}+\frac{\partial^2 V}{\partial z^2}. \quad (A.17)$$

It is convenient to write div grad as ∇^2, where

$$\nabla^2 = \frac{\partial^2}{\partial x^2}+\frac{\partial^2}{\partial y^2}+\frac{\partial^2}{\partial z^2} \quad (A.18)$$

is known as the Laplacian operator (pronounced 'del squared'). Expressions for this operator in spherical polar and cylindrical polar coordinates are given in eqns (2.4) and (2.5).

Other vector identities in which ∇ occurs twice are:

$$\text{curl grad } V = \nabla\wedge(\nabla V) = 0 \quad (A.19)$$
$$\text{div curl }\mathbf{P} = \nabla.(\nabla\wedge\mathbf{P}) = 0 \quad (A.20)$$
$$\text{grad div }\mathbf{P} = \nabla(\nabla.\mathbf{P}) \quad (A.21)$$
$$\text{curl curl }\mathbf{P} = \nabla\wedge(\nabla\wedge\mathbf{P}) = \text{grad div }\mathbf{P}-\nabla^2\mathbf{P}$$
$$= \nabla(\nabla.\mathbf{P})-\nabla^2\mathbf{P}. \quad (A.22)$$

The following formulae are useful in transformations from surface to volume integrals, and from line to surface integrals.

$$\int m\,\mathrm{d}\mathbf{S} = \int \text{grad } m\,\mathrm{d}\tau \quad (A.23)$$

$$\int \mathbf{P}\wedge\mathrm{d}\mathbf{S} = -\int \text{curl }\mathbf{P}\,\mathrm{d}\tau \quad (A.24)$$

$$\int \mathbf{P}.(\mathbf{Q}.\mathrm{d}\mathbf{S}) = \int \{(\mathbf{Q}.\text{grad})\mathbf{P}+\mathbf{P}\text{ div }\mathbf{Q}\}\,\mathrm{d}\tau \quad (A.25)$$

$$\int m\,\mathrm{d}\mathbf{s} = -\int (\text{grad } m)\wedge\mathrm{d}\mathbf{S} \quad (A.26)$$

$$\int \mathbf{P}\wedge\mathrm{d}\mathbf{s} = \int (\text{div }\mathbf{P})\,\mathrm{d}\mathbf{S}+\int (\text{curl }\mathbf{P})\wedge\mathrm{d}\mathbf{S}-\int (\mathrm{d}\mathbf{S}.\text{grad})\mathbf{P} \quad (A.27)$$

A.10. Transformation from a rotating coordinate system

When dealing with the effect of an applied magnetic field on an atomic system it is often convenient to transform to a rotating coordinate system. Vector methods make this transformation simple, as can be seen from the following treatment.

Suppose we are concerned with some vector quantity **A**, which to start with we will suppose to be fixed in the rotating coordinate system (a line on a spinning top is an example, but we do not have to restrict **A** to be simply a radius vector). The angular motion of the coordinate system is represented by a vector $\boldsymbol{\omega}$, whose magnitude is equal to the angular velocity and whose direction is parallel to the axis of rotation. Its sense is that in which a right-handed screw would advance if rotated in the same sense as the angular motion. If **A** is fixed in the rotating system, then in a time δt the end point of the vector is displaced by an amount $\delta\mathbf{A}$ relative to a fixed coordinate system, as shown in Fig. A.11. It is clear that the motion of the end point is a simple rotation about the axis defined by $\boldsymbol{\omega}$. Hence

$$\delta\mathbf{A} = (\omega\,\delta t)A\sin\theta = (\boldsymbol{\omega}\wedge\mathbf{A})\,\delta t.$$

Hence the velocity of **A** relative to the fixed system is

$$\lim_{\delta t\to 0}(\delta\mathbf{A}/\delta t) = (\mathrm{d}\mathbf{A}/\mathrm{d}t) = (\boldsymbol{\omega}\wedge\mathbf{A}). \tag{A.28}$$

If we now suppose that **A** is not fixed in the rotating system, but has a velocity $(\mathrm{D}\mathbf{A}/\mathrm{D}t)$ relative to that system, then we have, from the vector addition of the two velocities

$$\mathrm{d}\mathbf{A}/\mathrm{d}t = (\mathrm{D}\mathbf{A}/\mathrm{D}t)+(\boldsymbol{\omega}\wedge\mathbf{A}). \tag{A.29}$$

This relation may be applied to any vector **A**, and hence it may be applied to the vector $\mathrm{d}\mathbf{A}/\mathrm{d}t$ to find the second differential of **A**. Retaining the notation that $\mathrm{d}/\mathrm{d}t$ refers to rate of change in the fixed coordinate system, and $\mathrm{D}/\mathrm{D}t$ to rate of change relative to the rotating system, we have (since $\boldsymbol{\omega}$ is a constant)

$$\frac{\mathrm{d}^2\mathbf{A}}{\mathrm{d}t^2} = \frac{\mathrm{d}}{\mathrm{d}t}\left(\frac{\mathrm{d}\mathbf{A}}{\mathrm{d}t}\right) = \left(\frac{\mathrm{D}}{\mathrm{D}t}+\boldsymbol{\omega}\wedge\right)\left(\frac{\mathrm{d}\mathbf{A}}{\mathrm{d}t}\right) = \left(\frac{\mathrm{D}}{\mathrm{D}t}+\boldsymbol{\omega}\wedge\right)\left(\frac{\mathrm{D}\mathbf{A}}{\mathrm{D}\mathbf{t}}+\omega\wedge\mathbf{A}\right)$$

$$= \frac{\mathrm{D}^2\mathbf{A}}{\mathrm{D}t^2}+2\left(\boldsymbol{\omega}\wedge\frac{\mathrm{D}\mathbf{A}}{\mathrm{D}t}\right)+\boldsymbol{\omega}\wedge(\boldsymbol{\omega}\wedge\mathbf{A}). \tag{A.30}$$

A.11. Larmor's theorem

Suppose that a charge q is moving in a field of force (such as the attraction of a positively charged nucleus) whose value at any moment is described by the vector **F**. When a magnetic field is applied, the equation of motion is

$$m\frac{\mathrm{d}^2\mathbf{r}}{\mathrm{d}t^2} = \mathbf{F}+q\mathbf{v}\wedge\mathbf{B}, \tag{A.31}$$

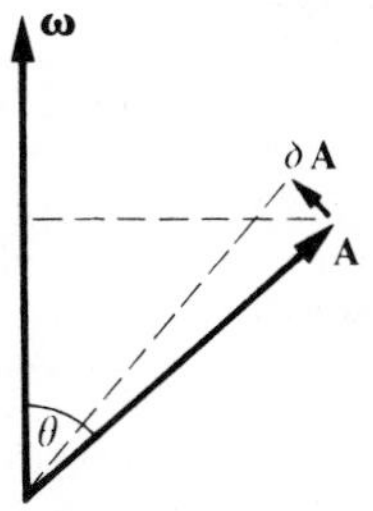

FIG. A.11. Change δA in time δt of a vector **A** rotating with velocity $\boldsymbol{\omega}$.

where m is the mass associated with the charge, and $\mathbf{v}$ is the instantaneous velocity ($= \mathrm{d}\mathbf{r}/\mathrm{d}t$). This is the equation of motion in vector form in a set of axes at rest with respect to the observer. Let us now change to a set of axes rotating with angular velocity $\boldsymbol{\omega}$ about the direction of $\mathbf{B}$. In transforming to rotating axes (see § A.10) we have the relation

$$m\frac{\mathrm{d}^2\mathbf{r}}{\mathrm{d}t^2} = m\frac{\mathrm{D}^2\mathbf{r}}{\mathrm{D}t^2} + 2m\left(\boldsymbol{\omega}\wedge\frac{\mathrm{D}\mathbf{r}}{\mathrm{D}t}\right) + m\{\boldsymbol{\omega}\wedge(\boldsymbol{\omega}\wedge\mathbf{r})\}, \tag{A.32}$$

where $\mathrm{D}^2\mathbf{r}/\mathrm{D}t^2$, $\mathrm{D}\mathbf{r}/\mathrm{D}t$ are the acceleration and velocity in the rotating coordinate frame, and $\boldsymbol{\omega}$ is the angular velocity expressed as a vector parallel to the axis of rotation, the direction of $\mathbf{B}$. The second term on the right-hand side is the 'Coriolis force' which appears if the particle is moving in the rotating system, and the last term is the centrifugal force normal to the axis of rotation. If $\boldsymbol{\omega}\wedge\mathbf{r}$ is small compared with $\mathrm{D}\mathbf{r}/\mathrm{D}t$ (as we shall show below to be the case), the centrifugal force will be small compared with the Coriolis force, and in the first approximation eqns (A.31) and (A.32) give

$$m\frac{\mathrm{D}^2\mathbf{r}}{\mathrm{D}t^2} = \mathbf{F} + q\mathbf{v}\wedge\mathbf{B} - 2m\boldsymbol{\omega}\wedge\frac{\mathrm{D}\mathbf{r}}{\mathrm{D}t} = \mathbf{F} + q\mathbf{v}\wedge\mathbf{B} + 2m\mathbf{v}\wedge\boldsymbol{\omega}, \tag{A.33}$$

where we have neglected the small difference between $\mathbf{v}$ and $\mathrm{D}\mathbf{r}/\mathrm{D}t$ (the velocity in the rotating frame) since they differ only by the quantity $\boldsymbol{\omega}\wedge\mathbf{r}$, which we have already assumed to be small in comparison. It is apparent that, if we choose the rate of rotation of the axes such that

$$\boldsymbol{\omega} = -(q/2m)\mathbf{B}, \tag{A.34}$$

the last two terms in (A.33) will vanish and the equation of motion is the same as if the magnetic field were absent. Thus to an observer rotating with the angular velocity given by (A.34) the motion of the charge appears to be the same as it would to a stationary observer in the absence of a magnetic field. Hence we may regard the motion of an electron of charge $-e$ in the field with flux density $\mathbf{B}$ as unchanged except for a precession with angular velocity $\boldsymbol{\omega} = +(e/2m)\mathbf{B}$ about the axis of $\mathbf{B}$. This is commonly known as the 'Larmor precession'.

The fact that it is justifiable to neglect the last term in eqn (A.32) can be seen as follows. When the electron is bound in the atom, it executes a periodic motion in its orbit whose frequency is of the same order as that of visible light. This corresponds to an angular frequency ω_0 of the order of 10^{15} rad s^{-1}. The terms $\mathrm{D}^2\mathbf{r}/\mathrm{D}t^2$ and $\mathrm{D}\mathbf{r}/\mathrm{D}t$ are then of order of magnitude $\omega_0^2 r$ and $\omega_0 r$ respectively, so that successive terms in eqn (A.32) decrease in magnitude by the ratio ω/ω_0. Since ω is only about 10^{11} rad s^{-1} even in a field of 1 T, the centrifugal force term is an order of magnitude smaller than the Coriolis force. In other words, the force on the electron due to the flux density $\mathbf{B}$ is small compared with the force exerted by the positively charged nucleus; if it were not, it would tear the atom apart.

The central force assumed above is that responsible for the orbital motion of an electron in an atom, and the angular velocity given by eqn (A.34) is identical with the angular velocity of precession (see § 14.1) of an electronic orbital magnetic moment in a magnetic flux density $\mathbf{B}$.

Appendix B: Depolarizing and demagnetizing factors

SUPPOSE a uniform field of strength $\mathbf{E}_0$ exists in a medium of relative permittivity ϵ_2. An ellipsoid with principal axes of length a, b, c and relative permittivity ϵ_1 is introduced with its principal a-axis parallel to $\mathbf{E}_0$ Then the field $\mathbf{E}_a$ inside the ellipsoid is also parallel to the a-axis; its value is

$$\mathbf{E}_a = \mathbf{E}_0 \frac{\epsilon_2}{\epsilon_2 + d_a(\epsilon_1 - \epsilon_2)}. \tag{B.1}$$

The quantity d_a is a pure number, given by the integral

$$d_a = \tfrac{1}{2}abc \int_0^\infty \frac{ds}{(s+a^2)R_s}, \tag{B.2}$$

where

$$R_s^2 = (s+a^2)(s+b^2)(s+c^2).$$

Values of d_a, d_b, d_c are tabulated by Osborn (1945) and Stoner (1945), and satisfy the identity

$$d_a + d_b + d_c = 1.$$

Two limiting sets of values are (for ellipsoids of circular cross-section)

	d_a	d_b	d_c
short, fat ellipsoid	1	0	0
long, thin ellipsoid	0	$\frac{1}{2}$	$\frac{1}{2}$

In the limit the short, fat ellipsoid is equivalent to a parallel-sided slab for which $\epsilon_1\mathbf{E}_a = \epsilon_2\mathbf{E}_0$; that is, the normal components of $\mathbf{D}$ are continuous, while the tangential components of $\mathbf{E}$ are continuous since $d_b = d_c = 0$. In the opposite limit the long, thin ellipsoid is equivalent to a circular cylinder; when $\mathbf{E}$ is parallel to the axis it is continuous, and when $\mathbf{E}$ is normal to the axis the formula agrees with the result of Problem 2.3. For a sphere $d_a = d_b = d_c = \frac{1}{3}$, and eqn (B.1) then reduces to eqn (2.43).

Eqn (B.1) cannot be used when a permanent moment is present, but it agrees with the definition of a 'depolarizing field' $\mathbf{E}_d = \mathbf{E}_a - \mathbf{E}_0$ given by eqn (2.48). The latter has the advantage that it is applicable when there also exists a permanent polarization parallel to the a-axis. This is an added complication, which is treated below for the magnetic case; the substitutions needed are $\mathbf{H}$ for $\mathbf{E}$; $\mathbf{M}$ for $\mathbf{P}/\epsilon_0$; μ for ϵ.

Suppose the ellipsoid has a permanent magnetization $\mathbf{M}_0$ parallel to the a-axis, as well as an induced moment. Then

$$\mathbf{M}_1 = (\mu_1 - 1)\mathbf{H}_a + \mathbf{M}_0,$$

and

$$\mathbf{M}_2 = (\mu_2 - 1)\mathbf{H}_0. \tag{B.3}$$

The treatment given by Lowes (1974) is equivalent to the presence of a demagnetizing field $\mathbf{H}_d = \mathbf{H}_a - \mathbf{H}_0$ given by

$$\mathbf{H}_d = -\frac{\mathbf{M}_1 - \mathbf{M}_2}{1+(d_a^{-1}-1)\mu_2}. \tag{B.4}$$

For a sphere $d_a = \frac{1}{3}$ and eqn (B.4) reduces to eqn (4.38c).

Lowes (1974) also derives the torque $\mathbf{T}$ on a permanent magnet in the form of an ellipsoid with magnetization $\mathbf{M}_0$ parallel to the a-axis in a flux density $\mathbf{B}_0$,

$$\mathbf{T} = V\frac{\mathbf{M}_0 \wedge \mathbf{B}_0}{\mu_2 + d_a(\mu_1 - \mu_2)} \tag{B.5}$$

where $V = \frac{4}{3}\pi abc$ is the volume of the magnet. For a long, thin magnet ($d_a \to 0$), the torque is simply $\mathbf{T} = \mu_0 V(\mathbf{M}_0 \wedge \mathbf{H}_0)$, in agreement with the experiment of Whitworth and Stopes-Roe (1971), but this is not true for other shapes. Eqn (B.5) gives only the torque associated with $\mathbf{M}_0$; there is also a torque associated with induced magnetization which tends to align a paramagnetic ellipsoid with its longest axis parallel to $\mathbf{H}_0$ (see Stratton 1941 (§ 3.29)).

References

Lowes, F. J. (1974). *Proc. R. Soc. Lond.* **A337,** 555.
Osborn, J. A. (1945). *Phys. Rev.* **67,** 351.
Stoner, E. C. (1945). *Phil. Mag.* **36,** 803.
Stratton, J. A. (1941). *Electromagnetic theory.* McGraw-Hill, New York.
Whitworth, R. W. and Stopes-Roe, H. V. (1971). *Nature, Lond.* **234,** 31.

Appendix C: Numerical values of the fundamental constants

c	velocity of light *in vacuo*	$2{\cdot}998\times10^{8}$ m s^{-1}
N_A	Avogadro's number	$6{\cdot}022\times10^{23}$ mol^{-1}
e	electronic charge	$1{\cdot}602\times10^{-19}$ C
m_e	electron rest mass	$9{\cdot}110\times0^{-31}$ kg
m_p	proton rest mass	$1{\cdot}673\times10^{-27}$ kg
m_p/m_e	ratio of proton to electron mass	$1{\cdot}836\times10^{3}$
h	Planck's constant	$6{\cdot}626\times10^{-34}$ J s
$\hbar$	Planck's constant/2π	$1{\cdot}055\times10^{-34}$ J s
$h/2e$		$2{\cdot}068\times10^{-15}$ J s C^{-1}
F	Faraday's constant (N_Ae)	$9{\cdot}648\times10^{4}$ C mol^{-1}
e/m	charge/mass for electron	$1{\cdot}759\times10^{11}$ C kg^{-1}
e^2/m		$2{\cdot}818\times10^{-8}$ C^2 kg^{-1}
a_0	Bohr radius $= 4\pi\epsilon_0\hbar^2/m_ee^2$	$5{\cdot}292\times10^{-11}$ m
$R_\infty c$	Rydberg constant$\times c = m_ee^4/8\epsilon_0^2h^3$	$3{\cdot}290\times10^{15}$ Hz
R_∞	Rydberg constant	$1{\cdot}097\times10^{7}$ m^{-1}
k	Boltzmann's constant	$1{\cdot}381\times10^{-23}$ J K^{-1}
R	gas constant $= N_Ak$	$8{\cdot}314$ J K^{-1} mol^{-1}
μ_B	Bohr magneton $= e\hbar/2m_e$	$9{\cdot}274\times10^{-24}$ A m^2
μ_N	nuclear magneton $= e\hbar/2m_p$	$5{\cdot}051\times10^{-27}$ A m^2
α	fine structure constant $= e^2/4\pi\epsilon_0\hbar c$	$(137{\cdot}0)^{-1}$
ϵ_0	permittivity of a vacuum $= (\mu_0c^2)^{-1}$	$8{\cdot}854\times10^{-12}$ F m^{-1}
$4\pi\epsilon_0$		$10^7/c^2 = 10^{-9}/9$ approximately
μ_0	permeability of a vacuum (by definition)	$4\pi\times10^{-7}$ H m^{-1}
Z_0	intrinsic impedance of free space	$3{\cdot}767\times10^{2}$ Ω
eV	electronvolt	$1{\cdot}602\times10^{-19}$ J
kT	energy for $T = 290$ K	$4{\cdot}004\times10^{-21}$ J

1 electronvolt is equivalent to:

- wavelength $\lambda = 1{\cdot}240\times10^{-6}$ m
- frequency $\nu = 2{\cdot}418\times10^{14}$ Hz
- wave number $\tilde{\nu} = 8{\cdot}065\times10^{5}$ m^{-1}
- temperature $T = 1{\cdot}160\times10^{4}$ K
- energy $W = 1{\cdot}602\times10^{-19}$ J

1 m^{-1} is equivalent to:

- wavelength $\lambda = 1$ m
- temperature $T = 1{\cdot}439\times10^{-2}$ K $= hc/k$

Appendix D: Some useful unit conversions

The system of units employed throughout this book is the International System (S.I.). However, other systems of units are still often used in the literature; these are usually m.k.s. (metre, kilogram, second) units or c.g.s. (centimetre, gram, second) units. In the theory of electricity and magnetism c.g.s. units may be either electrostatic (e.s.u.) or electromagnetic (e.m.u.).

The following is a brief list of equivalences which may be useful when reading the literature associated with topics in this book.

1 dyne = 10^{-5} newtons
1 erg = 10^{-7} joules
1 mm Hg = 1 torr ≈ 133·322 pascals
1 atmosphere = 101 325 pascals ≈ 10^5 Pa = 1 bar
1 calorie = 4·184 joules
1 e.m.u. of charge = 10 coulombs
1 e.s.u. of charge ≈ $3{\cdot}336 \times 10^{-10}$ coulombs
1 gauss = 10^{-4} teslas
1 oersted = $10^3/4\pi$ amperes per metre
1 maxwell = 10^{-8} webers
1 electronvolt ≈ $1{\cdot}602 \times 10^{-19}$ joules
1 micron = 1 micrometre = 10^{-6} m
1 ångström = 10^{-1} nanometres = 10^{-10} m
1 hertz = 1 cycle per second
1 Bohr magneton = $0{\cdot}9274 \times 10^{-20}$ e.m.u.
1 Faraday = $9{\cdot}649 \times 10^3$ e.m.u. mol^{-1}.

In S.I. units the volume susceptibility (per cubic metre) is a number larger by a factor 4π than the corresponding number (per cubic centimetre) in c.g.s. units, the latter being expressed in e.s.u. for electric susceptibility and in e.m.u. for magnetic susceptibility.

Index